Janus Particle Synthesis, Self-assembly and Applications

RSC Smart Materials

Series Editor:
Hans-Jörg Schneider, *Saarland University, Germany*
Mohsen Shahinpoor, *University of Maine, USA*

Titles in this Series:
1: Janus Particle Synthesis, Self-assembly and Applications

How to obtain future titles on publication:
A standing order plan is available for this series. A standing order will bring delivery of each new volume immediately on publication.

For further information please contact:
Book Sales Department, Royal Society of Chemistry, Thomas Graham House, Science Park, Milton Road, Cambridge CB4 0WF, UK
Telephone: +44 (0)1223 420066, Fax: +44 (0)1223 420247
Email: booksales@rsc.org
Visit our website at http://www.rsc.org/Shop/Books/

Janus Particle Synthesis, Self-assembly and Applications

Edited by

Shan Jiang
Massachusetts Institute of Technology, Cambridge, MA, USA
Email: sjiang2@mit.edu

Steve Granick
University of Illinois at Urbana-Champaign, IL, USA
Email: sgranick@illinois.edu

RSC Publishing

RSC Smart Materials No. 1

ISBN: 978-1-84973-423-3
ISSN: 2046-0066

A catalogue record for this book is available from the British Library

Published by The Royal Society of Chemistry,
Thomas Graham House, Science Park, Milton Road,
Cambridge CB4 0WF, UK

Registered Charity Number 207890

For further information see our website at www.rsc.org

Printed in the United Kingdom by Henry Ling Limited, Dorchester DT1 1HD, UK

Preface – An Introduction to Janus Particles

'Janus' is the name of an ancient Roman God, who has two faces peering into the past and the future, as shown in Figure 1a. Named after this Roman God, Janus particles have two distinguished surfaces/chemistries on the two sides, usually made of incompatible materials, such as hydrophobic *versus* hydrophilic and positive *versus* negative (charge). The idea of combining two incompatible properties into one object has precedent in an even earlier culture. In ancient Asian philosophy, it was believed that the seemingly opposite forces are interconnected in Nature and they give rise to the complicated change and transition in the world. The concept is called *Yin and Yang* (dark and bright) and symbolized in the classic Taoist Taijitu as shown in Figure 1b, which also looks like a Janus particle. So Janus particles can also be called '*Yin Yang*' particles. Just as how *Yin and Yang* sheds light on the underlying principles of Nature, this book will demonstrate how Janus particles offer insight into fundamental science and lead to invention of new materials.

1 A Brief History

The first publication about Janus particles was written by Casagrande and Veyssie in 1988.[1] Pierre-Gilles de Gennes, Nobel Prize in Physics winner in 1991, reiterated the concept in the context of *soft matter* and made it known to the scientific community in his Nobel Laureate speech.[2] In de Gennes's vision, Janus particles are the new materials that will adsorb on the water/air interface and form a monolayer with voids between them, which he called a 'skin that

RSC Smart Materials No. 1
Janus Particle Synthesis, Self-Assembly and Applications
Edited by Shan Jiang and Steve Granick
© The Royal Society of Chemistry 2012
Published by the Royal Society of Chemistry, www.rsc.org

(a)   (b)

Figure 1   (a) Janus God image engraved on an ancient Roman coin; (b) Taoist Taijitu symbol used on the national flag of South Korea.

can breathe.' It is interesting to note that in the same lecture, de Gennes identified fundamental commonality with polymers, liquid crystals and surfactants. He called attention to two characteristics shared by these vastly different systems: one is complexity, the other is flexibility. The Nobel Prize winner's remark will again be proved true in this book. The complexity lies not only in the geometry and chemistry of Janus particles, but also in the assembly structures induced either by the interactions among particles themselves or by the external field. The flexibility is shown by the dynamics of the structures formed by these particles, *i.e.* how they diffuse in the bulk and adsorb at the interface. In the context of soft matter, side-by-side with block copolymers, liquid crystals and surfactants, Janus particles fit in perfectly as a colloidal version of these fascinating systems. de Gennes was entranced by the connection and had a vision of the new physics that Janus particles are bound to offer.

Soft matter studies blossomed in the following years. Rapid progress was made in polymers. Block copolymers became a hot topic and the phase diagram was mapped out both theoretically and experimentally. Various patterns generated from assembly structures were studied extensively and proposed for applications for different purposes. Liquid crystal studies also gained great momentum driven by the huge profit from industry. Surfactant research expanded through studies of phospholipids and lipid-like molecules. Making floppy liposomes, big or small, to mimic the cell membrane became popular in the laboratory and lipid-based formulations for drug delivery and biomedication also became widespread in both academia and industry.

However, Janus particle research lagged behind while other areas in soft matter were thriving. During the 15 years following de Gennes's lecture, not much was done on Janus particles. The main obstacle in the field was the

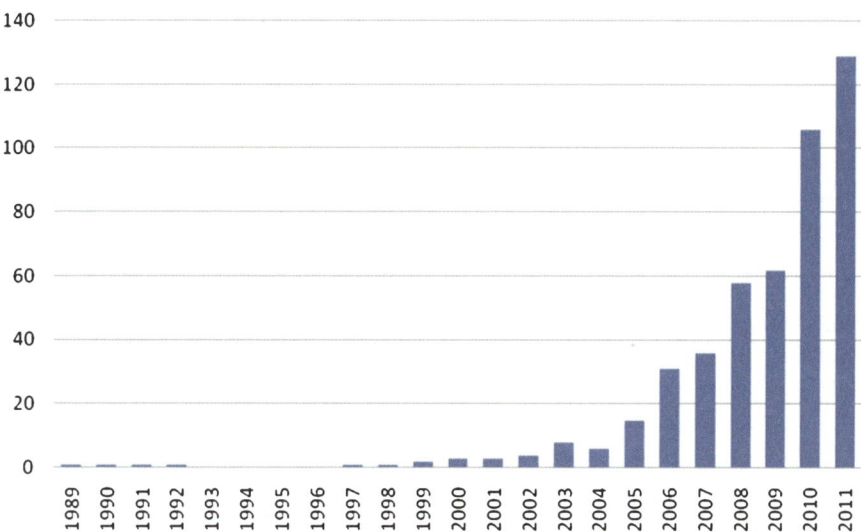

Figure 2 Statistics of publications on Janus particles from 1989 to 2011.

synthesis of these particles. People were searching for easy ways to fabricate Janus particles with well-defined geometry and chemistry. On the other hand, colloidal research has demonstrated many useful techniques to synthesize different particles in various sizes with homogeneous distributions. Enlightened by this progress, different methods for fabricating Janus particles started to emerge around 2003–2005. The momentum continued to build up in 2006 and the number of publications on Janus particles increased almost exponentially thereafter. The trend is still continuing today, as shown by Figure 2, based on article search in the database of Web of Science. The new methods of synthesizing Janus particles opened up the gate for further studies in the field.

2 How to Make Janus Particles

The first part of this book addresses the problem of synthesis. At de Gennes's time, the method was very limited in terms of control of size and geometry. One interesting suggestion is to modify the homogeneous particle surface to a different chemistry and then break the particles into fragments. In this way, all the fragments will adopt the Janus geometry automatically. This is an easy and low-cost approach. However, it only generates heterogeneous particles and offers no control over geometry. If the inner parts of the particles were fragmented, they would not be in Janus geometry and would be hard to separate out.

One of the early methods that successfully offered high-quality Janus particles is directional coating onto a monolayer of homogeneous particles deposited on a flat substrate.[3,4] The size and dispersity of the particles are determined solely by the starting particles. Directional coating generates Janus particles with perfect 50% coverage. If the metal coating is deposited on fluorescent particles, 'Moon particles' can be obtained. Studies described in Chapter 10 (Stephen Anthony and Minsu Kim) tracked the rotation and diffusion of the Moon particles and studied the details of dynamics using Janus particles as a probe.

The directional coating method can be viewed as the particles themselves protecting one side that is not coated. Inspired by the protection approach, many other methods were developed. One very versatile approach is the emulsion-based method,[5] taking advantage of the fact that particles were tightly adsorbed at the interface and partially protected by one phase. Janus geometry can be achieved by modifying the exposed side on the particles. This method can be simply scaled up since emulsions offer a huge amount of interface in a relatively small volume. What is more, the geometry of Janus particles can be fine tuned by adjusting how particles sit at the emulsion interface. The method is thoroughly reviewed in Chapter 4 (Chengliang Zhang, Wei Wei, Fuxin Liang and Zhenzhong Yang).

Protection and deprotection often require extra steps to achieve the Janus geometry. A more direct approach is simply to flow or stack two different materials together and form Janus particles. Chapter 3 (Tae-Hong Park and Joerg Lahann) describes the method of synthesizing Janus particles via electrohydrodynamic co-jetting. This method is similar to the microfluidic approach; however, the electric field induces particles of much smaller size. Chapter 5 (Joseph M. DeSimone, Jie-Yu Wang and Yapei Wang) describes a way to fabricate Janus particles using a special non-wetting mould. Both of these methods generate high-quality Janus particles with various geometries.

All the approaches mentioned above can generate micron or submicron particles; however, it is very challenging to fabricate small Janus particles below 100 nm. Synthesizing Janus particles of ultra-small size usually requires a special design or growing particles from scratch. Chapter 1 (Andreas Walther and Axel H. E. Müller) describes an ingenious way of synthesizing Janus particles using molecular assembly from block copolymers; Chapter 2 (Chao Wang and Chenjie Xu) discusses different ways of synthesizing fascinating dumbbell nanocrystals.

Table 1 summarizes some common methods used to synthesize Janus particles. However, this table is far from comprehensive and many more methods are being invented. Since details of these methods are discussed in the individual chapters, here we just give some guidelines for evaluation:

1. Homogeneity: whether particles are homogeneous in size and geometry.
2. Tunability: whether it is possible to change the shape and size of the particles and fine tune the Janus geometry.

Table 1 Common methods of synthesizing Janus particles.

Method	Special device required	Size/size distribution	Chemistry/materials	Production scale/ scalability
Molecular assembly	None	<100 nm monodisperse	Block copolymer assembly	mg easy to scale up
Nanocrystal dumbbell	High-temperature heating control device	<100 nm <10% dispersity	Metal or metal oxide coated with surfactants	mg easy to scale up
Microfluid	Microfluidic device	10–100 μm <5% dispersity	Photocurable monomers	mg hard to scale up
Electro co-jetting	Electrohydrodynamic co-jetting device	0.5–50 μm dispersity varies	Water-soluble polymers	mg difficult to scale up
Directional coating	Dielectric coating device	0.5–10 μm depends on starting particles	Metal and dielectric coatings	μg difficult to scale up
Emulsion	None	0.2–10 μm depends on starting particles	Particles with reactive surface	g easy to scale up
Lithography	Lithography device	0.5–100 μm monodisperse	Elastomer stamp and photocurable monomers	μg difficult to scale up

3. Functionality: whether it is easy to change materials of the particles and functionalize the surface.
4. Scalability: whether it is possible to scale up the procedure for large quantities.

Ultimately, ends dictate the means. For fundamental and assembly studies, we always want homogeneous particles with a tunable geometry; for emulsion stabilization, gram quantities of Janus particles are preferred; for drug delivery and biomedication, biocompatibility and biodegradability are the priorities.

3   What Can We Do With Janus Particles?

The second part of the book deals with this question – what can we do with Janus particles? Before answering the question, we can take a look at many fundamental biological structures that actually utilize the simple Janus motif. Nature has long been using the bottom up self-assembly approach to build up structures. For example, cell membranes are bilayers assembled from phospholipids, which are amphiphilic molecules with 'Janus geometry' of a charged hydrophilic head group and two hydrophobic tails. It has been observed in simulation that with the right geometry and interaction, Janus particles can self-assemble into bilayer structures.[6] Another good example is DNA, which is given integrity by thousands of nucleotides, each of which consists of phosphate and nucleobase structures. Again, the simple Janus geometry of the nucleotide determines the DNA structure: hydrogen bonding between the nucleobases constructs the core to store the genetic information, while phosphates act as the backbone to protect the information. Analogous structures were also observed in the assembly of Janus particles.[7] Chapter 9 (Yi Chen, Abigail K. R. Lytton-Jean and Hyuckjin Lee) shows beautiful examples of what self-assembly can achieve based on the specific binding of DNA molecules. Many elegant and unique structures were assembled, from crystals formed by Au nanoparticles with a DNA coating to crystal structures formed by DNA molecules themselves.

We can simply observe Janus particles. The first place in which people watched closely how Janus particles come together was not a bench-top experiment, but a computer simulation, in which perfect Janus particles can be created without too much trouble. Actually, simulation can create any geometry and shape of interest. The idea behind the simulation work is inspired by the concept of 'colloidal molecules,'[8] which first arose in the context of the packing of colloidal particles and the photonic crystal structures created by those particles. Clusters of simple colloids mimic those of small atoms and molecules with different shapes. Since the diffusion and assembly of colloidal particles are easily visible under an optical microscope, they can be used as a model system to study liquids, crystals, glasses and even atomic structures, which are hardly approachable in such detail at the equivalent single-atom or single-molecule level using conventional methods.

Janus particles can be viewed as one kind of fundamental colloidal molecule, whose molecular counterpart can be surfactant or dipolar molecules. Will Janus particles simply assemble in a way akin to how small molecules do? Can we create Janus particle micelles or even Janus particle liposomes? Theoretical work provides a good framework to solve these questions, as described in Chapter 6 (Achille Giacometti, Flavio Romano and Francesco Sciortino). To fully address the question also requires coordination between simulation and experiment. On the one hand, simulation can fully explore the possibility of Janus geometry and interactions that lead to different assembly structures. On the other hand, a simulation only makes good sense when the model can be realized in the experimental setup. There are also structures observed in experiments which can hardly be understood by just looking at them under the microscope. Only by combining experiment and simulation can we start to understand the underlying mechanisms. Chapter 7 (Liang Hong and Angelo Cacciuto) shows a perfect example of how experiment inspired simulation and simulation in turn helped understand experimental data. In that chapter, simulation actually further discovered new, previously unobserved interesting structures, based on the experimental feedback.

What we have learned is that some principles about assembly are indeed shared by both Janus particles and small molecules. Amphiphilic particles form clusters akin to micelles. However, there are also unique features which only exist in particle systems, such that the clusters adopt a specific geometry with a certain number of particles in the dilute regime. When the particle concentration increases, the clusters grow into long chains, with unique kinetics and stacking structures.[9]

Going beyond self-assembly, structures with different hierarchy can be forged when Janus particles are susceptible to the external field. Chapter 8 (Ilona Kretzschmar, Sumit Gangwal, Amar B. Pawar and Orlin D. Velev) discusses the details of how Janus particles respond and assemble under electric, magnetic and convective flow fields.

What else can we do with Janus particles besides assembly? Many hints can be taken from what is already known about small surfactant molecules and homogeneous colloids. Janus particles should stabilize emulsions as surfactant molecules can do. It is known that homogeneous particles can also stabilize emulsions with good stability. Theoretical calculation shows that the advantage of using Janus particles may show up when the particle size is ~ 10 nm, where the adsorption energy for homogeneous particles is close to the thermal fluctuation.[10] Chapter 11 (Shan Jiang and Steve Granick) presents detailed discussions of this. Chapter 1 also shows one example where Janus particles were used for emulsion polymerization and led to a very homogeneous particle size distribution, which cannot be achieved by a conventional surfactant-stabilized emulsion system.

Janus particles can assist catalysis, as demonstrated in Chapter 2. Gold–oxide hcterodimer crystals presented high catalytic activity over the traditional catalyst. The effect is due to electron transfer from the oxide to the Fermi level of the supported Au nanoparticles.[11] Janus particles can also assist to deliver drugs. Different drugs can be loaded together on to one particle and be released simultaneously to achieve synergistic effects. Imaging modality can also be combined with a drug to serve as a theranostics (therapeutics and diagnostics) agent. However, careful investigation is definitely needed to show the true advantage of using Janus particles over simply mixing things together. Chapter 3 and Chapter 12 (Zhiyong Poon and Paula Hammond) present some nice examples of how Janus particles have been applied for biomedication purposes.

4   Perspectives

As shown in Figure 3, the majority of current research on Janus particles has been focused on synthesis (statistics based on publications through 2011). At the same time, it is nice to see researchers from different backgrounds contributing together to the field. With many methods of making Janus at hand, maybe it is time to think more beyond synthesis. A balanced development from theory, simulation and experiment would be the optimal approach.

Many of the details about Janus particles are not yet clear. What is their phase diagram? Will they form different crystal phases? How do they arrange themselves at an oil/water interface? How do Janus particles interact with cells and proteins?

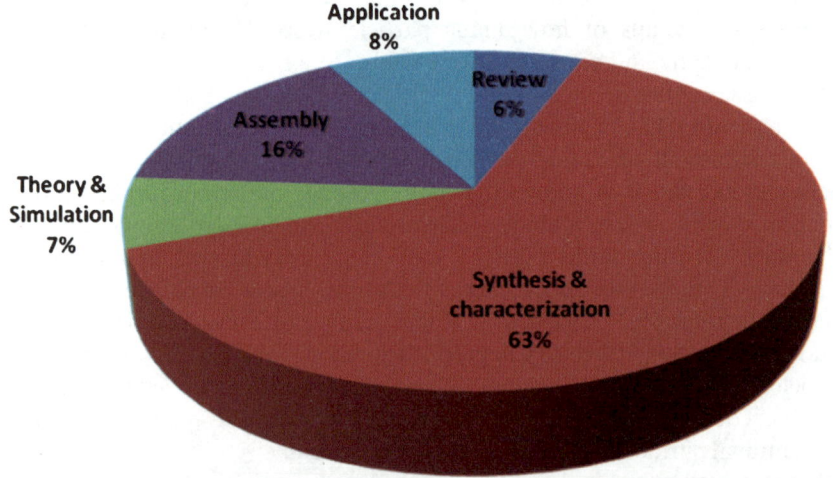

Figure 3   Statistics analysis of research topic distributed on Janus particles, based on the publications shown in Figure 2.

Again seeking analogy from what already happens in biology, is it possible to manufacture Janus particles like DNA molecules to store information or like proteins and enzymes to create lock-and-key structures, with more than just one patch on a single particle? How about Janus particles with different shapes? How about going beyond the simple Janus geometry and creating a triblock structure with patches on the two ends of the particle or at any location on the particle as we wish? Indeed, beautiful kagome lattice structures were created with the simple extension from two faces to three ('triblock').[12] The untapped possibilities with Janus and Janus-like particles are vast.

Wouldn't it be wonderful to have environmentally responsive Janus particles? Imagine if an assembled nanoparticle loaded with DNA, siRNA or drug molecules disassembles itself once the nanoparticle reaches inside a cell or a tumor site, in response to local change of pH or the presence of biomarkers. This may help kill cancer or cure diseases. Or imagine if a Janus particle can controllably stabilize and destabilize emulsions. This might help save the expense in harvesting oil from the deep sea.

Janus particles may not be the elixir to solve all these problems; however, the concept indeed offers an attractive new approach to solve many hard-to-tackle challenges. Janus particles may lead us to a whole new era of colloidal science research.

Acknowledgements

We would like to thank all of the contributors for their time and effort. We are also very grateful to the staff members at the RSC, especially Leanne Marle and Alice Toby-Brant, for their patient guidance through the production process.

Shan Jiang and Steve Granick
MIT and UIUC, April 2012

References

1. C. Casagrande and M. Veyssie, *C. R. Acad. Sci., Ser. Ii*, 1988, **306**, 1423.
2. P.-G. de Gennes, *Angew. Chem. Int. Ed. Engl.*, 1992, **31**, 842.
3. L. Petit, J. P. Manaud, C. Mingotaud, S. Ravaine and E. Duguet, *Mater. Lett.*, 2001, **51**, 478.
4. Z. N. Bao, L. Chen, M. Weldon, E. Chandross, O. Cherniavskaya, Y. Dai and J. B. H. Tok, *Chem. Mater.*, 2002, **14**, 24.
5. L. Hong, S. Jiang and S. Granick, *Langmuir*, 2006, **22**, 9495.
6. W. L. Miller and A. Cacciuto, *Phys. Rev. E*, 2009, **80**, 021404.
7. Y. C. Chen, V. Dimonie and M. S. Elaasser, *J. Appl. Polym. Sci.*, 1991, **42**, 1049.

8. M. Giersig, T. Ung, L. M. Liz-Marzan and P. Mulvaney, *Adv. Mater.*, 1997, **9**, 570.
9. Q. Chen, J. K. Whitmer, S. Jiang, S. C. Bae, E. Luijten and S. Granick, *Science*, 2011, **331**, 199.
10. B. P. Binks and P. D. I. Fletcher, *Langmuir*, 2001, **17**, 4708.
11. L. M. Molina and B. Hammer, *Phys. Rev. Lett.*, 2003, **90**, 206102.
12. Q. Chen, S. C. Bae and S. Granick, *Nature*, 2011, **469**, 381.

Contents

Chapter 1 **Soft, Nanoscale Janus Particles by Macromolecular Engineering and Molecular Assembly** **1**
Andreas Walther and Axel H. E. Müller

    1.1 Introduction 1
    1.2 Janus Particles *via* Direct Macromolecular Engineering 3
    1.3 Janus Particles *via* Direct Self-assembly and Transformations in Solution 9
    1.4 Janus Particles via Transformation of Self-assembled Polymer Bulk Structures 15
    1.5 Self-assembly Properties of Polymer based Janus Particles of Different Dimensionality 19
    1.6 Application as Structured Particulate Surfactants 23
    1.7 Summary and Outlook 25
    References 26

Chapter 2 **Design, Synthesis and Applications of Dumbbell-like Nanoparticles** **29**
Chao Wang and Chenjie Xu

    2.1 Introduction 29
    2.2 Synthesis 30
        2.2.1 DBNPs Containing Noble Metal and Transition Metal Oxide NPs 32
        2.2.2 DBNPs Containing Semiconductor NPs 36
        2.2.3 DBNPs Containing More Than Two Particles 38
    2.3 Functional Applications of DBNPs 40
        2.3.1 DBNPs as Heterogeneous Catalysts 40

RSC Smart Materials No. 1
Janus Particle Synthesis, Self-Assembly and Applications
Edited by Shan Jiang and Steve Granick
© The Royal Society of Chemistry 2012
Published by the Royal Society of Chemistry, www.rsc.org

　　　　2.3.2　DBNPs as a Multifunctional Platform for
　　　　　　　Biomedical Applications   41
　　2.4　Conclusion and Future Directions   48
　　References   50

Chapter 3　Janus Particles with Distinct Compartments *via*
Electrohydrodynamic Co-jetting   **54**
Tae-Hong Park and Joerg Lahann

　　3.1　Introduction   54
　　3.2　Compartmentalization of Nano- and Microparticles *via*
　　　　　Electrohydrodynamic Co-jetting   55
　　3.3　Microsectioning of Compartmentalized Fibers   63
　　3.4　Hybrid Janus Particles   64
　　3.5　Selective Surface Modification and Directional Self-
　　　　　assembly   66
　　3.6　Summary and Outlook   68
　　Acknowledgements   71
　　References   71

Chapter 4　Synthesis of Janus Particles by Emulsion-based Methods   74
Chengliang Zhang, Wei Wei, Fuxin Liang and
Zhenzhong Yang

　　4.1　Introduction   74
　　4.2　Synthesis at a Pickering Emulsion Interface   75
　　4.3　Synthesis in a Liquid Droplet   79
　　4.4　Synthesis upon Preformed Particles   83
　　4.5　Summary and Outlook   87
　　References   88

Chapter 5　Particle Replication in Non-wetting Templates: a Platform for
Engineering Shape- and Size-specific Janus Particles   **90**
Joseph M. DeSimone, Jie-Yu Wang and Yapei Wang

　　5.1　Introduction   90
　　5.2　PRINT Technique   91
　　5.3　Janus Particles Fabricated by the PRINT Technique   93
　　　　5.3.1　Stepwise Vertical Mold Filling   93
　　　　5.3.2　Horizontal Stepwise Mold Filling   95
　　5.4　Patchy PRINT Particles   100
　　　　5.4.1　Surface-modified Particles by Chemical
　　　　　　　Grafting   100

		5.4.2	Surface-functionalized Particles by Metal Deposition	101
	5.5		Self-assembly of Janus PRINT Particles	102
	5.6		Conclusion and Future Perspectives	105
			Acknowledgements	106
			References	106

Chapter 6 **Theoretical Calculations of Phase Diagrams and Self-assembly in Patchy Colloids** **108**
Achille Giacometti, Flavio Romano and Francesco Sciortino

	6.1	Introduction	108
	6.2	The Kern–Frenkel Model	110
	6.3	The Tools of Statistical Physics	112
	6.4	Monte Carlo Simulations	112
		6.4.1 Canonical *NVT* and *NPT* Methods	113
		6.4.2 Gibbs Ensemble Method	114
		6.4.3 Grand-canonical Ensemble μVT	114
		6.4.4 Fluid–Solid Coexistence: Thermodynamic Integration	115
	6.5	Integral Equation Theories	115
		6.5.1 General Scheme	115
		6.5.2 Iterative Procedure	117
		6.5.3 Thermodynamics	120
	6.6	Barker–Henderson Perturbation Theory	121
	6.7	Calculation of the Fluid–Fluid Coexistence Curves for the Integral Equation and Perturbation Theory	124
	6.8	Results	124
		6.8.1 Fluid–Fluid Coexistence Curves from the RHNC Integral Equation	124
		6.8.2 The Janus Limit	127
		6.8.3 One *Versus* Two Patches	129
		6.8.4 Evaluation of the Fluid–Fluid Coexistence Curves from Thermodynamic Perturbation Theory	129
		6.8.5 Fluid–Solid Coexistence	131
		6.8.6 Self-assembly in a Predefined Kagome Lattice	132
	6.9	Conclusions and Future Perspectives	134
		Acknowledgements	135
		References	135

Chapter 7   Self-assembly of Amphiphilic and Dipolar Janus Particles   138
              Liang Hong and Angelo Cacciuto

              7.1   Introduction   138
              7.2   Numerical Methods: Modeling of Janus Particles   141
                    7.2.1   Dipolar Janus Particles   141
                    7.2.2   Amphiphilic Janus Particles   145
              7.3   Experimental Methods   147
                    7.3.1   Dipolar Janus Particles   147
                    7.3.2   Amphiphilic Janus Particles   148
              7.4   Experiments on and Simulations of Janus Self-
                    assembly   149
                    7.4.1   Dipolar Janus Particles   150
                    7.4.2   Amphiphilic Janus Particles   154
              7.5   Off-balance Amphiphilic Janus Particles   161
              7.6   Conclusion   164
              Acknowledgements   166
              References   166

Chapter 8   Self-assembly of Janus Particles Under External Fields   168
              Ilona Kretzschmar, Sumit Gangwal, Amar B. Pawar and
              Orlin D. Velev

              8.1   Introduction   168
                    8.1.1   Convective Flow and Uniaxial Electric/
                            Magnetic Fields   169
                    8.1.2   Biaxial Combinations of Electric and Magnetic
                            Fields   174
              8.2   Janus Particle Preparation and Cell Set-up for Field
                    Assembly   176
                    8.2.1   Materials   176
                    8.2.2   Janus Particle Preparation   177
                    8.2.3   Assembly Cells for Field Assembly   178
              8.3   Field Assembly of Janus Particles   178
                    8.3.1   Janus Particles in Convective Flow Fields   179
                    8.3.2   Janus Particles in Electric Fields   183
                    8.3.3   Janus Particles in Magnetic Fields   191
                    8.3.4   Janus Particles in Biaxial Fields   195
              8.4   Future Outlook   197
              Acknowledgements   199
              References   199

Chapter 9   **DNA Self-Assembly: From Nanostructures to Macro Engineering 204**
Yi Chen, Abigail K. R. Lytton-Jean and Hyukjin Lee

9.1   Introduction   204
9.2   DNA Nanomachines   206
    9.2.1   Conformational Changes Induced by
           Environmental Changes   206
    9.2.2   Motions Fueled by Strand Displacement   208
    9.2.3   Autonomous Motion Powered by Enzymatic
           Activity   210
9.3   DNA-enabled Self-assembly of Inorganic/Organic
    Nanoparticles   211
    9.3.1   Properties of DNA-modified Gold
           Nanoparticles   212
    9.3.2   Directed Self-assembly of DNA-modified Gold
           Nanoparticles   214
9.4   Micro- to Macro-engineering by Self-assembly   217
9.5   Conclusion   220
References   220

Chapter 10   **Janus Particle Localization and Tracking for Studies of
Particle Dynamics**   **223**
Stephen M. Anthony and Minsu Kim

10.1   Introduction   223
10.2   Isolated Particle Localization   225
    10.2.1   Spatial Localization   225
    10.2.2   Angular Localization   227
    10.2.3   Experimental Validation   230
10.3   Optically Overlapping Particle Localization   232
    10.3.1   Image Preprocessing   235
    10.3.2   Overlapping Object Recognition   236
    10.3.3   Separation of Overlapping Janus Spheres   236
    10.3.4   Refining the Position and Extracting the
            Orientation   239
10.4   Probing Translational and Rotational Dynamics   239
10.5   Conclusion   242
Acknowledgements   242
References   242

Chapter 11   Janus Balance and Emulsions Stabilized by Janus Particles   244
Shan Jiang and Steve Granick

11.1   Introduction   244
11.2   Janus Particles at a Planar Interface   246
11.3   Janus Balance   247
    11.3.1   Contact Angle of Janus Particles at an
              Interface   248
    11.3.2   Adsorption Energy   249
    11.3.3   Quantification of Janus Balance   250
    11.3.4   An Example   251
    11.3.5   Outlook and Potential Implications   252
11.4   Emulsions Stabilized by Janus Particles   253
    11.4.1   An Example   253
    11.4.2   Other Progress   255
Acknowledgement   255
References   256

**Chapter 12   Applications of Janus and Anisotropic Particles for Drug
Delivery   257**
Zhiyong Poon and Paula T. Hammond

12.1   Overview   257
12.2   Nanoparticle Design to Overcome Barriers to Drug
       Delivery   258
12.3   Examples of Nanoparticle Systems: Liposomes,
       Micelles and Dendrimers   260
12.4   Anisotropic, Patchy and Janus Particles in Systemic
       Drug Delivery   262
    12.4.1   Particles with Anisotropic, Janus or Patchy
              Surfaces   262
    12.4.2   Interior Particle Compartmentalization   267
    12.4.3   Anisotropic Geometries   267
12.5   Conclusion   271
References   272

CHAPTER 1

Soft, Nanoscale Janus Particles by Macromolecular Engineering and Molecular Self-assembly

ANDREAS WALTHER*[a] AND AXEL H. E. MÜLLER*[b]

[a] DWI at RWTH Aachen University, 52056 Aachen, Germany;
[b] Macromolecular Chemistry II, University of Bayreuth, 95444 Bayreuth, Germany
*E-mail: walther@dwi.rwth-aachen.de or axel.mueller@uni-bayreuth.de

1.1 Introduction

Macromolecular engineering has evolved into a powerful toolbox for the preparation of complex polymer topologies with remarkable control over both the architecture and the distribution of monomer sequences into, *e.g.*, block-type structures or well-defined branched macromolecules. The rapid advances in controlled/living polymerization techniques during the last two decades have greatly facilitated this development. In the context of Janus particles, macromolecular engineering is interesting not only for the direct synthesis of phase-segregated unimolecular objects, but also for harnessing the self-assembly capabilities of tailor-made polymers owing to mutually incompatible polymer blocks, solvophobic effects or specific molecular interactions. Indeed, self-assembly of block copolymers has proven to be a remarkably elegant strategy to generate polymer-based nano-objects, where we have seen progress to increasingly sophisticated soft nanoparticles, from simple diblock copolymer micelles and vesicles, to multicompartment micelles (MCMs) with increasingly

RSC Smart Materials No. 1
Janus Particle Synthesis, Self-Assembly and Applications
Edited by Shan Jiang and Steve Granick
© Royal Society of Chemistry 2012
Published by the Royal Society of Chemistry, www.rsc.org

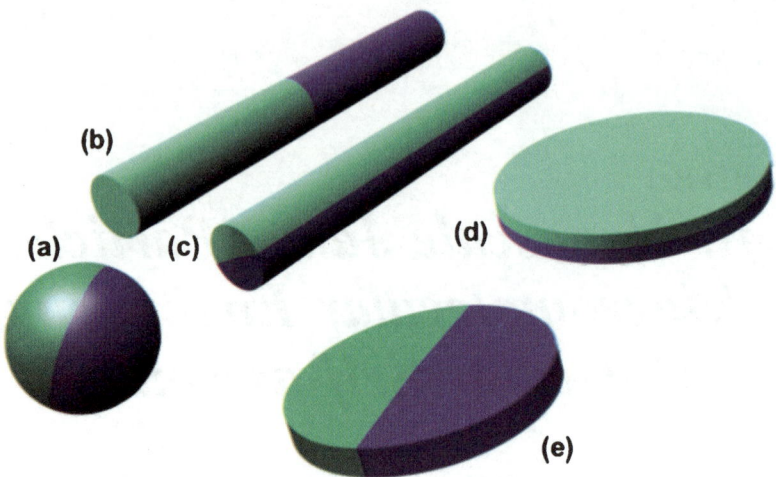

Figure 1.1   Different Janus particle topologies. Architectures (a)–(d) have so far been
realized by synthetic and self-assembly approaches for nanoscale
polymeric Janus particles.

complex geometries.[1–4] Still, directly breaking the symmetry into biphasic Janus
(Figure 1.1) particles or micelles has remained a considerable challenge.

Polymer-based Janus particles formed by direct synthesis or self-assembly of
block copolymers are unique among this class of non-centrosymmetric
colloids. First, truly nanoscale dimensions (i.e. <100 nm) can be approached
that are very difficult to tackle by, *e.g.*, common desymmetrization reactions at
interfaces or phase separation processes in emulsions, microfluidics or
electrohydrodynamic jetting. Second, smart polymer segments, able to respond
to environmental changes by phase transitions, impart a large-scale respon-
siveness to trigger superstructure formation or create strongly amphiphilic
particles relevant for surface nanostructuring and the stabilization of
interfaces. These properties render them a key building block for switchable
materials. As a third criterion, polymers are also the crucial soft materials to
communicate with the environment and to mediate interactions with cells,
proteins and other living matter when approaching the biological interface
with synthetic materials. Consequently, they are a valuable material class in the
multitude of Janus particles available nowadays.

In this chapter, we review and discuss recent developments towards
polymeric Janus particles. We place an emphasis on strategies specifically
involving advanced polymer synthesis to create unimolecular objects and on
methodologies utilizing self-assembly as well as post-transformations of self-
assembled structures to create biphasic particles. Thereby, we focus on the
small size regime and discuss particle architectures with different dimension-
alities, in which at least one dimension is truly nanoscale (*i.e.* <100 nm). It may
be noted that there are other approaches towards polymer-based Janus
particles on the (sub)micron scale, such as phase separation in emulsion

droplets, lithographic approaches in microfluidic channels and electrohydro-dynamic co-jetting, which are, however, beyond the focus of this contribution and are discussed in other chapters. This chapter is grouped into four topics. The first three are (a) Janus particles *via* direct macromolecular engineering, (b) Janus particles *via* direct self-assembly and/or transformations in solution and (c) Janus particles *via* transformation of self-assembled triblock terpolymer bulk structures. We finally discuss (d) self-assembly properties of the synthesized Janus particles and highlight some potential applications that have already been realized.

1.2 Janus Particles *via* Direct Macromolecular Engineering

The rapid advances in synthetic tools available to polymer chemists have triggered significant interest in the preparation of Janus particles. One of the earliest strategies involved the attachment or growth of different polymer chains to/from a single focal point or to/from a focal line with the aim of preparing spherical or cylindrical Janus particles, also known as heterografted star-shaped and cylindrical brush polymers. The resulting structures are outlined in Figure 1.2, which also highlights one of the major challenges for such nanoscale objects with high dynamics of the polymer chains. Phase separation of the chemically different polymer arms is required to realize a true Janus particle character in solution. However, phase separation for polymer arms emanating from a single focal point or from a dynamic micellar core – as will be discussed later – is not self-evident. It is governed by the interplay between entropy, favoring mixing of the polymer chains, and the enthalpic force of polymer chains to phase separate.[5,6] In solution, the latter is drastically

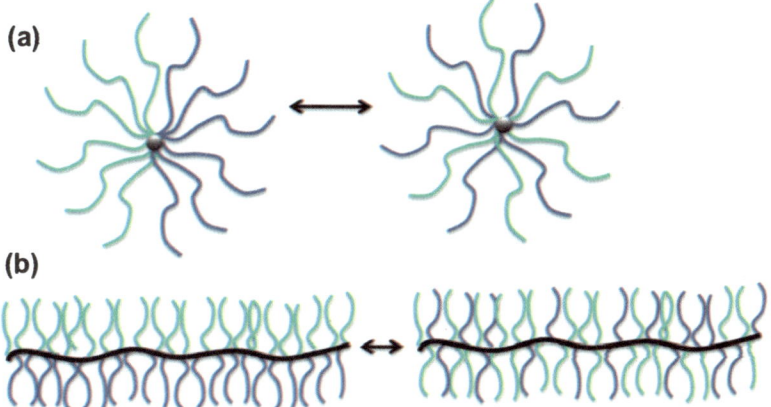

(a)

(b)

Figure 1.2   The interplay between entropy-favored chain mixing and polymer incompatibility-promoted phase separation: mixed and phase-segregated heteroarm grafted star-shaped (a) and cylindrical brush (b) polymers.

reduced compared with the bulk state and it has proven a challenging task to design systems that allow a freely occurring phase separation. In this context, it is also important to point to another major obstacle, namely the difficulty of providing solid *in situ* proof for corona segregation of polymer nano-objects in solution. The nanoscale dimensions and the often weak natural contrast of different organic parts for imaging are the main complications. This challenge can be best met by 2D ^{1}H–^{1}H NOESY NMR (NOE = nuclear Overhauser effect), an NMR technique probing intermolecular distances *via* through-space coupling, or by direct (cryogenic) transmission electron microscope (TEM) imaging using suitable staining methods to augment the natural contrast.

Some of the first evidence that a phase separation in heteroarm star polymers can indeed occur was delivered by Kiriy *et al.*[7] In their atomic force microscopy (AFM) investigations on a system of heteroarm star polymers composed of seven arms of polystyrene (PS) and seven arms of poly(2-vinylpyridine) (PS_7–$P2VP_7$), it was observed that different topologies of the molecules result upon deposition from different solvents on to mica (Figure 1.3). Chloroform ($CHCl_3$) led to a hat-shaped appearance, whereas tetrahydrofuran (THF) yielded a more globular shape. The dissimilar shapes were attributed to the adsorption of Janus-type conformations in the case of $CHCl_3$, whereas a mixed conformation was suggested for the molecules deposited from THF. These observations were supported by calculations of the solubility parameters, which confirmed that $CHCl_3$ is a more selective solvent for P2VP and thus can enforce intramolecular phase segregation and a Janus-type conformation in solution. One uncertainty related to the imaging of deposited molecules of dynamic species always lies in the unclear effect of the surface properties on the adsorption behavior, *i.e.* selective adsorption due to preferred adhesion – a problem hard to come by with *ex situ* techniques.

An improved focal point design was suggested by Ge *et al.*, who reported the synthesis and stimuli-responsive self-assembly of double-hydrophilic Janus-type A_7B_{14} heteroarm star copolymers with two types of water-soluble polymer arms, poly(*N*-isopropylacrylamide) (PNIPAAm) and poly(2-(diethylamino)ethyl methacrylate) (PDEAMA), emanating from the two opposing sides of a rigid toroidal β-cyclodextrin (β-CD) core.[8] Owing to the pre-encoded phase separation within the focal point, an enhanced tendency for phase separation of the two arms can be expected. The authors found an interesting schizophrenic self-assembly behavior. Depending on the conditions for triggering the solubility to insolubility phase transitions of the two polymer arms, which are high temperature for PNIPAAm and high pH for PDEAMA, it was possible to switch between two vesicle states by inverting the membrane structure. Such a vesicle inversion procedure is a highly unlikely scenario for simple coil–coil diblock copolymers and can serve as indirect evidence for the Janus character of these stars.

Zhu and co-workers described a facile and large-scale synthesis of possibly the smallest unimolecular Janus nanoparticles by intramolecular crosslinking of the inner P2VP block of a polystyrene-*block*-poly(2-vinylpyridine)-*block*-

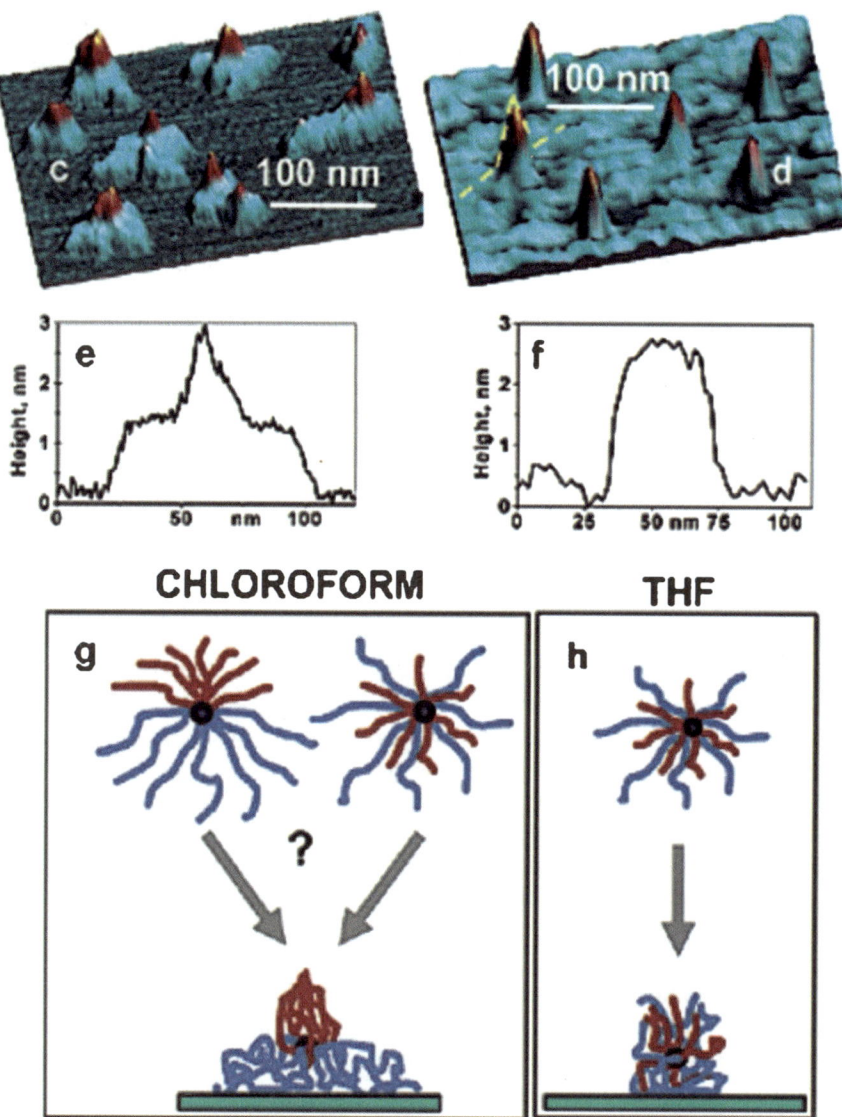

Figure 1.3 Solvent-induced transition from a mixed to a phase-separated structure in heteroarm grafted star polymers. Adapted with permission from *Macromolecules*, 2003, **36**, 8704.[7] Copyright 2003 American Chemical Society.

poly(ethylene oxide) (PS-*b*-P2VP-*b*-PEO or SVEO) triblock terpolymer (Figure 1.4).[9] Simple addition of an α,ω-dibromoalkane in a common solvent, dimethylformamide (DMF), resulted in nanoscale Janus particles with exactly one polymer arm of each end block attached to the central core. Interestingly, these Janus particles showed a concentration-dependent self-assembly beha-

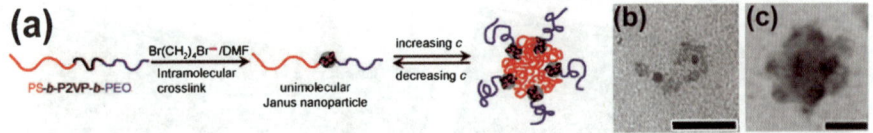

Figure 1.4   Ultrasmall Janus particles *via* intramolecular crosslinking of a polystyrene-*block*-poly(2-vinylpyridine)-*block*-poly(ethylene oxide) triblock terpolymer using a difunctional bromoalkane. TEM images of individual Janus nanoparticles and self-assembled aggregates resembling raspberry/football-shaped multicompartment micelles (scale bars = 50 nm). Reprinted and adapted with permission from *Macromolecules*, 2008, **41**, 8159.[9] Copyright 2008 American Chemical Society.

vior into supermicelles upon reaching a critical aggregation concentration, *cac*, of ∼2 mg mL^{-1}. Strikingly, this aggregation also took place in a good solvent (DMF) for both end blocks, PS and PEO, which is a first example of the unusual and intuitively unexpected self-assembly behavior of polymeric Janus particles in good solvents for both corona hemispheres.

Moving from these 3D systems with overall spherical character to 2D systems with cylindrical architectures, one can identify various synthetic efforts targeting different types of copolymer bottle brushes. Various groups have reported cylindrical copolymer brushes of poly(2-(dimethylamino)ethyl methacrylate) (PDMAEMA) and poly(ε-caprolactone) (PCL), PS and polylactide (PLA) or PS and PEO with a statistical distribution of side-chains along the backbone (Figure 1.2b).[10–12] Although these molecules showed clustering into some irregular aggregates upon exposure to selective solvents, there are no conclusive data and discussion on the potential Janus character of such structures. Simulations by de Jong and ten Brinke demonstrated that complete phase separation may only occur at very high incompatibilities of the two grafted polymers (similar to hetero-arm star-shaped polymers), as expressed by a large Flory–Huggins parameter, χ.[13] It was further suggested that a well-defined Janus cylinder may only be reached at theta conditions for both blocks and for rigid backbones (Figure 1.5). For good solvents and highly flexible backbones, common to most synthetic comb-shaped polymers, the extent of phase separation is reduced and the molecules undergo bending into different shapes.

Schmidt and co-workers found differently bent shapes for cylindrical brushes composed of P2VP and poly(methyl methacrylate) (PMMA), when imaged with AFM after deposition from different solvents.[14] After quaternization of parts of the P2VP segments and deposition from CHCl$_3$ and H$_2$O, strongly bent horseshoe and multiply bent, spiral/meander-type conformations were observed. The strong bending within the horseshoe structures was attributed to a basically quantitative phase separation along the main axis, whereas a patchy structure was ascribed to the meander-type patterns. Ishizu and co-workers reported imaging data on the side-by-side aggregation of cylindrical brushes obtained by polymerization of PS and PEO macromonomers during a very slow evaporation process of 1 week, starting from a THF–

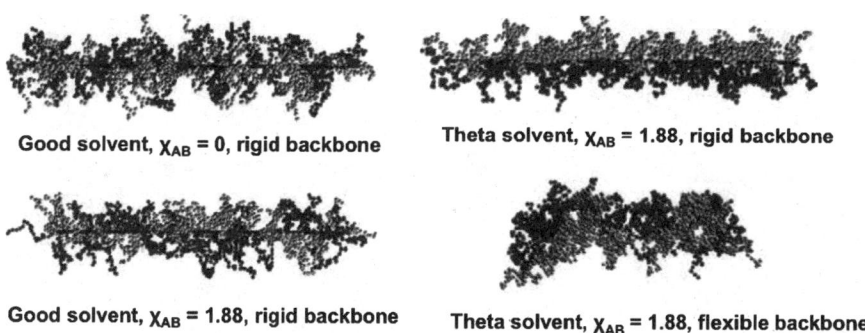

Good solvent, χ_{AB} = 0, rigid backbone

Theta solvent, χ_{AB} = 1.88, rigid backbone

Good solvent, χ_{AB} = 1.88, rigid backbone

Theta solvent, χ_{AB} = 1.88, flexible backbone

Figure 1.5 Snapshots of Monte Carlo simulations of hetero-grafted cylindrical copolymer brushes for high and low incompatibility of the side-chains, different solvent conditions and flexible *versus* rigid backbones. Reprinted and adapted with permission from *Macromol. Theory Simul.*, 2004, **13**, 318.[13] Copyright 2004 Wiley-VCH.

water solution.[15] These observations indicated that a reorientation can occur and that a biphasic character can develop if the correct solvent conditions (selectivity to drive self-assembly), concentration regime, and time scale are provided. These experimental systems and simulations, however, point to some limitations of this strategy when aiming at robustly phase-segregated two-dimensional cylindrical Janus brushes with corona segregation along the main axis. Further below, we will describe how to create similar Janus cylinders with very high precision using the controlled crosslinking of triblock terpolymer bulk phases.

Advances in polymer synthesis, however, have allowed the synthesis of a different type of cylindrical Janus particles. These are characterized by a phase separation perpendicular to the main axis into a cylindrical AB-type diblock brush.[16–22] Such structures are typically obtained by consecutive block polymerization of macromonomers or by the polymerization of AB diblock copolymers with orthogonally reactive moieties in both blocks permitting side-by-side polymer-analogous attachment of preformed polymer chains (grafting to) or the growth of polymers *via* (parallel) grafting from reactions. A variety of AB-type Janus cylinders have been reported with widely different physical properties. For instance, Rzyaev *et al.* described in great depth the synthesis and structure formation of PLA$_{comb}$-*b*-PS$_{comb}$ AB-type Janus brushes with total molecular weights exceeding 1 MDa.[21,22] Since the dimensions of the resulting lamellar bulk phases of the stiff AB cylinders approached the wavelength of visible light, a photonic bandgap behavior and an opalescent appearance could be observed for solid samples. Direct visualization by AFM was reported by Matyjaszwewski and Sheiko and co-workers for their poly(*n*-butyl acrylate)$_{comb}$-*block*-PCL$_{comb}$ (PnBA$_{comb}$-*b*-PCL$_{comb}$) AB diblock brushes (Figure 1.6) with amorphous and crystalline side-chains.[23]

In addition, Deffieux *et al.* employed a multi-step reaction scheme to create PS$_{comb}$-*block*-polyisoprene$_{comb}$ (PS$_{comb}$-*b*-PI$_{comb}$) with glassy and liquid-like

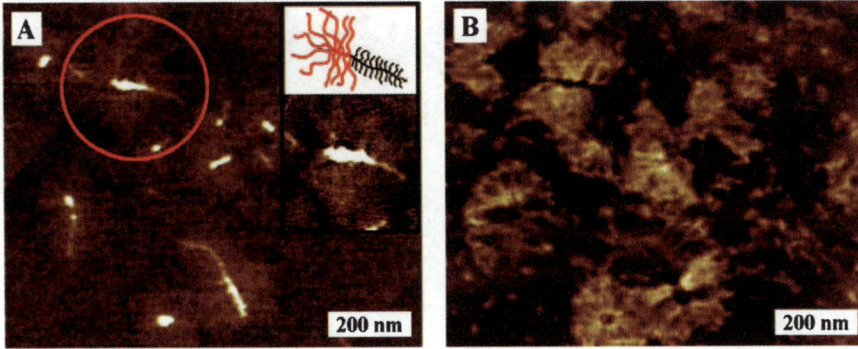

Figure 1.6   AFM images of heteroarm-grafted AB diblock Janus brushes with PnBA
and PCL side-chains. (a) The height image displays single bottle brushes
with a brighter PnBA head and a less distinct PCL tail. Verification of the
PnBA and PCL assignment can be seen in the corresponding phase image
(b). Reprinted with permission from *Macromolecules*, 2008, **41**, 6073.[23]
Copyright 2008 American Chemical Society.

side-chains, thus further broadening the property range.[16] Exposure to selective
solvents such as heptane induced self-assembly into homogeneous and nearly
spherical micelles. Similar self-assembly behavior was also reported for AB
diblock brushes composed of PS and poly(acrylic acid) (PAA) side-chains.[24]

In addition to the stepwise synthesis of unimolecular polymer objects *via*
classical polymer chemistry, the desymmetrization of particles at interfaces has
also reached the field of polymer-based nanoscale Janus particles. In general,
toposelective modifications of immobilized particles are widely used to break
the symmetry of inorganic particles and have greatly impacted the synthesis of
Janus objects.

In the context of soft nano-objects, Chen and co-workers reported a simple
one-pot process, in which they used PEO-*b*-P4VP-stabilized yttrium hydroxide
nanotubes (YNTs, diameter $\sim$200 nm and length 3–4 µm) as the interface for
symmetry breaking. After initial formation of the polymer-coated YNT rods due
to the adsorption of P4VP segments on the YNT surfaces *via* hydrogen bonding,
a mixture of a radical initiator [azobisisobutyronitrile (AIBN)] in divinylbenzene
(DVB) and additional NIPAAm was added to the dispersion (Figure 1.7).[25]
Because of the solubility characteristics of the various compounds, AIBN and
DVB accumulated in the P4VP phase, whereas NIPAAm remained dissolved in
the continuous phase. Subsequent heating induced polymerization of the DVB in
the confined space on the YNTs and nanoscopic, crosslinked polydivinylbenzene
(PDVB) beads were formed due to the increased incompatibility of P4VP and
PDVB polymers developing during the DVB polymerization. The radicals
reaching the outer surface also initiated the NIPAAm polymerization, which led
to the side-selective growth of PNIPAAm grafts, equaling an *in situ*
desymmetrization. The collapsed state of the PNIPAAm during the thermal
polymerization prevented the dissolution of the modified PDVB beads from the

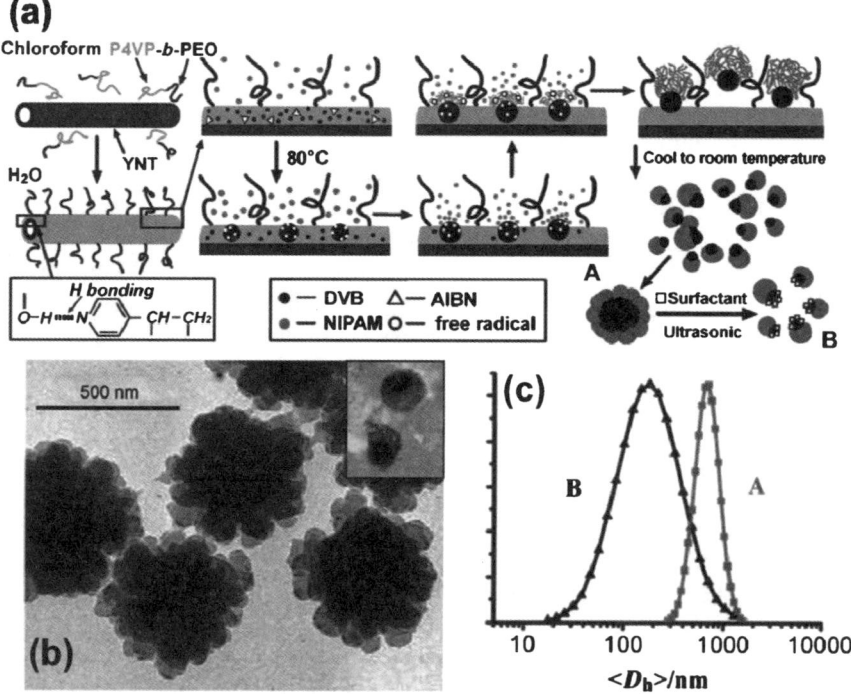

Figure 1.7 Reaction scheme illustrating the *in situ* desymmetrization of PDVB nanoparticles growing within the interfacial layer of water-dispersed polymer-coated YNTs. (b) TEM images of the supermicelles and of the individualized Janus particles (inset, same magnification). (c) DLS CONTIN plots of the supermicelles A and the individualized Janus particles after addition of surfactant. Reprinted with permission from *Angew. Chem. Int. Ed.* 2007, **46**, 6321.[25] Copyright 2007 Wiley-VCH.

YNTs. After resting at room temperature, the PNIPAAm/PDVB Janus particles separated from the polymer-coated inorganic rods. Dynamic light scattering (DLS) and TEM revealed flower-like aggregates of the strongly amphiphilic Janus particles with a hydrodynamic radius $<R_h>_z = 320$ nm, which could be dissociated by the addition of excess surfactant to yield isolated Janus particles with an average radius of 80 nm. Interestingly, the process might be suitable for different monomers and could potentially be cycled using the same polymer-coated YNTs.

1.3 Janus Particles *via* Direct Self-assembly and Transformations in Solution

Solution-based self-assembly of block copolymers into micellar aggregates has rapidly developed throughout recent decades. It originally appeared fairly straightforward to design systems capable of leading to a Janus-type conformation of micellar coronas. Systems conceived for that purpose involve ABC triblock terpolymers with an inner solvophobic block as micellar core and the A and C end blocks forming the corona. Additionally, a mixture of two diblock copolymers of the AB and BC type was expected to lead to feasible situations for Janus-type separation of the A/C corona and a common B core. Similar considerations for the phase separation of the polymer arms as discussed above for hetero-grafted polymers also apply. High incompatibility of the two A/C corona blocks and suitable solvent conditions are required (Figure 1.8a). Interestingly, however, Halperin calculated that a mixture of AB and BC will only lead to co-micellization into mixed micelles for low incompatibility of the two end blocks.[26] It was predicted that high incompatibility of both corona blocks would in fact lead to a set of two homogeneous populations of AB and CB micelles. Therefore, forced co-micellization of AB and CD diblock copolymers, carrying attractive interactions between the B and C blocks, was thought to be a viable alternative. The problem of uncertain co-micellization can be fully overcome for ABC triblock terpolymers possessing a chemical connectivity between the two corona blocks A and C.

A theoretical treatment of the corona segregation was presented by Charlaganov *et al.*, who used 2D self-consistent field theory (SCF) for an ABC triblock terpolymer system to calculate that corona phase segregation between A and C starts to occur at $\chi_{AC} \approx 0.5$ and leads to fully biphasic A/C coronas at $\chi_{AC} \approx 1$.[6] Such a high degree of incompatibility between the two corona-forming blocks is yet difficult to achieve in the case of dissolved polymer segments in solution.

A number of groups, including ourselves, devoted significant effort to understanding and eventually mastering the challenge of fully phase-separating coronas within micellar systems. For instance, Hu and Liu[27] demonstrated that introducing adenine and thymine into the PCEMA segments of two diblock copolymers, poly(*tert*-butyl acrylate)-*block*-poly{(2-cinnamoyloxyethyl methacrylate)-*ran*-[2-(1-thyminylacetoxyethyl methacrylate)]}, P*t*BA-P(CEMA-T) and PS-*block*-PCEMA-*ran*-[2-(1-adeninylacetoxyethyl methacrylate)]}, PS-P(CEMA-A), leads to enhanced mixing of both chains in micelles due to H-bonding of the nucleic acid pairs. The observation of better mixing due to secondary interactions between both solvophobic blocks points to their significant incompatibility, as predicted by theory. However, despite the existing repulsion of both corona blocks, full segregation could not be obtained and different populations of patchy multicompartment micelles (MCMs) were found. Similar results were obtained by Zheng *et al.*[28] for a system of two diblock copolymers, PCEMA-*block*-poly(glyceryl methacrylate) and PCEMA-*block*-PSGMA (PSGMA = succinated PGMA), with PCEMA forming the solvophobic core, and by Ma and co-workers[29,30] and Kim *et al.*[31] for a system using PNIPAm-*block*-P4VP and PEO-*block*-P4VP with a protonated P4VP as insoluble block or using stereo-complex formation of PNIPAAm-*block*-PLLA and PEO-*block*-PDLA, respectively.

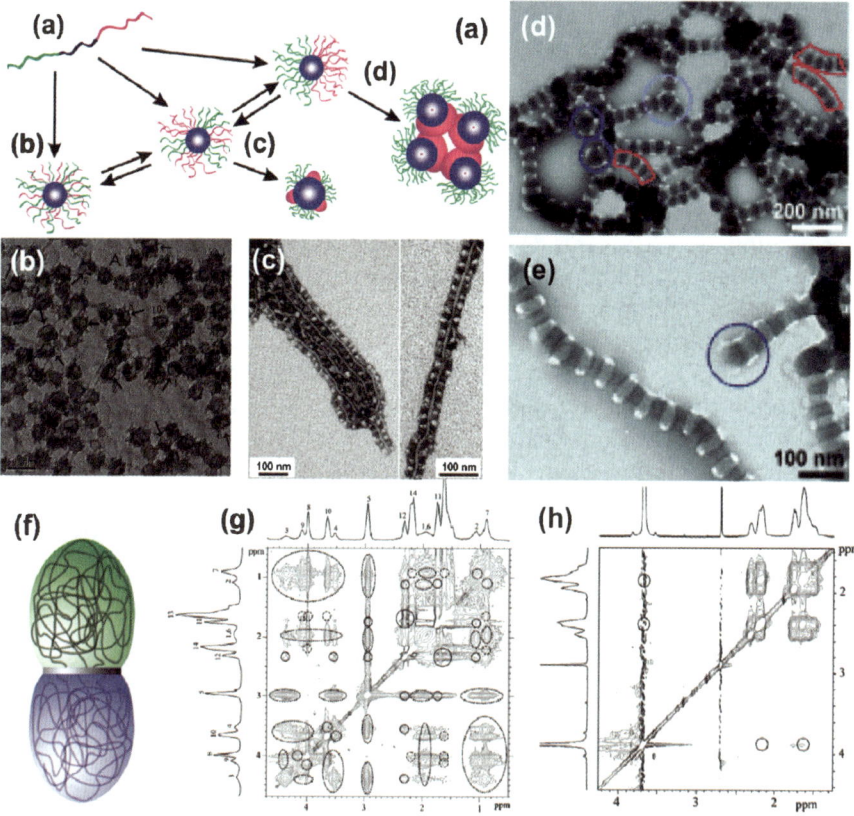

Figure 1.8 Direct self-assembly strategies for Janus micelles. (a) Possible corona configurations for triblock terpolymers with a solvophobic middle block and their superstructure formation upon triggering a solubility to insolubility transition on one of the end blocks. Reprinted with permission from *Langmuir*, 2010, **26**, 12237.[34] Copyright 2010 American Chemical Society. (b) Mixture of patchy multicompartment and Janus micelles obtained by direct dissolution of a poly(ethylene oxide)-*block*-poly(ε-caprolactone)-*block*-poly(2-aminoethyl methacrylate) (PEO-*b*-PCL-*b*-PAMA) triblock terpolymer in water, with PCL chains forming the micelle cores and the PEO and PAMA chains forming phase-segregated patchy or hemispherical coronas. The corona segregation was highlighted by selective silicification of the PAMA/PCL regions. Reprinted with permission from *Soft Matter*, 2010, **6**, 4851.[35] Copyright 2010 Royal Society of Chemistry. (c) Patchy worms formed by crystallization-driven self-assembly of a polystyrene-*block*-polyethylene-*block*-poly(methyl methacrylate) triblock terpolymer in organic media. PE forms the crystalline core and PS and PMMA phase segregate into a patchy corona. Reprinted with permission from *ACS Nano*, 2011, **5**, 9523.[37] Copyright 2011 American Chemical Society. (d, e) Superstructures formed in selective solvents (ethanol = non-solvent for P*t*BS) by stacking of patchy multicompartment and Janus micelles composed of a fluorinated polybutadiene core and a compartmentalized corona of poly(*tert*-butoxystyrene) and poly(*tert*-butyl methacrylate). Different amounts of

Fang *et al.*, in detailed TEM investigations, looked at the corona structure of micelles formed by ABC triblock terpolymers carrying a solvophobic semifluorinated middle block B and two corona blocks of poly[4-(*tert*-butoxy)styrene] (A, P*t*BS) and poly(*tert*-butyl methacrylate) (C, P*t*BMA).[32] A distribution of MCMs with few patches in the corona was found and the MCMs underwent supramicellar polymerization into extended, branched, mesoscale polymers upon exposure to a selective solvent inducing the precipitation of P*t*BS. This indicates a majority fraction of two opposing patches. Selective staining and subsequent imaging allowed the various building blocks and their patchiness to be resolved (Figure 1.8d, e).

In a refined system, Walther and co-workers investigated a range of bis-hydrophilic PEO-*b*-P*n*BA-*b*-PNIPAAm triblock terpolymers with respect to their thermo-reversible aggregation in water.[33,34] After micellization in water (PnBuA as cores), various heating cycles were used to induce the collapse of the PNIPAAm corona chains and to study its effect on the superstructure formation (Figure 1.8a). The expulsion of the solvent was used as means to raise artificially the χ-parameter in this system. Indeed, the formation of larger worm-like superstructures was seen with increasing heating cycles. This indicated a restructuring of the corona from a mixed to a phase-segregated morphology, thereby leading to larger sticky patches and enforced aggregation. A complete phase separation into Janus micelles was again not observed, as their supramicellar aggregation would have led to isolated clusters. Two to three corona patches of one block were the dominant corona configurations as judged by the cryo-TEM images obtained at high temperatures. Similar observations of patchy multicompartment micelles based on triblock terpolymers were also reported by Du and Armes for a PEO-*block*-poly(ε-caprolactone)-*block*-poly(2-aminoethyl methacrylate) triblock terpolymer in water, with PCL blocks forming the micelle cores (Figure 1.8b).[35]

Using a different driving force for self-assembly, Schmalz and co-workers reported on the thermo-reversible formation of worm-like micelles from a PS-

patches in the corona lead to linear segments, branching points and endcaps. Reprinted with permission from *Angew. Chem. Int. Ed.*, 2009, **48**, 2877.[32] Copyright 2009 Wiley-VCH. (f) Structure of Janus micelles formed by complex coacervation of two oppositely charged diblock copolymers (AB + CD). Reprinted with permission from *Soft Matter*, 2009, **5**, 999.[41] Copyright 2009 Royal Society of Chemistry. (g) 2D ^{1}H–^{1}H NOESY NMR contour plot of a 1:1 mixture of PDMAEMA45-*b*-PGMA90 and PAA42-*b*-PAAm417 in D_2O at 1 mM NaNO$_3$. Circles indicate intramolecular cross peaks within the corona blocks PAAm417 and PGMA90 (small circles, dotted lines). Subscripts denote the number-average degree of polymerization. Reprinted with permission from *Angew. Chem. Int. Ed.*, 2006, **45**, 6673.[38] Copyright 2006 Wiley-VCH. (h) 2D ^{1}H–^{1}H NOESY NMR contour plot of a 1:1 mixture of PAA42-*b*-PAAm417 and P2MVP42-*b*-PEO446 in D_2O at 1 mM NaNO$_3$. The circles indicate where cross peaks would appear in the case of close contact between PAAm and PEO chains. Significant cross peaks are absent. Reprinted with permission from *Angew. Chem. Int. Ed.*, 2006, **45**, 6673.[38] Copyright 2006 Wiley-VCH.

block-polyethylene-*block*-PMMA (SEM) triblock terpolymer with a crystallizable middle block in organic media.[36,37] TEM investigations revealed a core–corona structure for the worm-like micelles, in which the core was formed by crystalline polyethylene (PE) domains and the soluble corona exhibited a patched structure composed of microphase-separated PS and PMMA chains (Figure 1.8c), as proven by 2D ^{1}H–^{1}H NOESY NMR and TEM investigations of selectively stained samples. A fully biphasic corona could not yet be obtained.

Consequently, some generalities derive among the systems. Despite greatly different chemistries and also driving forces to generate the self-assembled, near solvent-free cores, it can be realized that success towards freely self-assembling Janus micelles is limited. This can be explained by considering that the gain in enthalpy by reducing the interface between a few patches is small compared with the accompanying loss of the entropy of the system due to full confinement and positional order of both sets of corona chains.

A breakthrough for self-assembled Janus micelles was reported by Voets and co-workers, who published very comprehensive studies on slightly anisometric Janus micelles using the forced co-assembly of two oppositely charged diblock copolymers (AB + CD) into a complex-core coacervate micelle with PEO and polyacrylamide (PAAm) as coronal blocks.[38–42] This system is different from the aforementioned approaches using solvophobic blocks or crystallization (solid cores) as it yields a hydrogel-like swollen core formed by an interpolyelectrolyte complex of B [poly(2-methylvinylpyrrolidinium iodide) (P2MVP) and C (PAA). In particular, the unique, slightly asymmetric shape, the absence of cross peaks in the 2D ^{1}H–^{1}H NOESY NMR spectra and additional self-consistent field modeling support the phase-separation into Janus micelles. Detailed NOESY NMR studies confirmed cross peaks for poly(glycidyl methacrylate) (PGMA) instead of PAAm, thus proving a mixed corona of PEO and PGMA, but showed no cross peaks for the PAAm–PEO pair, thus indicating a demixed corona (Figure 1.8f–h). The segregation was found to be fairly robust as changes in the ratio and degrees of polymerization of the corona blocks, different temperatures and salt-induced modification of the association numbers did not lead to any indications of mixed coronas. So far, this system represents the most remarkable success for freely self-assembling Janus micelles.

Although the commonly observed formation of MCMs (patchy micelles) with phase-segregated coronas and multiple patches in systems targeting Janus micelles may at first glance seem disappointing, Cheng *et al.*, for instance, demonstrated how to use those MCMs as intermediate templates to generate Janus particles by transformation reactions in solution (Figure 1.9).[43] They started with MCMs with PEO–poly(2-vinylnaphthalene) (P2VN) corona patches surrounding a non-covalently crosslinked PAA–diamine core, as formed by the co-assembly of PEO-*b*-PAA and P2VN-*b*-PAA diblock copolymers. Dialysis from DMF against water of pH 7 was used to trigger the collapse of the P2VN patches and a subsequent decrease in the pH value to 3.1 favored hydrogen bonding between PAA and PEO blocks, leading to the

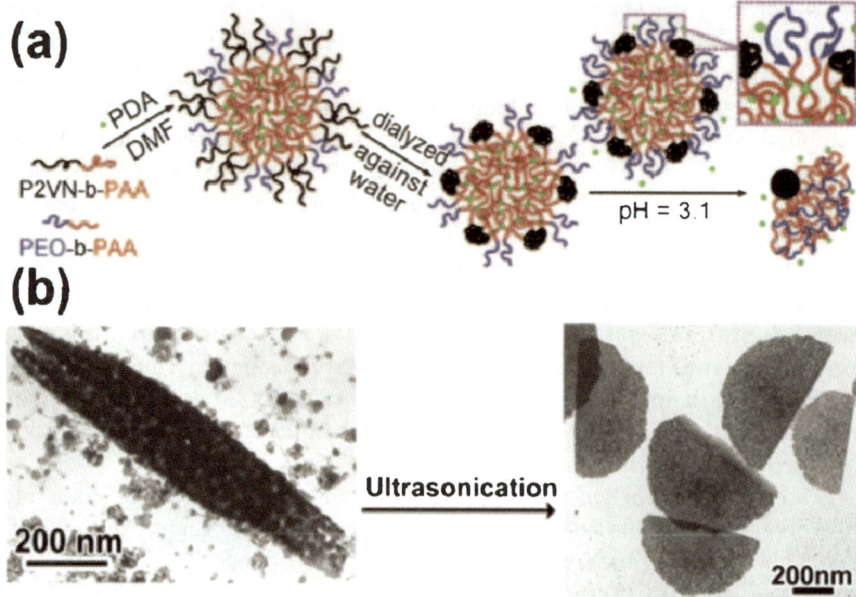

Figure 1.9   (a) Multistep reaction scheme illustrating the transformation of mixed shell micelles (patchy MCMs) into Janus particles. (b) The resulting Janus micelles assemble into tubes that unwind into trapezoids upon ultrasonication. Reprinted with permission from *Angew. Chem. Int. Ed.*, 2008, **47**, 10171.[43] Copyright 2008 Wiley-VCH.

formation of one P2VN and one PAA–PEO hemisphere. TEM images of the solutions at pH 3.1 display large submicron-sized tubular aggregates coexisting with Janus micelles. The tubular structures can unwind upon ultrasonication into trapezoidal and semicircular flat particles with similar extension. The thickness of the resulting trapezoids of ~30 nm suggests an alternating bilayer of the amphiphilic Janus particles. The complicated nature of the system using mixtures of polymers, and also various complex formations and solvent and solubility changes, do not yet allow a full picture to emerge of the unusual self-assembly processes. However, it is yet another example of the complex and often surprising self-assembly behavior of Janus particles.

Furthermore, Wooley and co-workers reported on the desymmetrization of shell-crosslinked polymer nanoparticles *via* a cyclic strategy to produce nanoscopic Janus particles that bear two kinds of clickable surface functional groups (thiol and azide).[44] The polymer nanoparticles were prepared through self-assembly of amphiphilic PAA-*b*-PS diblock copolymers in water, followed by azide functionalization and intramicellar crosslinking. The desymmetrization process involved the click conjugation of these small polymer nanoparticles with their azide functionalities on to comparably larger gold nanoparticles modified with heterotelechelic α-alkyne-ω-thiol-oligo(ethylene oxide) units. After separation of excess polymer nanoparticles, addition of more heterotelechelic ligands was used to liberate the Janus particles, now carrying both alkyne groups on the unmodified sides and thiol groups on the modified sides. The ligand exchange reactions also recovered the gold nanoparticle for further desymmetrization reactions. To confirm the anisotropic distribution of thiol groups on the resultant Janus particles, the thiol-functionalized regions were labeled with 2 nm citrate-stabilized gold nanoparticles and TEM imaging confirmed the biphasic characterization. This conjugation approach can be considered an advanced staining method to visualize compartmentalization in soft polymeric nanoparticles with weak contrast in electron microscopy.

1.4 Janus Particles *via* Transformation of Self-assembled Polymer Bulk Structures

As discussed in the previous section, it has remained very challenging to create Janus micelles *via* direct self-assembly in solution. However, prior to most of these investigations, the self-assembly of triblock terpolymers into a variety of bulk structures was used to design Janus particles.[45,46] Triblock terpolymers are a fascinating class of materials as they are able to form a wide variety of complex and highly defined microphase-segregated bulk morphologies.[47] The structure-guiding factors therein are the volume fractions of all components and the polymer incompatibilities as expressed by the mutual Flory–Huggins χ-parameters or interfacial tensions. Importantly, slight changes in the chemical compositions that lead to different interfacial tensions among the block segments can drastically influence the stability areas of certain morphologies. Therefore, not every triblock terpolymer displays the same amount of morphologies. Furthermore, slight external influences, *e.g.* the nature of the film-casting solvent, the addition of swelling solvents or crosslinking, may also trigger changes in the structure. This can, however, be turned into an advantage, as solvent annealing or film casting from selective solvents can be used to tailor the microphase-segregated structure into a specific, desirable non-equilibrium morphology.

Polystyrene-*block*-polybutadiene-*block*-poly(methyl methacrylate) (PS-*b*-PB-*b*-PMMA; SBM) is the most deeply studied triblock terpolymer system and its phase diagram is exemplarily depicted in Figure 1.10 (left). The lamellar structures were identified as suitable precursor morphologies for the fabrication of non-centrosymmetric Janus colloids. For instance, spherical Janus micelles can be prepared based on the selective crosslinking of spherical PB domains located at the interface of two lamellar domains of PS and PMMA in the so-called lamella-sphere (*ls*) morphology of SBM triblock terpolymers.[46] This methodology has been broadly and successfully developed by us and extended to the generation of Janus cylinders and Janus discs or Janus sheets (Figure 1.10, right).[48–51] Recently, Janus ribbons that form by a longitudinal connection of two Janus cylinders have also been synthesized due to the trapping of a metastable state during a phase transition.[52]

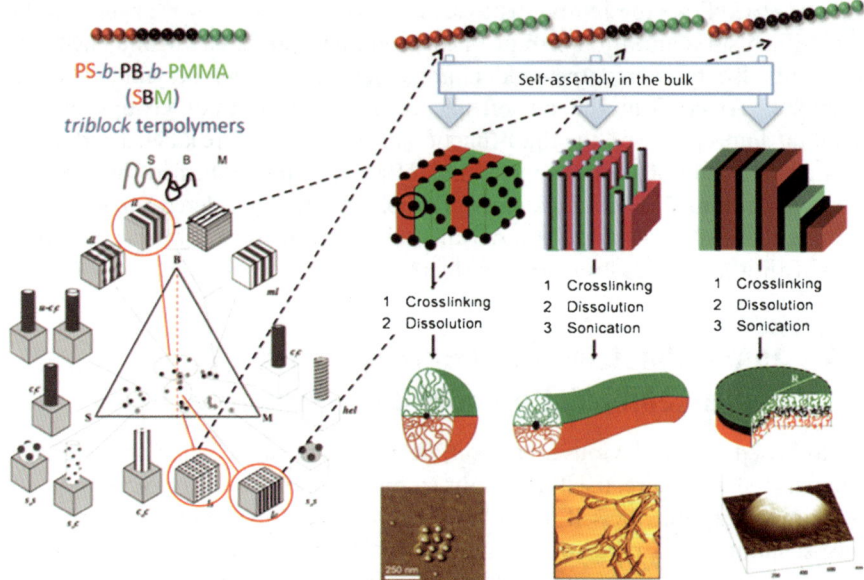

Figure 1.10 Janus particles of different architectures prepared *via* the selective cross-linking of triblock terpolymer bulk phases and subsequent dissolution. Left: phase diagram of SBM. Right: synthetic strategy for Janus micelles, Janus cylinders and Janus discs. Reprinted and adapted with permission from *Macromol. Rapid. Commun.*, 2000, **21**, 16, Copyright 2000 Wiley VCH, and *Macromolecules*, 2001, **34**, 1069,[46] *Macromolecules*, 2003, **36**, 7894[48] and *J. Am. Chem. Soc.*, 2007, **129**, 6187,[49] Copyright 2001, 2003 and 2007, American Chemical Society.

The underlying principle is to tailor the polymer structure (and also crosslinking procedures) in such a way that the geometry of the central part, B, within the microphase-segregated morphology of an ABC triblock terpolymer can be controlled in its dimensionality from spherical to cylindrical and then to a lamellar domain. This is typically achieved by increasing the weight fraction of the inner block, while keeping the weight fractions of the outer blocks symmetrical. The symmetrical volume fractions of the outer blocks maintain the overall lamellar structures, while the increase in the volume fraction of the inner block induces the phase transitions from lamella-sphere (*ls*) to lamella-cylinder (*lc*) and to fully lamellar *(ll)* morphology.

After a specific microphase-segregated morphology has been obtained, the non-centrosymmetric orientation of the terminal blocks, A and C, can be preserved by crosslinking the inner block, B. Thereafter, liberation of the core-crosslinked particle can be achieved *via* simple dissolution (in the case of spherical Janus micelles) or using an ultrasound-assisted dispersion of the crosslinked bulk films, necessary to cut down extremely long cylinders and/or large discs. A large variety of different chemical crosslinking methods, such as radical crosslinking, thermally or photo-induced dimerization or chemical crosslinking with additives, can be performed in the bulk phase. The choice of crosslinkable polymer segments is similarly wide. Some of the most frequently used include polydienes, cinnamoyl groups, polyacids/bases or gelable groups based on alkoxysilane motifs. This versatility in terms of crosslinking together with the rapid developments in controlled/living polymerization techniques over the past decade has greatly simplified the targeting of suitable polymer structures.

During the last decade, this method has been established as an extremely viable large-scale route towards the generation of nanoscopic Janus particles of different architectures. Due to the very well-defined nature of the microphase-segregated bulk structures, the resulting Janus particles also exhibit precise, near monodisperse cross-sections. Since the long period of the bulk structures depends on the overall molecular weight of the block terpolymer, the resulting dimension of the cross-section of the Janus colloid, *i.e.* radius, diameter and thickness of Janus micelles, Janus cylinder and Janus discs, respectively, can thereby be tuned. The sizes typically accessible *via* reasonable efforts in terms of accessible molecular weights of the precursor terpolymers are between 10 and 50 nm for the cross-section of the resulting colloids. Nowadays this approach allows the preparation of different Janus particles on the multigram scale (up to 100 g) with reasonable synthetic effort (a few days). The precision engineering of the particle shape, the versatility of the chemical composition and the scalability of the synthesis have proven very valuable in studying hierarchical self-assembly and potential fields of applications, as will be discussed later.

After the development of the phase diagram and the discovery of the lamella-sphere morphology in triblock terpolymers, Ishizu and co-workers[45] and Müller and co-workers[46] concurrently designed triblock terpolymer systems suitable for the preparation of Janus micelles. Ishizu and co-workers used a polystyrene-*block*-poly(2-vinylpyridine)-*block*-poly(*tert*-butyl methacrylate) ((PS-b-P2VP-PtBMA); SVT) polymer, which could be forced into a lamella-sphere bulk structure upon casting from toluene, being a poor solvent for P2VP.[45] The crosslinking of the central P2VP domains with 1,4-diiodobutane led to spherical-type Janus micelles with PS and PtBMA hemispheres. Light scattering analysis of the core-crosslinked Janus micelles revealed a weight-average of 47 polymers crosslinked within one biphasic particle. The Janus micelles exhibited an intensity-weighted hydrodynamic radius, $<R_h>_z$, of 38 nm in a good solvent (THF) for both corona hemispheres. Müller and co-workers reported detailed investigations of the synthesis and self-assembly behavior of Janus micelles based on SBM triblock terpolymers.[46] The PB center block was crosslinked by either 'cold vulcanization' using S_2Cl_2 or radical polymerization *via* a co-cast radical initiator. The resulting SBM Janus micelles, obtained after simple dispersion in organic solvents, could be converted into strongly amphiphilic and water-soluble SBMAA Janus micelles *via* alkaline hydrolysis of the PMMA part into poly(methacrylic acid) (PMAA).[53] Both species showed remarkable hierarchical self-assembly behavior in solution, as discussed below. The dedicated manipulation of the triblock terpolymer bulk phases also permits access to particle shapes of lower dimensionality, such as 1D Janus cylinders and 2D Janus discs. Both types of particles can hardly be afforded with a similar combination of precise control in cross-section, nanoscale dimensions and scalable production by any other technique.

Concerning cylindrical Janus particles, we reported their successful preparation *via* the selective crosslinking and subsequent dispersion of suitable lamella-cylinder bulk phase of SBM block terpolymers (Figure 1.12).[48,51] The process resulted in Janus cylinders with PS and PMMA hemicylinders that can be several micrometers in length. Liu *et al.* reported that the diameter of these cylinders can be adjusted *via* the molecular weight and resulting long period of the block terpolymer.[48] Walther *et al.* demonstrated that a simple sonication treatment can be used as a convenient tool to shorten the cylinders to the nanometer scale. The length was a function of both the energy and the duration of the sonication treatment.[51] Suitable staining of the corona hemicylinders also allowed the two corona compartments to be discerned in TEM imaging to give solid proof for the existence of corona segregation even after transfer of the crosslinked bulk structures in solution. In a recent extension of this concept, Wolf *et al.* synthesized the first fully water-soluble and pH-responsive Janus cylinders based on crosslinking and subsequent hydrolysis of a poly(*tert*-butoxystyrene)-*block*-polybutadiene-*block*-poly(*tert*-butyl methacrylate) triblock terpolymer forming a lamella-cylinder phase in the bulk.[52] After hydrolysis, the resulting Janus cylinders were composed of one hemicylinder of poly(4-hydroxystyrene) (PHS), which is soluble above pH 10, and a second hemicylinder of PMAA, soluble above pH 4. Detailed cryo-TEM investigations and the use of Cs counterions as staining agent for the charged PMAA side-chains allowed the Janus character to be directly visualized. This represents one of the most convincing, real space proofs of Janus particles in solution and may provide guidelines for the analysis of polymer-based Janus particles or other soft multicompartment polymer nano-objects in the future.

With respect to two-dimensional Janus discs, sheets or tiles, two different synthetic systems were used for the fabrication of planar Janus particles. We reported the preparation of Janus discs based on polystyrene-*block*-polybuta-diene-*block*-poly(*tert*-butyl methacrylate) (SBT) block terpolymers (Figure 1.13).[49] The chemical composition and incompatibility of the various components therein facilitated the formation of a fully lamellar (*ll*) morphology even at very low contents of PB. Although a low content of PB complicates tight crosslinking, a thin lamella of crosslinked PB simplified the dissolution of the crosslinked template, as less energy is necessary to disintegrate the crosslinked polymer structure. Owing to the low content of PB, the system was very susceptible to morphological changes upon exposure to solvents or chemicals required for the crosslinking.[54] The changes in bulk morphologies upon exposure to different crosslinking conditions were investigated in detail and revealed that careful control is required to achieve sufficient stabilization of the PB phase while maintaining the desired bulk morphology. Optimization led to the use of highly efficient thiol–polyene free radical polymerization reactions to achieve tight crosslinking of the thin PB domain.[54] After crosslinking, sonication-assisted dispersion was used to liberate the Janus discs/platelets. Similarly to the case of Janus cylinders, the particle size decayed in an exponential fashion with fragmentation of large Janus platelets into significantly smaller ones in the beginning. The average sizes could be tuned from the micrometer level down to the nanometer scale. The flat, disc-shaped character of the Janus particles was confirmed both by cryo-TEM and AFM (Figure 1.13a). Interestingly, the particles displayed a rather round-shaped appearance. This was attributed to the introduced ultrasound shock waves that tend to cut off protrusions and fragment Janus platelets along major existing crack tips, hence leading to a more circular appearance.

Recently, Zhang *et al.* reported the preparation of Janus discs based on the gelation crosslinking of a poly(2-vinylpyridine)-*block*-poly[3-(triethoxysilyl)-propyl methacrylate]-*block*-polystyrene triblock terpolymer containing siloxane moieties in the central block. Upon gelation into a silsesquioxane network, the terminal P2VP and PS blocks of the initial block terpolymer were compartmentalized on the two sides of the dispersed colloids, leading to hybrid Janus discs/platelets.[55]

1.5 Self-assembly Properties of Polymer-based Janus Particles of Different Dimensionality

In terms of self-assembly of completely hydrophobic SBM Janus micelles, a combination of various analytical tools, such as fluorescence correlation spectroscopy (FCS), small-angle neutron scattering (SANS) and static and dynamic light scattering (SLS and DLS), in addition to AFM, indicated the existence of an equilibrium between molecularly dissolved Janus micelles (unimers) and aggregates (multimers), so-called supermicelles.[46] The cluster formation into supermicelles was even observed in good solvents for both blocks (THF) and a surprisingly low critical aggregation concentration (*cac*) of ~ 7 mg L^{-1} was found with FCS (Figure 1.11a). Note that a similarly unexpected aggregation was later also observed in the case of ultrasmall Janus particles with only one polymer arm of PS and PEO tethered to a crosslinked P2VP core in SVEO triblock terpolymers (see above, Figure 1.4).[9] The individual SBM Janus micelles and their supermicelles had number-average hydrodynamic radii, $<R_h>_n$, of ~ 10 and 53 nm, respectively, as determined by FCS. These FCS results were corroborated by SANS and DLS and further analysis pointed to an average aggregation number of ~ 11 Janus micelle subunits within one supermicelle. When adsorbed on surfaces, AFM imaging revealed a height profile similar to a fried egg (Figure 1.11b, c). Well-ordered surface patterns could be observed due to the near monodisperse character of the Janus micelles and their supermicelles. The aggregates proved to be

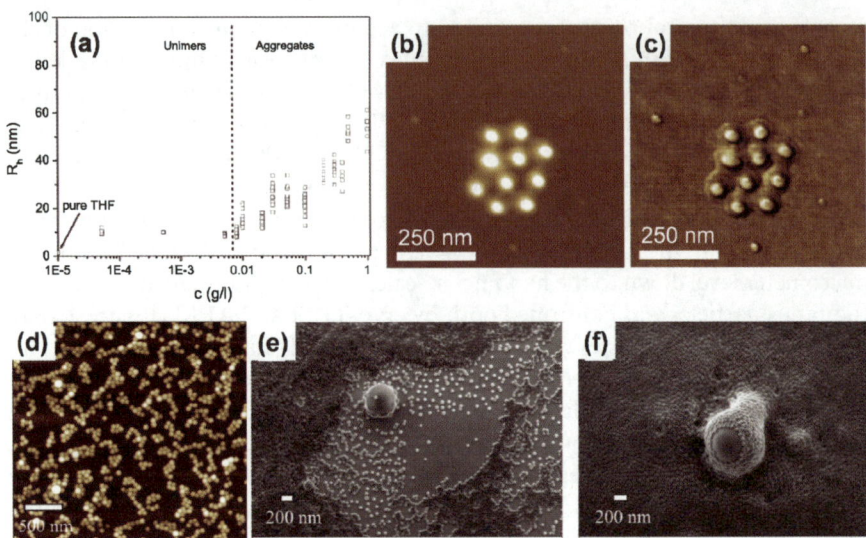

Figure 1.11 Superstructures formed by Janus micelles. (a) Fluorescence correlation spectroscopy data to identify superstructure formation of SBM Janus micelles. AFM height (b) and phase (b) images of superstructures of SBM Janus micelles and some isolated unimers. (d) AFM image of clusters (supermicelles) of strongly amphiphilic SBMAA Janus micelles. (e, f) SEM images of supermicelles and giant micelles formed by SBMAA Janus micelles. Reprinted and adapted with permission from *Macromolecules*, 2001, **34**, 1069[46] and *J. Am. Chem. Soc.*, 2003, **125**, 3260.[53] Copyright 2001 and 2003 American Chemical Society.

compact and stable and also did not show any tendency to open up into individual micelles when being spread as a Langmuir monolayer from CHCl$_3$ solution on to a water surface.[56] The number of Janus micelles that formed a domain in the spread monolayers was found to be similar to the number of Janus micelles associating into supermicelles in solution.

In contrast to the unexpected self-assembly of SBM Janus micelles in good solvents for both corona hemispheres, the SVT ones reported by Ishizu and co-workers required exposure to a selective solvent (acetone) to induce the collapse of the PS side and present a strongly solvophobic side to stimulate further aggregation.[45] Collapse of the PS side was accompanied by a diminished $<R_h>_z$ of 24 nm of the individual Janus micelles, compared with 38 nm before the collapse, and the occurrence of a small fraction of larger self-assemblies at 83 nm.

After hydrolysis of the PMMA segments of the SBM Janus micelles to PMAA, giving strongly amphiphilic SBMAA Janus micelles, detailed investigations of the solution properties revealed hierarchical self-assembly on two levels.[53] Similarly as in organic solution for the non-hydrolyzed analogues, self-assembly of single Janus particles into defined clusters took place above a *cac* of 0.03 mg mL^{-1} (Figure 1.11d). The resulting spherical

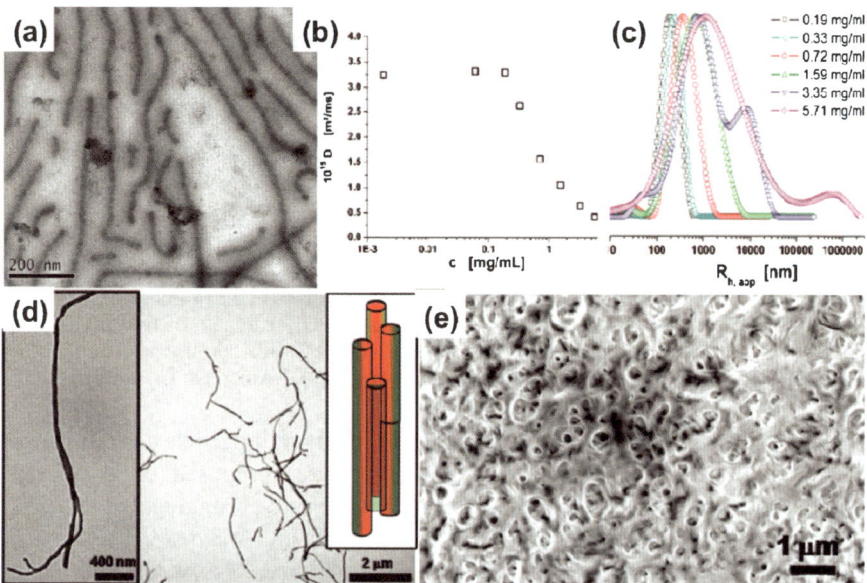

Figure 1.12   Overview of SBM Janus cylinders. (a) Cryo-TEM image depicting unimolecularly dissolved Janus cylinders in THF. (b) Evolution of the diffusion coefficient as a function of the concentration for SBM Janus cylinders in acetone. The *cac* is determined at the kink. (c) DLS CONTIN plots obtained at different concentrations as indicated within the figure. (d) TEM image of fibrillar aggregates obtained from Janus cylinders deposited at 0.5 g L^{-1} from acetone solution. (e) SEM image of network-like structures observed after deposition of a Janus cylinder dispersion at 5 g L^{-1} from THF. Reprinted and adapted with permission from *J. Am. Chem. Soc.*, 2009, **131**, 4720.[51] Copyright 2009 American Chemical Society.

supermicelles showed radii of 40–60 nm, which increased significantly upon ionization at higher pH values due to the stretching of the PMAA polyelectrolyte chains. About 30 SBMAA Janus micelles formed one supermicelle. In addition to these clusters, even larger aggregates, so-called giant micelles, with sizes up to 2 µm could be identified by imaging techniques (Figure 1.11e, f). Although the detailed structure of the very large giant micelles is unknown so far, it was suggested that the internal structure may be similar to that of multilamellar vesicles, albeit being composed of particles instead of surfactant molecules.

In contrast to spherical Janus micelles of the same chemical composition, the SBM Janus cylinders did not undergo supracolloidal aggregation in a good solvent for both hemicylinders.[51] DLS, SANS and cryo-TEM (in organic solvents) supported the presence of non-aggregated Janus cylinders at concentrations up to 50 g L^{-1} in THF (Figure 1.12a). Supracolloidal aggregation could be induced by transfer into selective solvents such as acetone (poor solvent for PS), which led to the observation of fiber-like

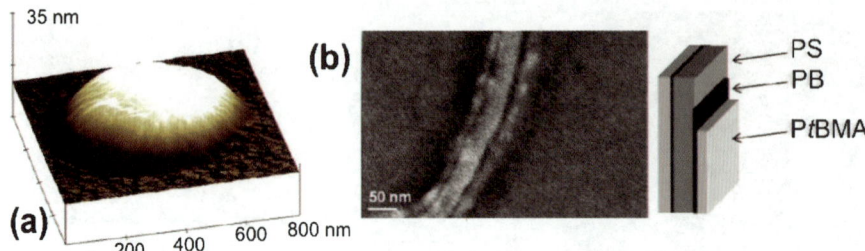

Figure 1.13 (a) 3D AFM height image of a flat Janus disc. (b) TEM image of the cross-section of a back-to-back stacked sandwich aggregate of two Janus discs formed in acetone. Reprinted and adapted with permission from *J. Am. Chem. Soc.*, 2007, **129**, 6187.[49] Copyright 2007 American Chemical Society.

aggregates, the length of which depended on the overall particle concentration (Figure 1.12b–e). The superstructured fibers were composed of 2–4 Janus cylinders at a given cross-section, that were toposelectively aggregated to shield the inner insoluble PS hemicylinder against the solvent. A *cac* of 0.2 g L^{-1} was found below which unimolecularly dissolved Janus cylinders were identified as stable particles. Below this *cac*, stabilization occurred *via* an intramolecular mechanism, in which the collapsed PS side was sufficiently protected by the PMMA arms extending around large parts of the cylinder to minimize the contact between PS and solvent. Aggregation on a second hierarchical level could be triggered when depositing Janus cylinders on surfaces from more concentrated solutions. This induced network-like patterns, in which different pore sizes and different surface compositions could be achieved simply by changing the concentration and the solvent quality.

In the case of SBT Janus discs, we identified their partial self-assembly in good solvents (THF) for both sides, PS and P*t*BMA.[49] Direct imaging *via* cryo-TEM in THF turned out to be particularly valuable in visualizing back-to-back stacking of Janus discs into sandwich-type structures. The supracolloidal aggregation could be enhanced by dispersion in selective solvents (*e.g.* acetone for PS). The detailed multicompartment cross-section and the inherent Janus character of the colloids could be demonstrated *via* embedding the aggregates into a photo-crosslinkable silicone oil, followed by microtome slicing and imaging of the ultrathin cross-sections by TEM (Figure 1.13b).

These SBT Janus discs were subsequently rendered strongly amphiphilic *via* acidic hydrolysis of P*t*BMA into PMAA, leading to SBMAA Janus discs consisting of one PS and one PMAA side.[50] Significant differences in the solution behavior were identified for differently sized Janus discs. Small discs (diameter <200 nm) showed only a minor amount of superstructures as seen by AFM and cryo-TEM and were thus mostly not aggregated *via* back-to-back stacking. The long PMAA chains, protruding out from the other side, partly shielded and sufficiently stabilized the fully collapsed, hydrophobic PS sides of the Janus discs against water. Significantly larger Janus discs were unable, however, to protect fully the PS side by expanded PMAA chains and displayed strong crumbling and bending and protected the PS side mostly *via* an intraparticle mechanism, *i.e.* by flipping over one part of the structure. Owing to the intrinsic stiffness of the crosslinked layer, efficient bending is favored with increasing size of the particle. Interestingly, large-scale back-to-back stacking was mostly not observed for this system and the presence of large hydrophobic faces exposed to water remains an unexpected observation.

1.6 Application as Structured Particulate Surfactants

Binks and Fletcher's first prediction of the enhanced interfacial adsorption capabilities of biphasic Janus particles spurred significant efforts to develop advanced particulate surfactants.[57] Janus particles uniquely combine amphiphilicity known from classical surfactants with the Pickering character that strongly holds solid particles at interfaces. Their calculations predicted an up to threefold stronger adsorption of Janus particles compared with particles of uniform wettability. This effect is most relevant for nanoscale particles, as the dynamics on this length scale are much higher than for micron-scale particles, whose adsorption strength is already very large simply due to the Pickering effect. Therefore, breaking the symmetry of nanoscale particles can lead to substantial and crucial improvements of the interfacial desorption energy compared with thermal energy. Moreover, since the engineering capabilities of particle synthesis nowadays allow precise tailoring of the particle architecture, it is also possible to nanostructure the interface laterally or impart additional properties, such as side-selective reflectivity or catalysis.

A first proof of principle for the enhanced interfacial activity of Janus nanoparticles was delivered by Glaser *et al.*, who verified that bimetallic slightly amphiphilic Janus particles indeed induce a larger decrease in the quasi-equilibrium interfacial tension of liquid/liquid interfaces as compared with homogeneous particles.[58] Following this demonstration, we investigated the size-dependent effects of disc-shaped SBT[49] and cylindrical SBM Janus particles[59] on the interfacial tensions of liquid/liquid interfaces. In both cases, progressively enhanced adsorption was found for Janus particles when increasing their dimensions, *i.e.* Janus disc diameter and Janus cylinder length, respectively (Figure 1.14a, b). Nonetheless, one has to keep in mind that these results are expected based on simple considerations of the Pickering effect, which basically states that the maximum of the desorption energy of a solid homogeneous sphere located at a liquid/liquid interface is proportional to the square of its radius ($E \sim R^2$). However, more important were the observations that Janus discs are significantly more powerful in reducing the interfacial tension as compared with the linear, non-crosslinked terpolymer, serving as precursor polymer for the synthesis of the Janus particles. Moreover, a comparison of the effectiveness of Janus cylinders of a given length with that of a homogeneous cylinder (*e.g.* core–shell PB–PS) clearly demonstrated a significantly higher interfacial activity, thereby further justifying research on advanced Janus-type surfactant particles (Figure 1.14b). Additionally, Ruhland *et al.* also characterized the time-

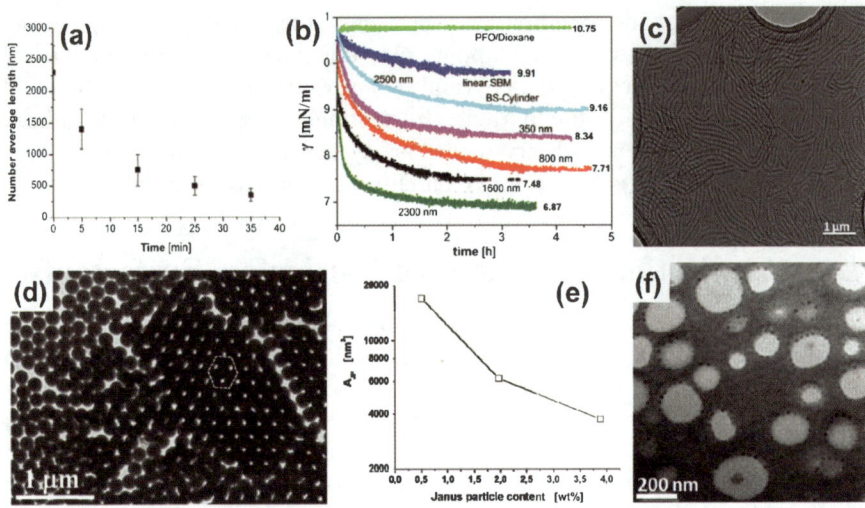

Figure 1.14 Interfacial activity of Janus particles. (a) Size evolution of Janus cylinders, used for interfacial tension measurements, as a function of sonication time. (b) Interfacial tension isotherms of Janus cylinders of different length at the perfluorooctane/dioxane interface measured with a pendant drop tensiometer. The non-crosslinked linear SBM triblock terpolymer used for the synthesis and a homogeneous BS core–shell cylinder with a PB core and a PS corona are shown for comparison. (c) TEM image of the interfacially adsorbed Janus cylinders (length 2300 nm) after transfer to a lacey carbon-coated grid. A local mesoscale order can be identified at longer adsorption times (here 2 h). (a–c) Reprinted and adapted with permission from *Langmuir*, 2011, **27**, 9807.[59] Copyright 2011 American Chemical Society. (d) TEM image of latex particles obtained by emulsion polymerization using SBMAA Janus particles as stabilizers. (e) Interfacial area of a latex particle stabilized by a single Janus particle. (d, e) Reprinted and adapted with permission from *Angew. Chem. Int. Ed.*, 2008, **47**, 711.[60] Copyright 2008 Wiley VCH. (f) TEM image of a nanostructured PS–PMMA (PMMA = white domains) blend stabilized by SBM Janus particles. The Janus particle can be identified as black dots at the interface. Reprinted and adapted with permission from *ACS Nano*, 2008, **2**, 1167.[61] Copyright 2008 American Chemical Society.

dependent evolution of the surface structures formed during the adsorption of Janus cylinders at interface.[59] Initially, only isolated cylinders were observed in *ex situ* TEM images of the deposited interfacial areas. After prolonged adsorption, an increasingly better order with a local mesoscale liquid crystalline arrangement of the cylinders could be found, indicating pathways towards nanostructuring of droplet interfaces (Figure 1.14c). Thereafter, loosely attached multilayers were observed near the quasi-equilibrium interfacial tension at which the interface was nearly completely covered.

In terms of real-life and industrially relevant applications, we employed amphiphilic SBMAA and SBM Janus micelles (diameter $\sim$20 nm) as stabilizer for emulsion polymerization[60] and compatibilizer for polymer blends,[61] respectively.

Emulsion polymerizations of styrene and *n*-butyl acrylate could be conducted in a facile batch process and did not require additives or miniemulsion polymerization techniques, as do other Pickering emulsion polymerizations using non-amphiphilic particles. The resulting latex dispersions displayed very well-controlled particle sizes with extremely low polydispersities, typically below 1.02 (Figure 1.14d). The particle size was controlled by the amount of particulate stabilizer added to the system. A detailed analysis of the surface coverage of the latex particles revealed a loose coverage of the latex surface by the Janus particles, in which the surface area stabilized by one Janus particle exceeded its cross-section several times (Figure 1.14e). A comparison with block copolymer systems or standard Pickering particles known from the literature strongly indicates a superior performance of the Janus particles and thus renders this material and simple process highly interesting for further fundamental studies and also industrial applications.

In a second application study, we used SBM Janus particles on a multigram scale for the blend compatibilization of a PS–PMMA polymer blend system in a twin-screw mini-mixer.[61] It was shown that the Janus particles locate exclusively at the interface of the two polymer phases despite the high temperature and shear conditions (Figure 1.14f). Constant decay of the domain size of the dispersed phase could be observed, independent of the blend composition used. The performance of the Janus particles in compatibilizing the polymer blend was found to be significantly superior to other state-of-the-art compatibilizers, such as linear triblock terpolymers of the same composition. Common problems such as micellization of the stabilizer and insufficient adsorption at the interface were absent to a major extent. The origin of the continuous decrease lies in the enhanced adsorption of the Janus particles at the interface, which is in turn caused by their biphasic particle character. In contrast to block copolymers or homogeneous particles, the Janus particles are located at the interface, even at high temperature and shear, because the desorption energy of a Janus particle from the interface under processing conditions was calculated to be almost as high as for a homogeneous particle at room temperature. In addition, the Janus particles exhibited an ordered arrangement at the polymer blend interface. Therefore, they provide efficient means for the nanoscopic engineering of polymer blend systems while matching macroscopic processing constraints.

1.7 Summary and Outlook

The toolbox of polymer chemistry and the self-assembly capabilities of block copolymers provide powerful strategies for the preparation of nanoscopic, responsive and precisely structured Janus particles with a wide range of architectures. Although some limitations clearly remain to be overcome as in the case of the direct self-assembly and spontaneous symmetry breaking of block copolymers in solution, other approaches have evolved into large-scale synthetic routes for precisely engineered soft nano-objects. This has not only allowed the detailed study of the unusual self-assembly behavior of different Janus particles as a function of the architecture, but also permitted prototype application studies employing these particles as advanced particulate surfactants.

What could be some of the next challenges to master? There is still a need to develop even simpler, direct and large-scale access routes to soft nanoscopic Janus particles. Attracting further attention from industry requires finding solution-based strategies that are feasible at high concentrations during particle synthesis and allow easily tunable chemistry. Progress in this direction could push the developments in the direction of interfacial compatibilizers further from laboratory-scale models into technological processes for high-end applications.

On a more fundamental level, there is still a lot of unexplored territory with respect to controlled stimulus-responsive self-assembly, in which simple triggers achieve large rearrangements of structures and properties. Encoding the ability for programmable and on-demand structure formation based on such 'intelligent' Janus particles may be formulated as one of the ultimate goals. Furthermore, hardly any application studies have addressed fields beyond the stabilization of interfaces. Just to name one field of interest, such small compartmentalized particles with well-defined functionalities located inside polymer coronas are, for instance, ideal candidates for biomedical applications and the amphiphilic character may well lead to fundamentally different interactions with cell membranes and opens up possibilities for spatially separated biosensing/ biotargeting and detection. Therefore, interfacing the progress on the side of macromolecular engineering and polymer self-assembly further with application specialists would span some crucial gaps and aid in the focused development of new generations of tailored and responsive soft Janus nano-objects.

References

1. C. A. Fustin, V. Abetz and J. F. Gohy, *Eur. Phys. J. E* 2005, **16**, 291–302.
2. A. O. Moughton, M. A. Hillmyer and T. P. Lodge, *Macromolecules*, 2011, **45**, 2–19.
3. S. J. Holder and N. A. J. M. Sommerdijk, *Polym. Chem.*, 2011, **2**, 1018–1028.
4. A. H. Gröschel, F. H. Schacher, H. Schmalz, O. V. Borisov, E. B. Zhulina, A. Walther and A. H. E. Müller, *Nat. Commun.*, 2012, **3**, 710.
5. Y. Chang, W. C. Chen, Y. J. Sheng, S. Y. Jiang and H. K. Tsao, *Macromolecules*, 2005, **38**, 6201–6209.
6. M. Charlaganov, O. V. Borisov and F. A. M. Leermakers, *Macromolecules*, 2008, **41**, 3668–3677.
7. A. Kiriy, G. Gorodyska, S. Minko, M. Stamm and C. Tsitsilianis, *Macromolecules*, 2003, **36**, 8704–8711.
8. Z. S. Ge, J. Xu, J. M. Hu, Y. F. Zhang and S. Y. Liu, *Soft Matter*, 2009, **5**, 3932–3939.
9. L. Cheng, G. Hou, J. Miao, D. Chen, M. Jiang and L. Zhu, *Macromolecules*, 2008, **41**, 8159–8166.
10. D. X. Wu, Y. F. Yang, X. H. Cheng, L. Liu, J. Tian and H. Y. Zhao, *Macromolecules*, 2006, **39**, 7513–7519.

11. L. N. Gu, Z. Shen, S. Zhang, G. L. Lu, X. H. Zhang and X. Y. Huang, *Macromolecules*, 2007, **40**, 4486–4493.
12. M. R. Xie, J. Y. Dang, H. J. Han, W. Z. Wang, J. W. Liu, X. H. He and Y. Q. Zhang, *Macromolecules*, 2008, **41**, 9004–9010.
13. J. de Jong and G. ten Brinke, *Macromol. Theory Simul.*, 2004, **13**, 318–327.
14. T. Stephan, S. Muth and M. Schmidt, *Macromolecules*, 2002, **35**, 9857–9860.
15. K. Tsubaki, H. Kobayashi, J. Sato and K. Ishizu, *J. Colloid Interface Sci.*, 2001, **241**, 275–279.
16. A. Deffieux, D. Lanson, M. Schappacher and R. Borsali, *Macromolecules*, 2006, **39**, 7107–7114.
17. R. Borsali, D. Lanson, F. Ariura, M. Schappacher and A. Deffieux, *Macromolecules*, 2009, **42**, 3942–3950.
18. R. Borsali, D. Lanson, M. Schappacher and A. Deffieux, *Macromolecules*, 2007, **40**, 9503–9509.
19. R. Borsali, D. Lanson, M. Schappacher and A. Deffieux, *Macromolecules*, 2007, **40**, 5559–5565.
20. A. Deffieux, F. Ariura, M. Schappacher and R. Borsali, *React. Funct. Polym.*, 2009, **69**, 402–408.
21. J. Rzayev, *Macromolecules*, 2009, **42**, 2135–2141.
22. J. Rzayev, J. Bolton and T. S. Bailey, *Nano Lett.*, 2011, **11**, 998–1001.
23. H.-i. Lee, K. Matyjaszewski, S. Yu-Su and S. S. Sheiko, *Macromolecules*, 2008, **41**, 6073–6080.
24. Z. Li, J. Ma, C. Cheng, K. Zhang and K. L. Wooley, *Macromolecules*, 2010, **43**, 1182–1184.
25. L. Nie, S. Liu, W. Shen, D. Chen and M. Jiang, *Angew. Chem. Int. Ed.*, 2007, **46**, 6321–6324.
26. A. Halperin, *J. Phys. (Paris)*, 1989, **49**, 131–137.
27. J. Hu and G. Liu, *Macromolecules*, 2005, **38**, 8058–8065.
28. R. Zheng, G. Liu and X. Yan, *J. Am. Chem. Soc.*, 2005, **127**, 15358–15359.
29. R. Ma, B. Wang, Y. Xu, Y. An, W. Zhang, G. Li and L. Shi, *Macromol. Rapid Commun.*, 2007, **28**, 1062–1069.
30. G. Li, L. Shi, R. Ma, Y. An and N. Huang, *Angew. Chem. Int. Ed.*, 2006, **45**, 4959–4962.
31. S. H. Kim, J. P. K. Tan, F. Nederberg, K. Fukushima, Y. Y. Yang, R. M. Waymouth and J. L. Hedrick, *Macromolecules*, 2008, **42**, 25–29.
32. B. Fang, A. Walther, A. Wolf, Y. Y. Xu, J. Y. Yuan and A. H. E. Müller, *Angew. Chem. Int. Ed.*, 2009, **48**, 2877–2880.
33. A. Walther, P. E. Millard, A. S. Goldmann, T. M. Lovestead, F. Schacher, C. Barner-Kowollik and A. H. E. Müller, *Macromolecules*, 2008, **41**, 8608–8619.
34. A. Walther, C. Barner-Kowollik and A. H. E. Müller, *Langmuir*, 2010, **26**, 12237–12246.
35. J. Du and S. P. Armes, *Soft Matter*, 2010, **6**, 4851–4857.
36. H. Schmalz, J. Schmelz, M. Drechsler, J. Yuan, A. Walther, K. Schweimer and A. M. Mihut, *Macromolecules*, 2008, **41**, 3235–3242.

37. J. Schmelz, M. Karg, T. Hellweg and H. Schmalz, *ACS Nano*, 2011, **5**, 9523–9534.
38. I. K. Voets, A. de Keizer, P. de Waard, P. M. Frederik, P. H. H. Bomans, H. Schmalz, A. Walther, S. M. King, F. A. M. Leermakers and S. M. A. Cohen, *Angew. Chem. Int. Ed.*, 2006, **45**, 6673–6676.
39. I. K. Voets, A. de Keizer, M. A. C. Stuart and P. de Waard, *Macromolecules*, 2006, **39**, 5952–5955.
40. I. K. Voets, R. Fokkink, A. de Keizer, R. P. May, P. de Waard and M. A. C. Stuart, *Langmuir*, 2008, **24**, 12221–12227.
41. I. K. Voets, R. Fokkink, T. Hellweg, S. M. King, P. de Waard, A. de Keizer and M. A. C. Stuart, *Soft Matter*, 2009, **5**, 999–1005.
42. I. Voets, F. Leermakers, A. de Keizer, M. Charlaganov and M. Stuart, *Adv. Polym. Sci.*, 2011, **241**, 163–185.
43. L. Cheng, G. Zhang, L. Zhu, D. Chen and M. Jiang, *Angew. Chem. Int. Ed.*, 2008, **47**, 10171–10174.
44. S. Zhang, Z. Li, S. Samarajeewa, G. Sun, C. Yang and K. L. Wooley, *J. Am. Chem. Soc.*, 2011, **133**, 11046–11049.
45. R. Saito, A. Fujita, A. Ichimura and K. Ishizu, *J. Polym. Sci. Part A: Polym. Chem.*, 2000, **38**, 2091–2097.
46. R. Erhardt, A. Böker, H. Zettl, H. Kaya, W. Pyckhout-Hintzen, G. Krausch, V. Abetz and A. H. E. Müller, *Macromolecules*, 2001, **34**, 1069–1075.
47. F. S. Bates and G. H. Fredrickson, *Phys. Today*, 1999, **52**, 32–38.
48. Y. F. Liu, V. Abetz and A. H. E. Müller, *Macromolecules*, 2003, **36**, 7894–7898.
49. A. Walther, X. Andre, M. Drechsler, V. Abetz and A. H. E. Müller, *J. Am. Chem. Soc.*, 2007, **129**, 6187–6198.
50. A. Walther, M. Drechsler and A. H. E. Müller, *Soft Matter*, 2009, **5**, 385–390.
51. A. Walther, M. Drechsler, S. Rosenfeldt, L. Harnau, M. Ballauff, V. Abetz and A. H. E. Müller, *J. Am. Chem. Soc.*, 2009, **131**, 4720–4728.
52. A. Wolf, A. Walther and A. H. E. Müller, *Macromolecules*, 2011, **44**, 9221–9229.
53. R. Erhardt, M. F. Zhang, A. Böker, H. Zettl, C. Abetz, P. Frederik, G. Krausch, V. Abetz and A. H. E. Müller, *J. Am. Chem. Soc.*, 2003, **125**, 3260–3267.
54. A. Walther, A. Göldel and A. H. E. Müller, *Polymer*, 2008, **49**, 3217–3227.
55. K. Zhang, L. Gao and Y. Chen, *Polymer*, 2010, **51**, 2809–2817.
56. H. Xu, R. Erhardt, V. Abetz, A. H. E. Müller and W. A. Gödel, *Langmuir*, 2001, **17**, 6787–6793.
57. B. P. Binks and P. D. I. Fletcher, *Langmuir*, 2001, **17**, 4708–4710.
58. N. Glaser, D. J. Adams, A. Böker and G. Krausch, *Langmuir*, 2006, **22**, 5227–5229.
59. T. M. Ruhland, A. H. Gröschel, A. Walther and A. H. E. Müller, *Langmuir*, 2011, **27**, 9807–9814.
60. A. Walther, M. Hoffmann and A. H. E. Müller, *Angew. Chem. Int. Ed.*, 2008, **47**, 711–714.
61. A. Walther, K. Matussek and A. H. E. Müller, *ACS Nano*, 2008, **2**, 1167–1178.

CHAPTER 2
Design, Synthesis and Applications of Dumbbell-like Nanoparticles

CHAO WANG*[a] AND CHENJIE XU*[b]

[a] Department of Chemical and Biomolecular Engineering, Johns Hopkins University, Baltimore, MD 21218; [b] Division of Bioengineering, School of Chemical and Biomedical Engineering, Nanyang Technological University, Singapore 637457
*E-mail: chaowang@jhu.edu or xuchenjie@gmail.com

2.1 Introduction

The enormous potential of nanomaterials with engineered properties and functions has been driving scientists into the era of nanotechnology over the past 20 years.[1,2] One of the most exciting topics in the field is to design materials through bottom-up approaches, in which the engineered building units are utilized to create new materials for applications in electronics, catalysis, energy and medicine.[2] Given their unique properties and small sizes, nanoparticles (NPs) are one of the preferred units for assembling new materials with multiple functions.[3] Among the diverse platforms assembled through NPs with different properties, heterodimer or dumbbell-like nanoparticles (DBNPs) are of particular interest.[4]

DBNPs refer to those with two or more different nanoparticulated components in intimate contact, as shown in Figure 2.1. As DBNPs are a combination of NPs with different properties, it is natural to speculate that

RSC Smart Materials No. 1
Janus Particle Synthesis, Self-Assembly and Applications
Edited by Shan Jiang and Steve Granick
© The Royal Society of Chemistry 2012
Published by the Royal Society of Chemistry, www.rsc.org

DBNPs would possess the properties of each component. The incorporation of different properties in one system thus enables the DBNPs to achieve multiple tasks at the same time. For example, gold–iron oxide (Au–Fe$_3$O$_4$) NPs have been developed as contrast agents for both X-ray computed tomography (CT) and magnetic resonance imaging (MRI),[5] which integrates the advantages of CT (fast scanning speed) and MRI (high spatial resolution) in one contrast agent.

While the NP components keep their original properties, the interfacial interaction originating from electron transfer across the nanoscale contact at the interface of these components can further induce new properties that are not present in the single-component NPs.[4] For example, Au NPs are chemically inert but Au NPs deposited on a metal–oxide support, a structure similar to that shown in Figure 2.1, have shown high catalytic activity for CO oxidation.[6] This high activity of Au–oxide catalysts has been rationalized in terms of a junction effect, arising from the transfer of electrons from the oxide to the Fermi level of the supported Au NPs.[7–9]

This chapter discusses the current efforts in the design, preparation and application of DBNPs. In particular, the focus is placed on the DBNPs containing noble metal, magnetic NPs and/or quantum dots (QDs). The nanoscale junctions present in these structures facilitate electron transfer across the interface, changing the local electronic structures and therefore their physical and chemical properties. Being composed of two or more functional components, they are also ideal platforms for the integration of different biomedical functions, including molecular imaging, drug delivery and magnetic control, in a single system.

2.2   Synthesis

DBNPs can be prepared through several mechanisms, including direct heterogeneous nucleation, liquid–liquid phase transfer and non-epitaxial

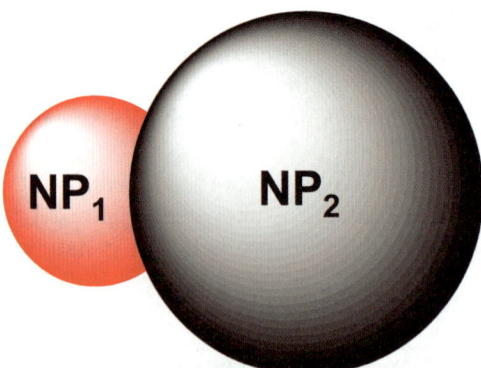

Figure 2.1   Schematic illustration of a dumbbell-like NP containing two different components.

deposition followed by coalescence/crystallization.[3,10] In all cases, the formation of dumbbell structures instead of separate individual NPs is achieved by overgrowth of a second component on the preformed seeds. This is similar to what has been known as seed-mediated growth for the synthesis of core–shell NPs,[11] but the nucleation and growth are anisotropic and preferential along one specific direction on the seeding NPs. Therefore, for the synthesis of DBNPs it is critical to promote the heterogeneous nucleation on seeding NPs while suppressing the homogeneous nucleation (the formation of separate NPs of the second component) and isotropic growth (into core–shell NPs).

In the classical nucleation theory, the driving force for nucleation and growth is described by the excess Gibbs free energy (ΔG_r) of the system:[12]

$$\Delta G_r = \Delta G_S + \Delta G_V = 4\pi r^2 \gamma - \frac{4}{3}\pi r^3 \times \frac{RT \ln S}{V_m} \tag{2.1}$$

where r is radius of the nuclei, γ the surface free energy per unit area, R the ideal gas constant, T the temperature of the solution, S the normalized concentration of reaction species under equilibrium conditions ($S = C/C_{eq}$) and V_m the molar volume of the material. ΔG_S and ΔG_V, correspond to the contributions from the surface and bulk free energies, respectively. If the solution is undersaturated ($S \leq 1$), ΔG_r is positive and nucleation is not favored. When the solution concentration goes beyond the equilibrium, namely supersaturated ($S > 1$), ΔG_r decreases with increase in radius and nucleation and growth can occur.

For heterogeneous nucleation on seeding NPs, eqn (2.1) becomes

$$\Delta G_r = \Delta G_S + \Delta G_V + \Delta G_{inter} \tag{2.2}$$

where the additional term ΔG_{inter} accounts for the adhesive energy at the interface between the seeds and the overgrown particles.[13] In order to promote nucleation on existing seeds, ΔG_{inter} must be negative so that the system energy is lower compared with separate nucleation and formation of free particles of the second component. This is why epitaxial growth or chemical bond formation at the interface is found to be important for the growth of DBNPs.[4,14] The interfacial energy (ΔG_{inter}) usually decreases with increase in particle size:

$$\Delta G_{inter} = \gamma' S_{inter} \approx \gamma' \pi r^2 \tag{2.3}$$

where $\gamma' < 0$ is the interfacial energy per unit area and S_{inter} is the area of the interface. From eqns (2.1)–(2.3), it can be seen that heterogeneous nucleation can start when the solution concentration is below the critical concentration for homogeneous nucleation ($C < C_{eq}$), if $\Delta G_S + \Delta G_{inter}$ decreases with increase in r. Therefore, heterogeneous nucleation can be reinforced in the synthesis by controlling the concentration of the precursor to be below the homogeneous nucleation threshold for the second component.

Figure 2.2 Schematic illustration of the growth of DBNPs *via* heterogeneous nucleation and Ostwald ripening process.

With heterogeneous nucleation, the composite NPs can grow into core–shell or dumbbell-like morphologies. The product morphology depends on the wetting properties of the second component over the seeds,[15] but is mainly determined by the lattice mismatch between the two materials.[16] In the case of epitaxial growth, the interfacial energy increases with the lattice mismatch due to the accumulation of strain energy at the interface. When the difference in lattice constant is small, the nucleation and growth of the second component can go in different directions by epitaxial growth. However, when the lattice mismatch is relatively large, empirically $\sigma \geq 3\%$, epitaxial growth is confined on certain crystal planes [*e.g.* (111) for FCC crystals] where the difference in atomic distances is minimized. Depending on the interplay between ΔG_S and ΔG_{inter}, the overgrowth can result in clover-like NPs with several particles on one seed or one-to-one DBNPs after an Ostwald ripening process (Figure 2.2).[14]

With the general growth mechanisms in mind, we now present several examples including noble metal–transition metal oxide DBNPs, DBNPs containing semiconductor NPs and DBNPs with multiple cores.

2.2.1 DBNPs Containing Noble Metal and Transition Metal Oxide NPs

The first type of DBNPs is the noble metal–transition metal oxide composite, including Au–Fe$_3$O$_4$[14] and Ag–Fe$_3$O$_4$[17] DBNPs, with a plasmonic–magnetic bifunctionality. In both cases the DBNPs were synthesized by organic solution approaches through seed-mediated growth and Au–Fe$_3$O$_4$ NPs were made with Au NPs as seeds whereas Ag–Fe$_3$O$_4$ NPs were made with Fe$_3$O$_4$ NPs as seeds.

The Au–Fe$_3$O$_4$ DBNPs were prepared *via* the decomposition of iron pentacarbonyl, Fe(CO)$_5$, over the surface of the preformed Au NPs, followed by oxidation in air, as illustrated in Figure 2.3a (direct heterogeneous nucleation).[14] Specifically, Au NPs were first prepared by the injection of a chloroauric acid (HAuCl$_4$) solution into the organic solution in the presence of oleylamine as surfactants. The size of the Au NPs was tuned by controlling the temperature at which the HAuCl$_4$ solution was injected, with larger NPs obtained at higher temperatures. Subsequently, the prepared and purified Au seeds were mixed with Fe(CO)$_5$ in 1-octadecene solvent in the presence of oleic acid and oleylamine, refluxed at 300 °C and followed by room temperature air oxidation (Figure 2.3b). The size of the Fe$_3$O$_4$ NPs was controlled by adjusting

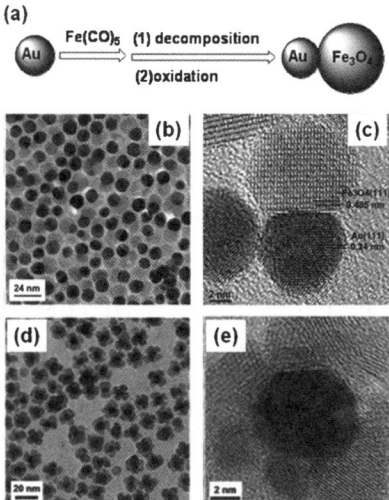

Figure 2.3 (a) Schematic illustration of the growth of Au–Fe$_3$O$_4$ DBNPs. (b, d) transmission electron microscopy (TEM) and (c, e) high-resolution TEM (HRTEM) images of the Au–Fe$_3$O$_4$ DBNPs with one-to-one (b, c) and clove-like (d, e) morphologies. Adapted with permission from reference 14. Copyright 2005 American Chemical Society.

the ratio between Fe(CO)$_5$ and Au and larger Fe$_3$O$_4$ NPs were obtained with more Fe(CO)$_5$ precursor. The overgrowth for Fe$_3$O$_4$ over Au seeds was found to be epitaxial (Figure 2.3c). The lattice constant for Au in the FCC phase is 4.080 Å, and 8.394 Å for inverse spinel structured magnetite. The size of the Fe$_3$O$_4$ lattice is almost double that for Au with a mismatch of $\sim$3%, which facilitates epitaxial growth.[14] The epitaxial growth was also found to be preferentially along <111> directions (Figure 2.3c).

The interesting story here is that the overgrowth of Fe$_3$O$_4$ on Au forms one-to-one DBNPs instead of one Au-to-several Fe$_3$O$_4$ DBNPs. This was attributed to the plausible electron transfer between Au and Fe.[14] Fe(CO)$_5$ decomposes into Fe, which nucleates on Au NPs. Once the Fe nucleus is formed, the free electrons from Fe tend to flow across the junction to Au. As a result, Au in Au–Fe composites becomes electron 'rich,' unsuitable for multinucleation of Fe and giving only the Au–Fe dumbbell structure. When exposed to air, Fe is oxidized to Fe$_3$O$_4$, forming Au–Fe$_3$O$_4$ DBNPs. From the above discussion, it can be considered that the electron coupling effect caused an increase in the interfacial energy per unit area [γ' in eqn (2.3)] and hence nucleation and growth on multiple crystal planes became energetically unfavorable. Interestingly, changing the solvent from octadecene to diphenyl ether resulted in clover-like Au–Fe$_3$O$_4$ NPs[14] (Figure 2.3d, e). The employment of polar solvents, such as ethers, is believed to compensate the charge accumulation on the Au seeds and thus the nucleation and growth of multiple Fe$_3$O$_4$ particles on each Au seeds becomes possible.

Another example is Ag–Fe$_3$O$_4$ NPs that could be made by controlled nucleation of Ag on the preformed Fe$_3$O$_4$ NPs at liquid/liquid interfaces (Figure 2.4).[17] Simply, an organic solution (*e.g.* hexane) containing Fe$_3$O$_4$ NPs was mixed with an aqueous solution of AgNO$_3$ and agitated by ultrasonication. As hexane and water are immiscible, the sonication broke the hexane layer into small droplets that formed a microemulsion in the AgNO$_3$ solution. The Fe$_3$O$_4$ NPs assembling at the liquid/liquid interface acted as catalytic centers for the reduction of Ag$^+$ and nucleation/growth of Ag NPs. This method has been further generalized to the synthesis of Ag–FePt and other Ag-based DBNPs.[16] A detailed examination by HRTEM revealed that the inter-particle relationship in these Ag-based DBNPs is not epitaxial, which indicated that the growth mechanism leading to one-to-one Ag–Fe$_3$O$_4$ DBNPs was not heterogeneous nucleation. Instead, at the liquid/liquid interface, the partial exposure of the Fe$_3$O$_4$ NPs to the AgNO$_3$ aqueous phase limited the nucleation of multiple Ag particles on each Fe$_3$O$_4$ seed. Once an Ag nucleus had been formed on an Fe$_3$O$_4$ seed, further deposition and growth of the existing Ag seed were more favorable than nucleation on another site of this seed owing to the high mobility and low surface energy of Ag.

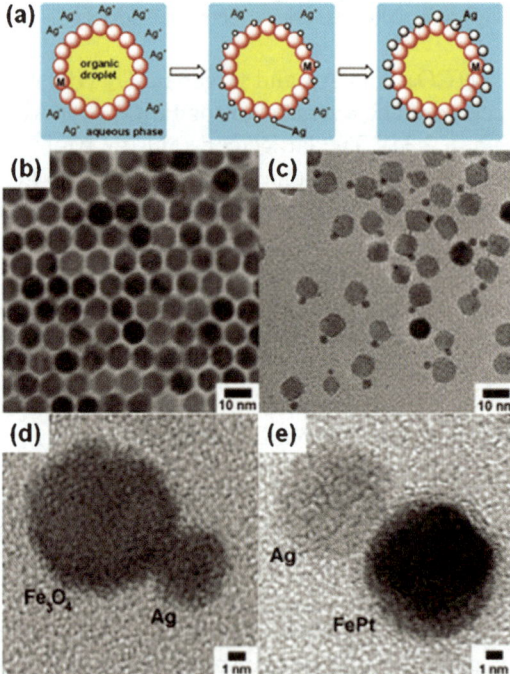

Figure 2.4   (a) Schematic illustration of the formation of DBNPs at the liquid/liquid interface. TEM images of (b, d) Ag–Fe$_3$O$_4$ and (c, e) Ag–FePt DBNPs prepared by this method are also presented. Adapted with permission from reference 17. Copyright 2005 American Chemical Society.

In addition to liquid–liquid phase transfer, $Ag-Fe_3O_4$ DBNPs were also made by nucleation and growth of Ag on sulfur-modified Fe_3O_4 NPs (Figure 2.5).[18] Fe_3O_4 NPs were first made by reductive decomposition of $Fe(acac)_3$ in diphenyl ether and then reacted with hexadecanethiol (1-$C_{16}H_{33}SH$) at reflux (265 °C) in the presence of oleic acid and oleylamine. The HS group attached to the iron oxide surface and, at the high reaction temperature, the SH–Fe linkage decomposed to give a thin layer of FeS on the particle surface. The FeS-modified Fe_3O_4 NPs were then mixed with $AgNO_3$ in tetralin (1,2,3,4-tetrahydronaphthalene) in the presence of oleylamine. The mixture was heated at 100 °C for 1 h, giving the $Ag-Fe_3O_4$ DBNPs. In this case, the amorphous nature of the $Ag-S-Fe_3O_4$ junction substantially reduced the interfacial energy (ΔG_{inter}) for the overgrowth of Ag on the Fe_3O_4 seeds and the surface energy (ΔG_S) became dominant in the system energy [eqn (2.2)], making it more favorable to form a single Ag particle on each seed in order to lower the surface energy.

The organic solution approach to $Au-Fe_3O_4$ NPs[14] by using Au NPs as seeds discussed above has also been generalized to the synthesis of other types of DBNPs.[6,19] By this method, a range of DBNPs, with the noble metal being Ag, Pt, Pd or Ag–Ag alloy and the other component being iron or cobalt oxides, was produced (Figure 2.6).

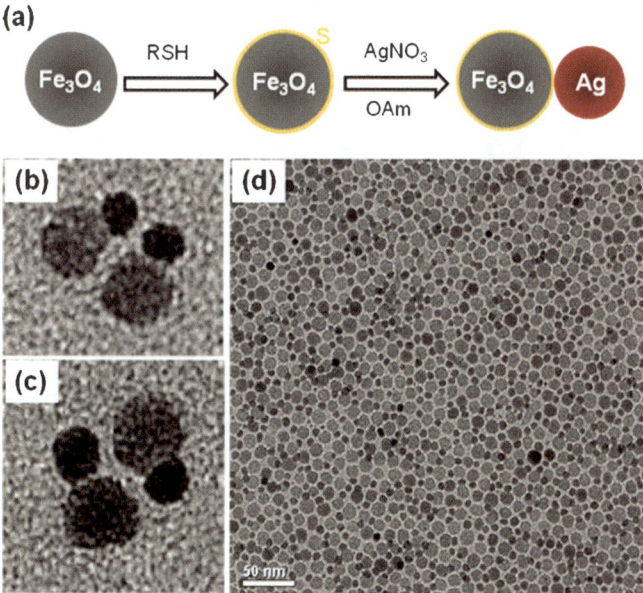

Figure 2.5 (a) Schematic illustration of the synthesis of $Ag-Fe_3O_4$ DBNPs and TEM images of the two $Ag-Fe_3O_4$ DBNPs aligned parallel (b) and antiparallel (c) and an assembly (d) of the $Ag-Fe_3O_4$ DBNPs. Adapted with permission from reference 18. Copyright 2005 American Chemical Society.

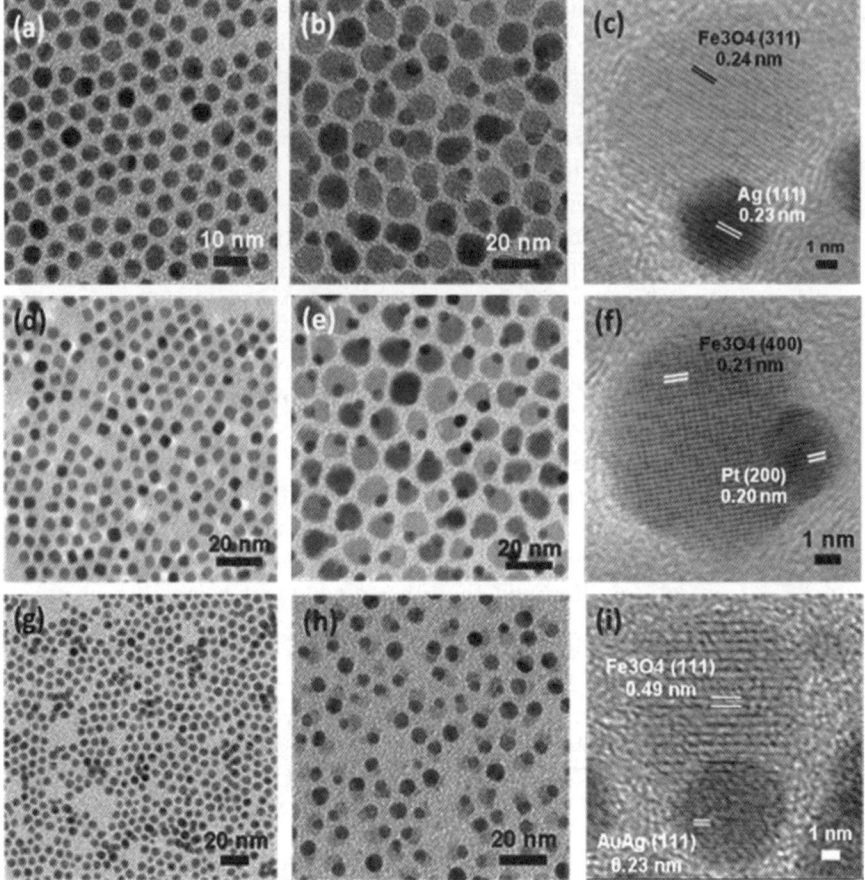

Figure 2.6   Noble metal–magnetic DBNPs from organic solution synthesis: (a–c) Ag–
Fe_3O_4, (d–f) Pt–Fe_3O_4 and (g–i) AuAg–Fe_3O_4. Adapted with permission
from reference 6. Copyright 2010 American Chemical Society.

2.2.2   DBNPs Containing Semiconductor NPs

Semiconductor nanocrystals or QDs are another important type of nanomaterials which have attracted considerable interest due to their unique electronic
and optical properties.[20–22] The incorporation of QDs into a composite
nanomaterial has long been pursued and intensively studied.[23,24] As the
synthetic conditions for QDs, which usually involve chalcogenides, are usually
not compatible with the synthesis of common metal and metal oxide NPs,
DBNPs containing QDs are usually prepared by seed-mediated growth with
metal or metal oxide NPs as seeds.

The first example of DBNPs containing QDs was demonstrated in FePt–
CdS NPs. These DBNPs were made from the FePt–CdS core–shell NPs
followed by dewetting of the CdS shell (Figure 2.7).[15] In the organic solution
after the growth of FePt NPs, elemental sulfur was added in an inert

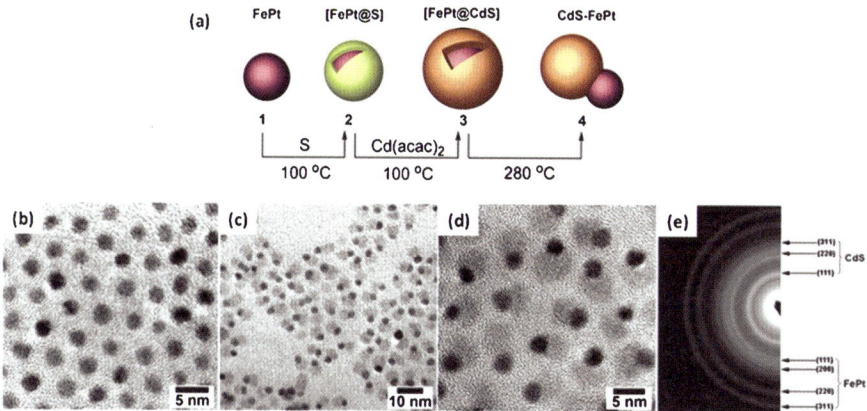

Figure 2.7 (a) Schematic illustration of the formation of FePt–CdS DBNPs via sulfur dewetting over FePt seeds. (b) TEM image of the FePt seeds. (c) TEM, (d) HRTEM images and (e) selected area electron diffraction pattern of the FePt–CdS DBNPs. Reprinted with permission from reference 15. Copyright 2004 American Chemical Society.

atmosphere. The high affinity between FePt and sulfur allowed sulfur to deposit on the surface of FePt seeds, forming FePt–S core–shell structures. The subsequent addition of cadmium acetylacetonate [$Cd(acac)_2$], 1,2-hexadecane-diol and trioctylphosphine oxide (surfactant) produced metastable FePt–CdS core–shell NPs, where CdS was amorphous. Upon further heating to an elevated temperature ($\sim$280 °C), the amorphous CdS started to crystallize and the dewetting of CdS over FePt led to the formation of FePt–CdS DBNPs.

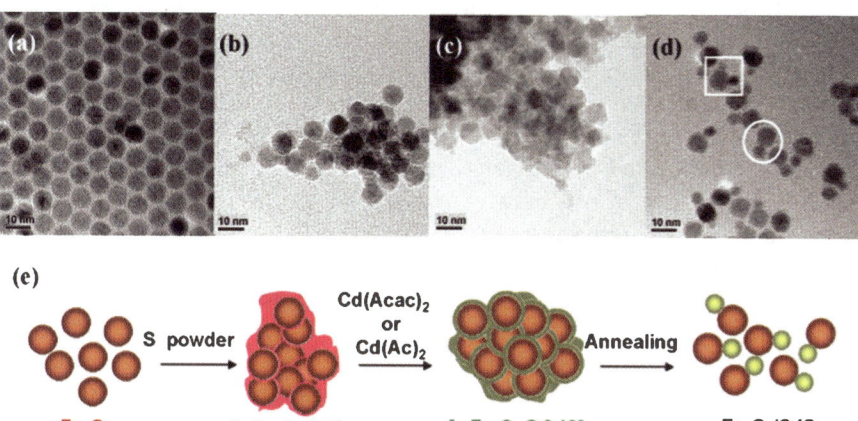

Figure 2.8 TEM images of the reaction mixture at different stages of the reaction. (a) γ-Fe_2O_3 NPs, (b) γ-Fe_2O_3 NPs after addition of sulfur powder, (c) γ-Fe_2O_3 NPs after addition of sulfur and Cd precursors and (d) the final γ-Fe_2O_3–CdS DBNPs. (e) Schematic illustration of the reaction pathway. Reprinted with permission from reference 25. Copyright 2005 American Chemical Society.

This method was later extended to the synthesis of γ-Fe$_2$O$_3$–MS (M = Zn, Cd, Hg) DBNPs where γ-Fe$_2$O$_3$ NPs were used as seeds (Figure 2.8).[25] The intermediates, in the form of Fe$_2$O$_3$ NPs enclosed by sulfur or amorphous chalcogenides, were clearly observed *via* TEM studies (Figure 2.8b, c). Unlike the case with FePt–CdS where the inter-particle interface was found not to be epitaxial, the Fe$_2$O$_3$–MS heterojunction was found to be epitaxial. Nonetheless, in both systems the driving force towards one-to-one DBNPs is the tendency for crystallization and reducing surface energy for the semiconductor components.

2.2.3 DBNPs Containing More Than Two Particles

Based on the chemistry described above, more exotic DBNPs containing metal, metal–metal oxide NPs and/or QDs have been made. For example, PbS–Au–Fe$_3$O$_4$ NPs were obtained by mixing Au–Fe$_3$O$_4$ NPs with Pb–oleate complex and sulfur.[26] Once sulfur had been added to the reaction flask containing the Pb–oleate complex and Au–Fe$_3$O$_4$ NPs, there was a competition between the adsorption of sulfur on the Au surface and reaction with the Pb–oleate to form PbS NPs. The competitive reaction led to the formation of a heterojunction between PbS and Au. The final product obtained adopted a ternary structure (Figure 2.9).

Even more complicated nanostructures have been produced by seed-mediated approaches. Ternary Au–Au–Fe$_3$O$_4$ NPs were synthesized by overgrowing additional Au particles (Au$_2$) on to Au$_1$–Fe$_3$O$_4$ DBNPs.[16] The competitive growth of Au$_2$ on Au$_1$ in the Au–Fe$_3$O$_4$ seeds induced a 'tug of war' between Au$_2$ and Fe$_3$O$_4$ in the Au$_2$–Au$_1$–Fe$_3$O$_4$ ternary nanostructures formed. As a result, Au$_2$ extracted Au$_1$ from the Au$_1$–Fe$_3$O$_4$ conjugation, generating a new dumbbell-like Au–Au and a dented Fe$_3$O$_4$ NP (Figure 2.10). Particle size-controlled growth and modeling analysis of the stress and strain distribution across the NPs revealed that the observed detachment was due to the stress accumulated at the heterogeneous interface in Au$_1$–Fe$_3$O$_4$, which increased as the size of Au$_2$ grew.

In another study, ternary Fe$_3$O$_4$–Au–Fe$_3$O$_4$ NPs were obtained by fusing and merging the Au particles from two Au–Fe$_3$O$_4$ DBNP NPs (Figure 2.11).[26] In this synthesis, Au–Fe$_3$O$_4$ NPs in solution were heated in the presence of sulfur. The addition of sulfur was found to be critical for the formation of ternary DBNPs as it binds Au strongly and served as glue between two Au NPs. Heating the non-S-treated Au–Fe$_3$O$_4$ DBNPs under the same conditions did not yield the Fe$_3$O$_4$–Au–Fe$_3$O$_4$ dumbbell structure.

It is also interesting that in the synthesis of ternary DBNPs, the connection is always through the metal particle, namely the central particle is metallic (*e.g.* Au in the above three examples). Besides the possible restrictions of lattice mismatch, it has to be related to the unique properties of the metals. Noble metals such as gold are soft and surface atoms have high mobility. Compared with metal oxides and semiconductor chalcogenides, they can accommodate

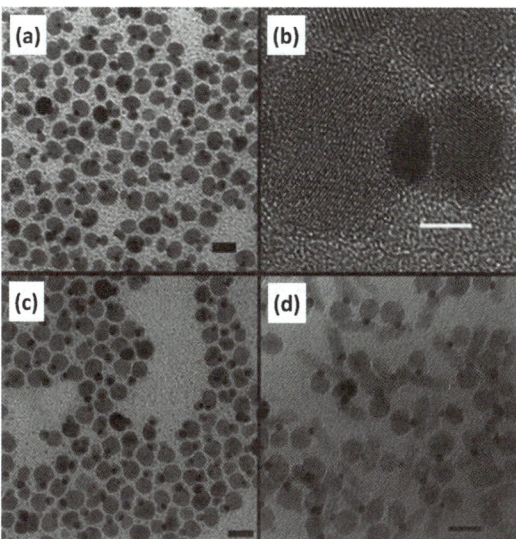

Figure 2.9 TEM images (a, c and d) and HRTEM image (b) of Fe_3O_4–Au–PbSe (a, b) and Fe_3O_4–Au–PbS (c, d) ternary DBNPs. The scale bars are 20 nm in (a), (c) and (d) and 4 nm in (b. Reprinted with permission from reference 26. Copyright 2006 American Chemical Society.

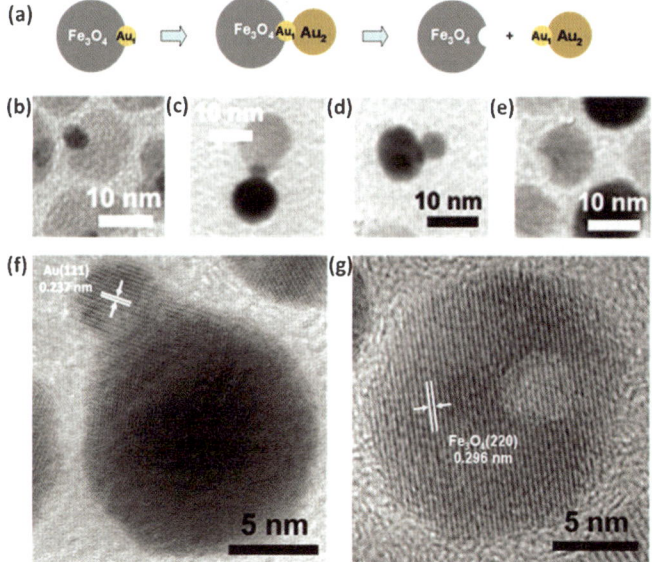

Figure 2.10 'Tug of war' in Au_2–Au_1–Fe_3O_4 DBNPs. (a) Schematic illustration of the Au_2 overgrowth on Au_1 NP and Au_1 NP detachment from the Fe_3O_4 NP, forming the new dumbbell-like Au_1–Au_2 and the dented Fe_3O_4 NPs. TEM images of an individual Au–Fe_3O_4 (b), Au_2–Au_1–Fe_3O_4 (c), Au_2–Au_1 (d) and dented Fe_3O_4 NP (e) and HRTEM images of an Au_2–Au_1–Fe_3O_4 (f) and a dented Fe_3O_4 NP (g). Reprinted with permission from reference 16. Copyright 2009 American Chemical Society.

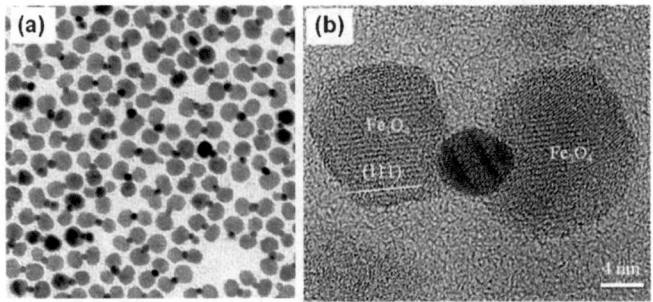

Figure 2.11   (a) TEM and (b) HRTEM images of Fe_3O_4–Au–Fe_3O_4 DBNPs obtained by fusion of the Au component of two Au–Fe_3O_4 NPs. Reprinted with permission from reference 26. Copyright 2006 American Chemical Society.

more strain energy at the interfaces and thus allow for the formation of multiple junctions on each particle. Similar argument may hold true in other DBNPs containing multiple components.[27]

2.3   Functional Applications of DBNPs

2.3.1   DBNPs as Heterogeneous Catalysts

The DBNPs have structural features similar to those of the supported heterogeneous catalysts except that the support itself is nanoscale. It is known that the catalytic behavior of the supported NPs is usually different from that of their individual counterparts. For example, Au NPs supported on transition metal oxides can show high catalytic activity for CO oxidation at low temperatures,[28–30] and also many other chemical reactions.[31–33] Despite the debates on the detailed mechanisms and active sites, the enhanced Au catalysis is usually ascribed to the local modification of its electronic structures by the oxide support.[8,9,34] Such electronic modification requires intimate contact between the metal and metal oxide, but this is usually hard to control by conventional impregnation or co-deposition synthesis. Therefore, the DBNPs from organic solution synthesis, in which the epitaxial relationship between metal and metal oxide particles facilitates the electronic interaction at the interface, are advantageous for heterogeneous catalysis.

Recently, DBNPs have been demonstrated to be highly active catalysts for CO oxidation (Figure 2.12).[6] Au– and Pt–Fe_3O_4 DBNPs obtained by epitaxial growth of Fe_3O_4 on Au or Pt NPs were deposited on amorphous carbon and annealed at 300 °C in 8% O_2–He forming gas to remove the residues of organic surfactants on the particle surface. The treated NPs were then applied as catalysts for CO oxidation. It was found that both the Au– and Pt–Fe_3O_4 NPs were more active than their counterparts prepared by conventional methods. The Au–Fe_3O_4 NPs exhibit a $T_{1/2}$ (the temperature at which CO conversion reaches 50%) of −25 °C, compared with 30 °C for the conventional Au–Fe_2O_3, and 67 °C for Pt–Fe$_3$O$_4$ NPs *versus* $\sim$100 °C for conventional Pt–Fe$_2$O$_3$. The enhanced activities thus unambiguously reveal the superior supporting effect of metal oxide in the solution-processed DBNPs with intrinsic epitaxial particle interconnection, which may not only facilitate electronic coupling, but also enhance the thermal stability and catalyst durability of the noble metal NPs.[35]

The epitaxial heterojunctions present in DBNPs have also been found to be beneficial for charge transfer in electrocatalysis. Pt–Fe$_3$O$_4$ DBNPs synthesized by epitaxial growth of Fe on to Pt nanoparticles followed by Fe oxidation in air (Figure 2.6d–f) were demonstrated to show a 20-fold improvement in catalytic activity for the oxygen reduction reaction compared with plain Pt catalysts (Figure 2.13). The introduction of a heterogeneous junction effect thus represents a novel approach in tailoring advanced catalysts for electrochemical and energy conversion reactions.

2.3.2 DBNPs as a Multifunctional Platform for Biomedical Applications

In addition to being a catalyst, DBNPs are also attractive for the biomedical field, given the multifunctionalities coming from the different components of DBNPs.[36] Before being applied for biomedical applications, including

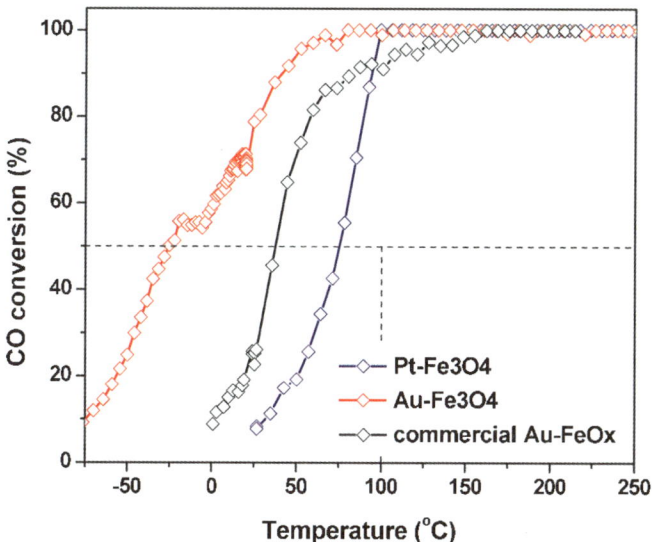

Figure 2.12 CO oxidation light-off curves recorded with Au–Fe$_3$O$_4$ and Pt–Fe$_3$O$_4$ DBNPs and commercial Au–Fe$_2$O$_3$ (World Gold Council) catalyst. The $T_{1/2}$ (the temperature at which CO conversion reaches 50%) value for conventional Pt–Fe$_2$O$_3$ catalysts is marked by dashed lines. Note that isolated Au or Pt NPs were not active in the corresponding temperature range. Reprinted with permission from reference 6. Copyright 2010 American Chemical Society.

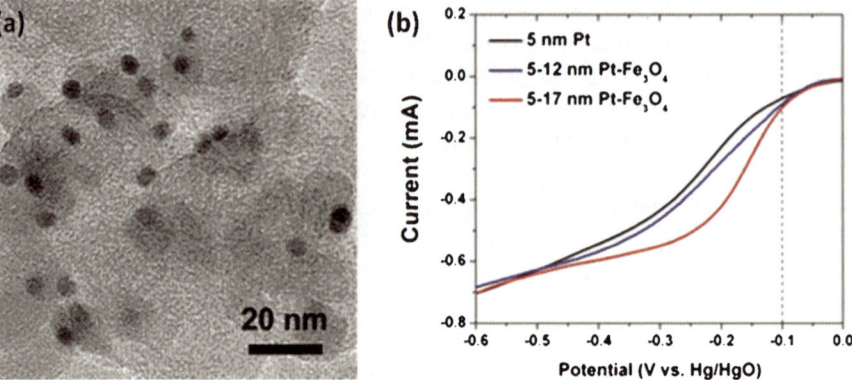

Figure 2.13 (a) TEM image of carbon-supported Pt–Fe$_3$O$_4$ DBNP catalyst. (b) Polarization curves for the oxygen reduction reaction recorded for the Pt–Fe$_3$O$_4$/C catalysts with different particle sizes. The data were obtained from rotating disk electrode (RDE) measurements in 0.5 M KOH with a rotation speed of 1600 rpm and a sweep rate of 10 mV s^{-1}. Reprinted with permission from reference 19. Copyright 2009 American Chemical Society.

diagnosis/detection,[37,38] molecular imaging,[5,39] magnetic control[40] and drug delivery,[41] the particles have to be stabilized in physiological buffers to allow a thorough interaction between biological molecules and particles.

2.3.2.1 Surface Modification

For any type of NPs, including DBNPs in solutions, the main forces affecting their stability are attractive van der Waals forces (also magnetic attraction in the case of magnetic NPs) and repulsive double-layer and steric interactions.[42] Such forces have been extensively discussed in the literature[43,44] and are outside the scope of this chapter. Nevertheless, in general, in the organic phase, steric forces from the absorbed organic coating drive DBNPs apart. In the aqueous phase, double-layer interactions are the main contributor for the separation of DBNPs from each other. Thus, to enable the DBNPs to be dispersible in aqueous media, they should be coated with a layer of dispersing surfactants or dispersants, which provides both stability and functional features to DBNPs.

There are various kinds of materials that can act as dispersants for DBNPs in aqueous media, including 2,3-dimercaptosuccinic acids, catechol derivatives such as dopamine, silanes, lipids, proteins, dendrimers, gelatin, polysaccharides, cellulose, chitosan, poly(ethylene glycol) (PEG), polyvinylpyridine (PVP) and poly(vinyl alcohol) (PVA) (more details could be found in a recent review by Reimhult *et al.*[45]). They could be grafted on to the surface of DBNPs through either ligand addition or ligand exchange (Figure 2.14).[46]

Ligand addition does not need to remove the original protecting ligands and usually generates a core–shell or layer-by-layer structure.[47] NPs modified in this strategy usually have better dispersibility and chemical stability. For example, Jon and co-workers recently showed that Au–Fe$_3$O$_4$ DBNPs covered with hydrophobic oleylamine/oleic acid could be rendered hydrophilic through coating another layer of amphiphilic poly(DMA-*r*-mPEGMA-*r*-MA).[5] The polymer contains the hydrophilic PEG moiety, enabling the particles to disperse in physiological buffers. Also, the long hydrophobic alkyl chain insects into the original hydrophobic layer of DBNPs via van der Waals interaction for the encapsulation of DBNPs.

The second method, surfactant exchange, replaces the original surfactants with the dispersants (*e.g.* catechol derivatives[48]) that have stronger affinity to the surface of DBNPs. It is both cost-effective and time-efficient, which makes these dispersants the top choice for the proof-of-concept test of self-made DBNPs in the laboratory. Furthermore, as different components of DBNPs have their unique surface chemistry properties, this permits the individual replacement of the original surfactants on each component. For example, we demonstrated that the Fe$_3$O$_4$ core and Au core in Au–Fe$_3$O$_4$ DBNPs could be individually modified with dopamine and thiol-containing molecules.[39,41] As revealed by our early work, the specific coordination bond between Fe and dopamine enabled us to replace the original oleic acid on the Fe$_3$O$_4$ core.[48] Thus, by using dopamine as an anchor, we connected the functional PEG to the surface of Fe$_3$O$_4$, which linked the functional groups [*e.g.* epidermal growth factor receptor (EGFR) antibody, a monoclonal antibody targeting EGFR on cancer cells). To modify the gold core, the thiol-containing molecules were incubated with now water-soluble DBNPs. The high affinity of thiol for gold tripped off the original oleylamine on the gold core and put the second functional groups (*e.g.* the therapeutic agent cisplatin) in their place.[41]

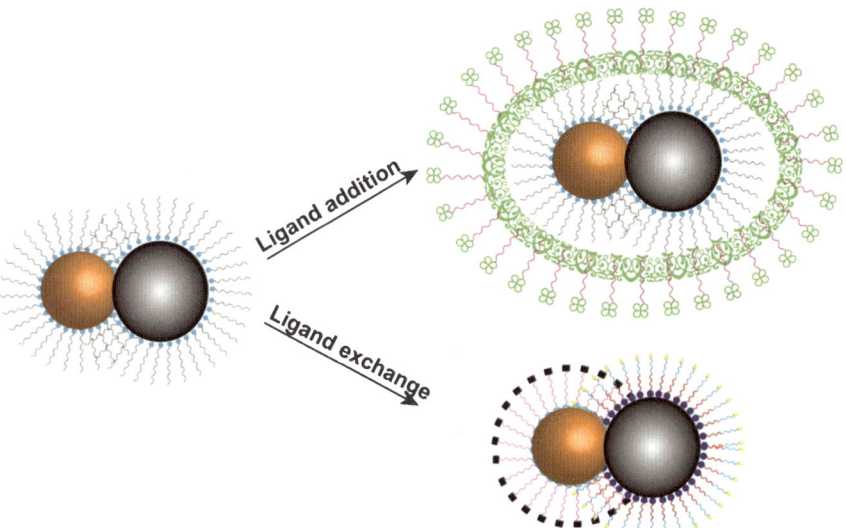

Figure 2.14 Two strategies for the surface modification of DBNPs.

After modification, there are three parameters which normally need to be characterized: the morphology, hydrodynamic diameter and zeta potential. Morphology provides evidence for the integrity of DBNPs during the modification. The size and charge of DBNPs greatly influence the recognition and internalization by cells, the circulation time *in vivo* and the distribution in organs.[49,50] For example, when DBNPs are delivered into the circulation through systematic administration and have a hydrodynamic diameter larger than 200 nm, DBNPs could be easily sequestered by the reticuloendothelial system (RES) of the spleen and liver. Again, if their size is less than 10 nm, they will be subject to rapid renal clearance.[51] Thus, to ensure the maximum circulation time, the optical hydrodynamic diameter is between 10 and 200 nm.

2.3.2.2   Molecular Imaging

Molecular imaging is a biomedical research discipline dealing with the visualization, characterization and quantification of biological processes at the cellular and subcellular levels. The images produced through molecular imaging (usually non-invasive) would help reflect the cellular and molecular pathways and mechanism of diseases in the subjects. Existing imaging techniques include MRI, optical imaging, CT and photoacoustic imaging, and each of them has their own advantages and disadvantages. For example, Fe_3O_4 NP-based MRI is advantageous for a large penetration depth and high spatial resolution (10–100 μm), but suffers from lower sensitivity, the high background coming from 'field disturber' and long acquisition time.[52,53] On the other side, although suffering from a small penetration depth, optical imaging based on Au NPs or QDs is time- and cost-efficient, which facilitates rapid testing of biological hypotheses and proof-of-principles in living experimental models.[54] Therefore, in the case of cell tracking with NP labeling,[53] using DBNPs comprising Fe_3O_4 and Au NPs permits the combination of the advantages of MRI and optical imaging, which would allow the prompt optical examination of cells right after labeling and longitudinal tracking *in vivo* following administration.

The potential candidates for two or more imaging modalities reported thus far include, but are not limited to, Au–Fe_3O_4,[14] Ag–Fe_3O_4,[17] Au–$FePt$[55] and CdSe–Fe_2O_3 NPs.[56] Here we use the Au–Fe_3O_4 NPs as an example to demonstrate DBNPs as multifunctional probes.

As synthesized, Au–Fe_3O_4 DBNPs are coated with a hydrophobic layer composed of oleate and oleylamine. To make them hydrophilic, we adopted the ligand-exchange method (Figure 2.15a). Specifically, the original oleate/oleylamine coating on Fe_3O_4 core was replaced by a catechol unit present in the dopamine molecule that is linked to PEG (M_r = 3000), while oleylamine around the Au core was replaced by HS–PEG–NH_2 (M_r = 2204), with thiol (HS–) attaching to Au. Finally, the targeting agent (*e.g.* EGFR monoclonal antibody)[57] was connected to PEG via ethyl(dimethylaminopropyl)carbodiimide/N-hydroxysuccinimide (EDC/NHS) chemistry on the Fe_3O_4 side.[58]

EGFR is well known to be highly expressed on cancer cells, including breast cancer cell line A431 cells. Through grafting EGFR antibody on the surface of DBNPs, we could label/detect EGFR-positive cancer cells with Au–Fe$_3$O$_4$ DBNPs, which might potentially be useful in the early diagnostics and localization of cancers through MRI[59] and for visual location of tumor borders and spread during surgery.[60] Despite multiple steps of chemical reaction, there was no noticeable change in the morphology of NPs (Figure 2.15b).

To demonstrate the ability of Au–Fe$_3$O$_4$ DBNPs as contrast agents for multiple imaging modalities, we examined their potential in T_2-based MRI and reflection imaging (Figure 2.15c, d). It is well known that Fe$_3$O$_4$ NPs provide negative contrast by creating a large dipolar magnetic field gradient that is experienced by protons in close proximity to the particle.[61] One of the effects is the signal dephasing due to the local field inhomogeneity induced in water molecules near Fe$_3$O$_4$ NPs.[62] This effect is seen as hypointensity or negative contrast on T_2-weighted and T_2*-weighted images due to shortening of T_2 relaxation times (Figure 2.15c, row i). Also, the higher Fe concentration, the higher the negative contrast will be. Now, since Au–Fe$_3$O$_4$ DBNPs have the Fe$_3$O$_4$ component, it is natural to predict and prove that Au–Fe$_3$O$_4$ DBNPs also causes negative contrast in the T_2-weighted images (Figure 2.15c, rows ii–iv). However, as a result of the junction effect in the dumbbell structure,[14] the signal dephasing ability of Fe$_3$O$_4$ was weakened in Au–Fe$_3$O$_4$ DBNPs.[63] As shown in Figure 2.15c, rows i–iii, at the same iron concentration, Fe$_3$O$_4$ NPs caused higher negative contrast than Au–Fe$_3$O$_4$ DBNPs. Also, the larger the Au core in DBNPs, the weaker the signal dephasing ability was.

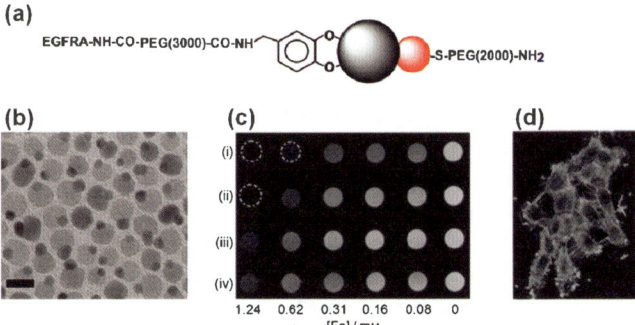

Figure 2.15 Au–Fe$_3$O$_4$ DBNPs as dual imaging agents. (a) Schematic illustration of surface functionalization of the Au–Fe$_3$O$_4$ DBNPs, (b) TEM image of 8–20 nm Au–Fe$_3$O$_4$ dumbbell MNPs. Scale bar is 20 nm. (c) T_2-weighted MRI images of (i) 20 nm Fe$_3$O$_4$, (ii) 3–20 nm Au–Fe$_3$O$_4$, (iii) 8–20 nm Au–Fe$_3$O$_4$ MNPs and (iv) A 431 cells labeled with 8–20 nm Au–Fe$_3$O$_4$ MNPs. (d) Reflection images of the A431 cells labeled with 8–20 nm Au–Fe$_3$O$_4$ MNPs. Reprinted with permission from reference 39. Copyright © 2008 John Wiley and Sons, Inc.

After establishing their ability as contrast agents for MRI, we then started to examine whether the optical properties of Au were preserved. As a result of surface plasmon resonance, Au NPs are optically active and have been used in a variety of optical imaging methods, including light-scattering imaging, two-photon luminescence imaging and surface-enhanced Raman scattering.[64] We discovered that it also presented enhanced light reflection between 500 and 800 nm, but unfortunately the reflection ability of Au NPs was decreased in Au–Fe_3O_4 DBNPs owing to the junction effect.[39] Nevertheless, the partially reserved optical properties of the Au core still allowed us to visualize A431 cells labeled with Au–Fe_3O_4 DBNPs with a scanning confocal microscope at 594 nm (Figure 2.15d). The signal reflected the typical morphology of epithelial cells and was much stronger on the plasma membrane of cells due to the preferred binding between EGFR and EGFR antibody.

In addition to being a platform to combine MRI and optical imaging, Au–Fe_3O_4 DBNPs are also suitable for both MRI and CT imaging,[5] given that the Au core has strong X-ray absorption.[65] As a proof-of-concept, Jon and co-workers examined Au–Fe_3O_4 DBNPs for their X-ray absorption ability that is attributed to the Au core and T_2-weighted contrast enhancement that comes from the Fe_3O_4 core first.[5] They then explored the dual imaging ability in an orthotopic hepatoma mouse model. More specifically, Au–Fe_3O_4 DBNPs were injected intravenously into a mouse that had an artificial tumor in the liver. After only 1 h, the accumulation of Au–Fe_3O_4 DBNPs allowed the authors to distinguish the hepatoma from the surrounding normal liver tissue through both CT and MRI.

2.3.2.3   *Magnetic Control*

A unique feature of magnetic NPs is to respond well to magnetic field, which allows us to manipulate and remotely control specific cellular components *in vitro* and, more importantly, *in vivo*, providing clinicians and scientists with a powerful tool for investigating cell function and molecular signaling pathways. So far, it has been utilized for cellular manipulation,[66] protein purification,[67] bacteria detection,[68] drug delivery[69] and stem cell engraftment.[70] Traditionally, the evaluation of the success of magnetic control relies on outcomes such as the purification efficiency, therapeutic effects and the number of engrafted stem cells. This could take time and increase costs. One solution is to visualize the magnetic control *in situ* with the help of optical tools. To achieve this goal, DBNPs comprising a magnetic core and optically active core have been tested.[36,40,71,72]

Xu and co-workers demonstrated the magnetic-controlled aggregation of glutathione-modified Fe_3O_4–CdSe DBNPs in HEK293T cells[40] (Figure 2.16a–c). Specially, they prepared Fe_3O_4–CdSe DBNPs that had a 1.1 emu g^{-1} magnetic moment and fluorescence emission at 610 nm. After ligand exchange, the Fe_3O_4–CdSe DBNPs became hydrophilic and were internalized by cells. To confirm that the change of signal location during the magnetic control came from the movement of DBNPs instead of cells, the authors genetically expressed green fluorescent protein (GFP) in the cellular plasma as the background. Before applying a magnetic field, the fluorescent signal from both Fe_3O_4–CdSe DBNPs and GFP distributed evenly through the cytoplasm (Figure 2.16d). Then a magnet was placed beside the cells and induced the movement of Fe_3O_4–CdSe DBNPs along the direction of the magnetic field, which finally accumulated beside the magnet (Figure 2.16e).

2.3.2.4   Drug Delivery

A further bioapplication of DBNPs is the delivery of therapeutics. Compared with single-core NPs, DBNPs provide two or more types of surface chemistry that allow the functionalization of each core individually. The advantage of individual functionalization instead of one-step conjugation is the opportunity to separate two interfering ligands into different parts of one system. Furthermore, as discussed above, DBNPs can act as contrast agents for

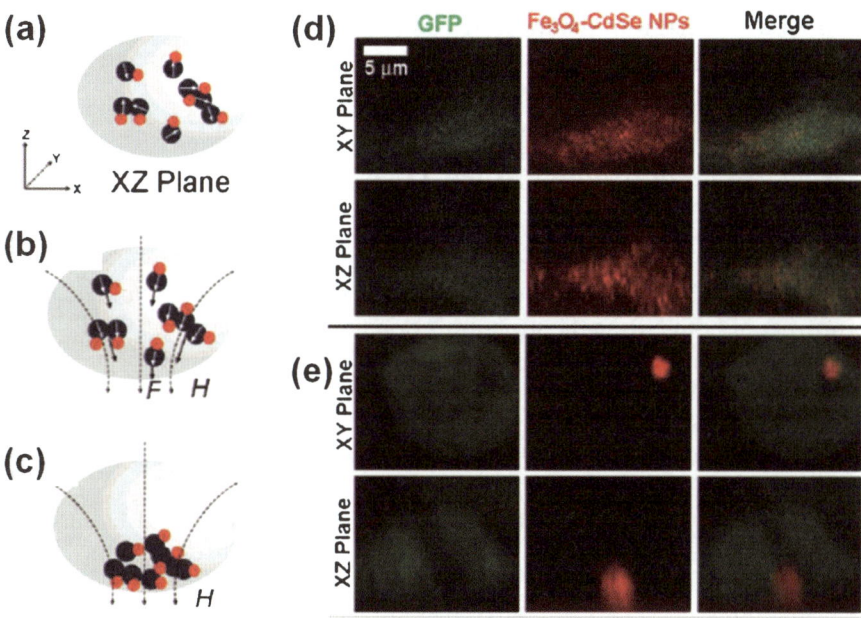

Figure 2.16   Visualization of magnetic control with Fe_3O_4–CdSe DBNPs. Schematic illustration of magnetic control: (a) DBNPs distributed randomly in the cell cytoplasm without a magnetic field; (b, c) once the magnetic field was applied, the DBNPs moved towards and aggregated at one side of the cell close to the magnet; (d) an HEK293T cell expressing GFP was incubated with the Fe3O4-CdSe@GSH DBNPs and had DBNPs distributed evenly in the cytoplasm; (e) DBNPs aggregated at one side of the cell close to the magnet after 8 h. Adapted with permission from reference 40. Copyright 2008 American Chemical Society.

multiple imaging modalities, which permits the sensitive examination of the distribution of therapeutics following administration.

As an example, we utilized Au–Fe$_3$O$_4$ DBNPs to deliver the chemotherapeutic agent cisplatin to Her2-positive breast cancer cells while visualizing the delivery through the reflection imaging.[41] As mentioned earlier, Au–Fe$_3$O$_4$ DBNPs are composed of an Au core and Fe$_3$O$_4$ core with different surface chemistry. The Fe$_3$O$_4$ core has high affinity to the catechol-derived molecule dopamine whereas the Au core prefers to bind with thiol groups. Thus, with dopamine, we linked the targeting agent for Her2-positive cancer cells (Herceptin) on to the Fe$_3$O$_4$ core (Figure 2.17a). Subsequently, cisplatin was anchored on the Au surface through a thiolated bidentate dicarboxylate ligand that is similar to the structure of carboplatin.[73] Through controlling the size of the Au core, we were able to tune the loading efficiency of cisplatin. Herceptin allowed the final conjugates to target Her2-positive SkBr3 breast cancer cells, not the Her2-negative MCF-7 breast cancer cells. More importantly, this binding could be visualized through reflection imaging (Figure 2.17b). In DBNP-labeled SkBr3 cells, it is clear to see the DBNP signal from both the plasma membrane and cytoplasm. As a result of platinum accumulation, SkBr3 cells labeled with platinum–Au–Fe$_3$O$_4$–Herceptin DBNPs were subject to the increasing expression of p53 (a tumor-suppressor protein) and massive apoptosis (Figure 2.17c, d).

2.4   Conclusion and Future Directions

Recent synthetic progress has led to the controlled formation of DBNPs with two or more NPs in intimate contact. The interfacial interaction existing between two different NPs can induce changes in the physical and/or chemical properties of both NPs in the structure. Owing to this interaction, DBNPs have shown beneficial properties for catalytic applications. The presence of two or more types of distinctive surfaces and functional components in the DBNPs also facilitates selective surface modification and multiple functionalization, making them useful tools for molecular probing, medical diagnosis and chemotherapy.

Despite the mentioned studies using DBNPs for catalytic and biomedical applications, the DBNP system has not been intensively investigated as much as the traditional Fe$_3$O$_4$ NPs, QDs and Au NPs. This might be due to the concerns about the stability of the system. As shown in recent work by Sun and co-workers,[74] different components in the dual-core system (*i.e.* Ag–Fe$_3$O$_4$ DBNPs) easily dissociated under extensive heating. This suggests that the stability of the dumbbell-like structures should be considered more in the future design and application of DBNP systems.

Further research on DBNPs should also address the growth mechanism, in particular the control over the morphology of the composite nanostructures.[13] The target is to expand our understanding of the thermodynamic and kinetic parameters governing the heterogeneous nucleation and anisotropic growth in

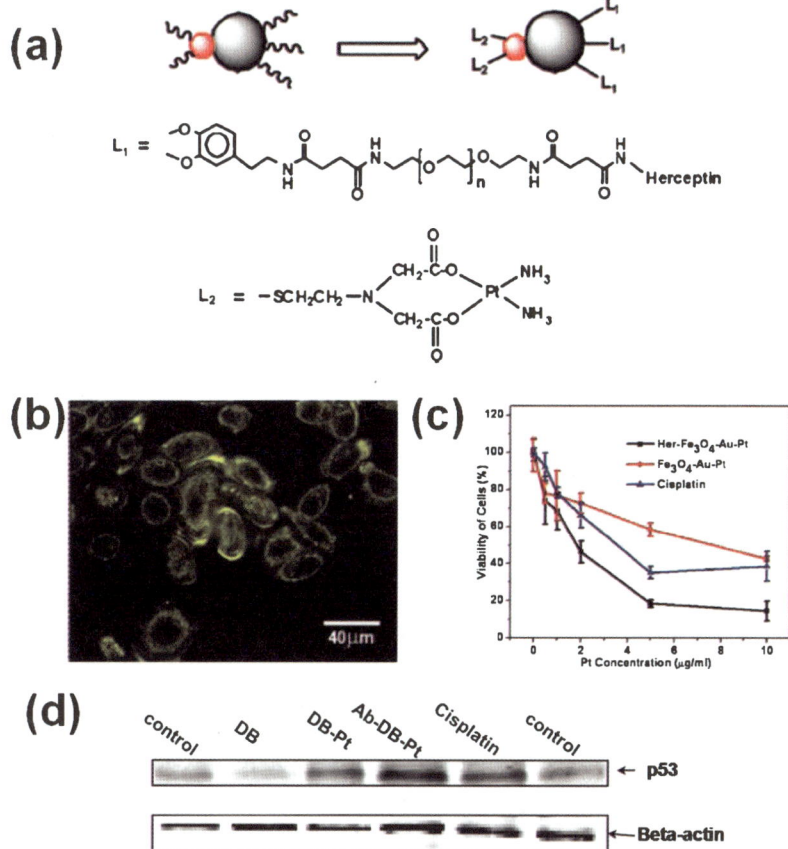

Figure 2.17 Au–Fe_3O_4 DBNPs as drug carriers. (a) Schematic illustration of Au–Fe_3O_4 DBNPs coupled with Herceptin and a platinum complex. (b) Reflection image of Her2-positive SkBr3 cells after incubation with platinum–Au–Fe_3O_4–Herceptin DBNPs. (c) Viability of SkBr3 cells after incubation with platinum-Au– Fe_3O_4 DBNPs, platinum–Au–Fe_3O_4–Herceptin NPs and free cisplatin. (d) p53 expression in SkBr3 cells after incubation with Au–Fe_3O_4, Fe_3O_4–Au–platinum, Herceptin–Fe_3O_4–Au–platinum or cisplatin at the same Pt concentration or Fe concentration, where a higher platinum accumulation in cells resulted in higher expression of p53. Reprinted with permission from reference 41. Copyright 2009 American Chemical Society.

homogeneous solutions and then to achieve the rational design and synthesis of DBNPs with the desired structures and properties.

Finally, other applications of DBNPs are yet to be explored. By tailoring the electronic coupling between the two components, DBNPs are expected to be of interest for the fabrication of electronic and optical devices.[14,18] The multifunctionality of DBNPs may also see its role in the development of electrochemical materials for energy conversion and storage.[19,75] Exploration of these aspects, plus continual investigations of catalytic and biomedical applications, will lead to a wide range of novel materials with advanced properties and functional performance.

References

1. M. R. Mozafari, *Nanomaterials and Nanosystems for Biomedical Applications*, Springer, Dordrecht, 2007.
2. V. M. Rotello, *Nanoparticles: Building Blocks for Nanotechnology*, Kluwer Academic/Plenum Publishers, New York, 2004.
3. M. Lattuada and T. A. Hatton, *Nano Today*, 2011, **6**, 286–308.
4. C. Wang, C. J. Xu, H. Zeng and S. H. Sun, *Adv. Mater.*, 2009, **21**, 3045–3052.
5. D. Kim, M. K. Yu, T. S. Lee, J. J. Park, Y. Y. Jeong and S. Jon, *Nanotechnology*, 2011, **22**, 155101.
6. C. Wang, H. F. Yin, S. Dai and S. H. Sun, *Chem. Mater.*, 2010, **22**, 3277–3282.
7. M. Haruta, T. Kobayashi, H. Sano and N. Yamada, *Chem. Lett.*, 1987, 405–408.
8. L. M. Molina and B. Hammer, *Phys. Rev. Lett.*, 2003, **90**., 206102.
9. S. Laursen and S. Linic, *Phys. Rev. Lett.*, 2006, **97**, 026101.
10. L. Carbone and P. D. Cozzoli, *Nano Today*, 2010, **5**, 449–493.
11. B. L. Cushing, V. L. Kolesnichenko and C. J. O'Connor, *Chem. Rev.*, 2004, **104**, 3893–3946.
12. M. Volmer and A. Weber, *Z. Phys. C*, 1926, **119**, 227.
13. C. Wang, W. D. Tian, Y. Ding, Y. Q. Ma, Z. L. Wang, N. M. Markovic, V. R. Stamenkovic, H. Daimon and S. H. Sun, *J. Am. Chem. Soc.*, 2010, **132**, 6524–6529.
14. H. Yu, M. Chen, P. M. Rice, S. X. Wang, R. L. White and S. H. Sun, *Nano Lett.*, 2005, **5**, 379–382.
15. H. W. Gu, R. K. Zheng, X. X. Zhang and B. Xu, *J. Am. Chem. Soc.*, 2004, **126**, 5664–5665.
16. C. Wang, Y. J. Wei, H. Y. Jiang and S. H. Sun, *Nano Lett.*, 2009, **9**, 4544–4547.
17. H. W. Gu, Z. M. Yang, J. H. Gao, C. K. Chang and B. Xu, *J. Am. Chem. Soc.*, 2005, **127**, 34–35.
18. Y. Q. Li, G. Zhang, A. V. Nurmikko and S. H. Sun, *Nano Lett.*, 2005, **5**, 1689–1692.
19. C. Wang, H. Daimon and S. H. Sun, *Nano Lett.*, 2009, **9**, 1493–1496.
20. M. G. Bawendi, M. L. Steigerwald and L. E. Brus, *Annu. Rev. Phys. Chem.*, 1990, **41**, 477–496.
21. A. J. Nozik, *Annu. Rev. Phys. Chem.*, 2001, **52**, 193–231.
22. A. J. Nozik, M. C. Beard, J. M. Luther, M. Law, R. J. Ellingson and J. C. Johnson, *Chem. Rev.*, 2010, **110**, 6873–6890.
23. T. Mokari, E. Rothenberg, I. Popov, R. Costi and U. Banin, *Science*, 2004, **304**, 1787–1790.
24. D. K. Yi, S. T. Selvan, S. S. Lee, G. C. Papaefthymiou, D. Kundaliya and J. Y. Ying, *J. Am. Chem. Soc.*, 2005, **127**, 4990–4991.
25. K. W. Kwon and M. Shim, *J. Am. Chem. Soc.*, 2005, **127**, 10269–10275.

26. W. L. Shi, H. Zeng, Y. Sahoo, T. Y. Ohulchanskyy, Y. Ding, Z. L. Wang, M. Swihart and P. N. Prasad, *Nano Lett.*, 2006, **6**, 875–881.
27. M. R. Buck, J. F. Bondi and R. E. Schaak, *Nat. Chem.*, 2012, **4**, 37–44.
28. M. Haruta, N. Yamada, T. Kobayashi and S. Iijima, *J. Catal.*, 1989, **115**, 301–309.
29. M. S. Chen and D. W. Goodman, *Science*, 2004, **306**, 252–255.
30. A. A. Herzing, C. J. Kiely, A. F. Carley, P. Landon and G. J. Hutchings, *Science*, 2008, **321**, 1331–1335.
31. D. I. Enache, J. K. Edwards, P. Landon, B. Solsona-Espriu, A. F. Carley, A. A. Herzing, M. Watanabe, C. J. Kiely, D. W. Knight and G. J. Hutchings, *Science*, 2006, **311**, 362–365.
32. Q. Fu, H. Saltsburg and M. Flytzani-Stephanopoulos, *Science*, 2003, **301**, 935–938.
33. L. Kesavan, R. Tiruvalam, M. H. Ab Rahim, M. I. bin Saiman, D. I. Enache, R. L. Jenkins, N. Dimitratos, J. A. Lopez-Sanchez, S. H. Taylor, D. W. Knight, C. J. Kiely and G. J. Hutchings, *Science*, 2011, **331**, 195–199.
34. Z. P. Liu, X. Q. Gong, J. Kohanoff, C. Sanchez and P. Hu, *Phys. Rev. Lett.*, 2003, **91**, 266102.
35. H. F. Yin, C. Wang, H. G. Zhu, S. H. Overbury, S. H. Sun and S. Dai, *Chem. Commun.*, 2008, 4357–4359.
36. D. Jańczewski, Y. Zhang, G. K. Das, D. K. Yi, P. Padmanabhan, K. K. Bhakoo, T. T. Y. Tan and S. T. Selvan, *Microsc. Res. Tech.*, 2011, **74**, 563–576.
37. J.-S. Choi, Y.-W. Jun, S.-I. Yeon, H. C. Kim, J.-S. Shin and J. Cheon, *J. Am. Chem. Soc.*, 2006, **128**, 15982–15983.
38. S.-H. Choi, H. B. Na, Y. I. Park, K. An, S. G. Kwon, Y. Jang, M.-H. Park, J. Moon, J. S. Son, I. C. Song, W. K. Moon and T. Hyeon, *J. Am. Chem. Soc.*, 2008, **130**, 15573–15580.
39. C. Xu, J. Xie, D. Ho, C. Wang, N. Kohler, E. G. Walsh, J. R. Morgan, Y. E. Chin and S. Sun, *Angew. Chem. Int. Ed.*, 2008, **47**, 173–176.
40. J. Gao, W. Zhang, P. Huang, B. Zhang, X. Zhang and B. Xu, *J. Am. Chem. Soc.*, 2008, **130**, 3710–3711.
41. C. J. Xu, B. D. Wang and S. H. Sun, *J. Am. Chem. Soc.*, 2009, **131**, 4216–4217.
42. F. Caruso, *Colloids and Colloid Assemblies: Synthesis, Modification, Organization and Utilization of Colloid Particles*, Wiley-VCH, Weinheim, 2004.
43. A. Stradner, H. Sedgwick, F. Cardinaux, W. C. K. Poon, S. U. Egelhaaf and P. Schurtenberger, *Nature*, 2004, **432**, 492–495.
44. I. Capek, *Nanocomposite Structures and Dispersions: Science and Nanotechnology – Fundamental Principles and Colloidal Particles*, Elsevier, Amsterdam,, 2006.
45. E. Amstad, M. Textor and E. Reimhult, *Nanoscale*, 2011, **3**, 2819–2843.
46. C. J. Xu and S. Sun, *Dalton Trans.*, 2009, 5583–5591.
47. Y. Tai, L. Wang, G. Yan, J.-M. Gao, H. Yu and L. Zhang, *Polym. Int.*, 2011, **60**, 976–994.
48. C. J. Xu, K. M. Xu, H. W. Gu, R. K. Zheng, H. Liu, X. X. Zhang, Z. H. Guo and B. Xu, *J. Am. Chem. Soc.*, 2004, **126**, 9938–9939.

49. V. Mailander and K. Landfester, *Biomacromolecules*, 2009, **10**, 2379–2400.
50. S. E. A. Gratton, P. A. Ropp, P. D. Pohlhaus, J. C. Luft, V. J. Madden, M. E. Napier and J. M. DeSimone, *Proc. Natl. Acad. Sci. U. S. A.*, 2008, **105**, 11613–11618.
51. A. K. Gupta and M. Gupta, *Biomaterials*, 2005, **26**, 3995–4021.
52. R. Guzman, N. Uchida, T. M. Bliss, D. He, K. K. Christopherson, D. Stellwagen, A. Capela, J. Greve, R. C. Malenka, M. E. Moseley, T. D. Palmer and G. K. Steinberg, *Proc. Natl. Acad. Sci. U. S. A.*, 2007, **104**, 10211–10216.
53. C. Xu, L. Mu, I. Roes, D. Miranda-Nieves, M. Nahrendorf, J. A. Ankrum, W. Zhao and J. M. Karp, *Nanotechnology*, 2011, **22**, 494001.
54. R. Weissleder and M. J. Pittet, *Nature*, 2008, **452**, 580–589.
55. J. S. Choi, Y. W. Jun, S. I. Yeon, H. C. Kim, J. S. Shin and J. Cheon, *J. Am. Chem. Soc.*, 2006, **128**, 15982–15983.
56. S. T. Selvan, P. K. Patra, C. Y. Ang and J. Y. Ying, *Angew. Chem. Int. Ed.*, 2007, **46**, 2448–2452.
57. R. I. Nicholson, J. M. W. Gee and M. E. Harper, *Eur. J. Cancer*, 2001, **37**, 9–15.
58. J. R. McCarthy and R. Weissleder, *Adv. Drug. Deliv. Rev.*, 2008, **60**, 1241–1251.
59. E. C. Cho, C. Glaus, J. Chen, M. J. Welch and Y. Xia, *Trends Mol. Med.*, 2010, **16**, 561–573.
60. M. Patlak, *J. Natl. Cancer Inst.*, 2011, **103**, 173–174.
61. R. Hao, R. Xing, Z. Xu, Y. Hou, S. Gao and S. Sun, *Adv. Mater.*, 2010, **22**, 2729–2742.
62. W. J. Rogers, C. H. Meyer and C. M. Kramer, *Nat. Clin. Pract. Cardiovasc. Med.*, 2006, **3**, 554–562.
63. H. Zeng and S. Sun, *Adv. Funct. Mater.*, 2008, **18**, 391–400.
64. X. Huang, P. K. Jain, I. H. El-Sayed and M. A. El-Sayed, *Nanomedicine*, 2007, **2**, 681–693.
65. J. F. Hainfeld, D. N. Slatkin, T. M. Focella and H. M. Smilowitz, *Br. J. Radiol.*, 2006, **79**, 248–253.
66. J. Dobson, *Nat. Nanotechnol.*, 2008, **3**, 139–143.
67. C. Xu, K. Xu, H. Gu, X. Zhong, Z. Guo, R. Zheng, X. Zhang and B. Xu, *J. Am. Chem. Soc.*, 2004, **126**, 3392–3393.
68. H. Gu, K. Xu, C. Xu and B. Xu, *Chem. Commun.*, 2006, 941–949.
69. Y. Namiki, T. Namiki, H. Yoshida, Y. Ishii, A. Tsubota, S. Koido, K. Nariai, M. Mitsunaga, S. Yanagisawa, H. Kashiwagi, Y. Mabashi, Y. Yumoto, S. Hoshina, K. Fujise and N. Tada, *Nat. Nanotechnol.*, 2009, **4**, 598–606.
70. K. Cheng, T. S. Li, K. Malliaras, D. R. Davis, Y. Zhang and E. Marban, *Circ. Res.*, 2010, **106**, 1570–1581.
71. A. Kale, S. Kale, P. Yadav, H. Gholap, R. Pasricha, J. P. Jog, B. Lefez, B. Hannoyer, P. Shastry and S. Ogale, *Nanotechnology*, 2011, **22**, 225101.
72. Q. Wei, H.-M. Song, A. P. Leonov, J. A. Hale, D. Oh, Q. K. Ong, K. Ritchie and A. Wei, *J. Am. Chem. Soc.*, 2009, **131**, 9728–9734.

73. N. J. Wheate, S. Walker, G. E. Craig and R. Oun, *Dalton Trans.*, 2010, **39**, 8113–8127.

74. S. Peng, C. H. Lei, Y. Ren, R. E. Cook and Y. G. Sun, *Angew. Chem. Int. Ed.*, 2011, **50**, 3158–3163.

75. Y. C. Lu, Z. C. Xu, H. A. Gasteiger, S. Chen, K. Hamad-Schifferli and Y. Shao-Horn, *J. Am. Chem. Soc.*, 2010, **132**, 12170–12171.

CHAPTER 3

Janus Particles with Distinct Compartments via Electrohydrodynamic Co-jetting

TAE-HONG PARK†[a] AND JOERG LAHANN*[a,b]

[a] Department of Chemical Engineering, University of Michigan, Ann Arbor, MI 48109, USA; [b] Departments of Materials Science and Engineering, Macromolecular Science and Engineering and Biomedical Engineering, University of Michigan, Ann Arbor, MI 48109, USA
*E-mail: lahann@umich.edu
†Present address: Nuclear Chemistry Research Division, Korea Atomic Energy Research Institute, Daejeon 305-353, Korea

3.1 Introduction

Nano- and microparticles have been successfully utilized in a variety of research areas, such as diagnostics, therapeutics and photonic crystals. Structural and compositional flexibility, tunability and versatility of colloidal particles have been accomplished *via* rapid and continuous progress in particle fabrication methods, which allows for the precise engineering of nano- and microparticles with respect to shape, size, surface structures and mechanical properties.[1,2] In particular, the complementary control of internal (bulk) and external (surface) features has been increasingly pursued as an important design parameter for multifunctional particles. Anisotropic particles possessing two or more distinct shapes, surface patterns or compositions in a single entity may allow fundamentally new design elements needed for building complex hierarchical assemblies.[3-13] Materials comprised of self-assembled

RSC Smart Materials No. 1
Janus Particle Synthesis, Self-Assembly and Applications
Edited by Shan Jiang and Steve Granick

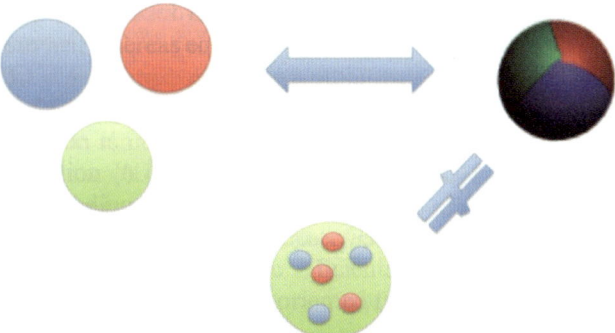

Figure 3.1   Multicompartmental particles (right) can display the combined set of properties associated with the individual components. This architecture is distinctly different from geometries based on simple mixing of materials (bottom). Reproduced with permission from reference 10.

particles may find diverse applications in multiplexed bioassays,[14] vehicles for multiple therapeutic modalities,[15] biohybrid materials[16] and microactuators.[17]

Anisotropic surface functionalization, in part, can generate patchy particles with orthogonal surface properties. Surface anisotropy can exhibit bipolar characteristics, such as simultaneously being hydrophilic and hydrophobic, or can be used for self-assembly, photonic materials or drug delivery.[3] On the other hand, compositional bulk anisotropy may permit additional Janus particle properties (Figure 3.1). By recognizing that compartmentalized material distribution in particles may, in some cases, be as important as their actual chemical make-up, the versatile utilities of such compositionally anisotropic particles may be broad, going well beyond the initial patchy particles.[8,10] In particular, the independent control of individual compartments within a single particle allows for inclusion of multiple materials or material combinations with dissimilar sets of properties, which are able to present controlled release of multiple drugs with independent release kinetics (provided that each drug is loaded in another compartment) or combination of fully decoupled modalities for combined imaging and therapy. In addition, the selective surface functionalization of multicompartmental particles may lead to more versatile and multifunctional particles.

3.2   Compartmentalization of Nano- and Microparticles *via* Electrohydrodynamic Co-jetting

Compartmentalized particles have been fabricated with several distinct methods. Traditionally, the self-assembly of block copolymers[18,19] and phase separation induced during seeded polymerizations[20,21] lead to diverse architectures of anisotropic particles. Lithographic and template-assisted methods often show great flexibility in constructing non-spherical Janus particles in large quantities.[22–24] In addition, a growing body of research has emphasized the potential of droplet-based strategies for the preparation of multifunctional particles. Microfluidic techniques combined with photolithography provide a promising route towards compartmentalized particles with respect to highly uniform size and shape in addition to more complex inner and outer geometries.[14,25–27] However, a narrow choice of materials is available and relatively large particle sizes are typically reported. Further progress will benefit from diversification of materials and the design of widely applicable processing devices.

Electrospraying and electrospinning processes of polymer solutions are straightforward methods for fabricating nano-/microparticles and also fibers by applying a high electrical voltage to polymeric solutions. Control of the polymer solution properties such as material selection, concentration, viscosity and conductivity can result in a diverse range of sizes and shapes in the nano- and micrometer range.[28] Building on these processes, electrohydrodynamic (EHD) co-jetting involves two or more capillary needles in a side-by-side configuration that allow different polymer solutions to be processed in parallel. Under the laminar flow regimen encountered in these systems, biphasic droplets are formed at the outlet point of the adjacent needles (Figure 3.2).[10,29] Application of an electric field to the nozzle leads to a stable Taylor cone at the tip of the biphasic droplet. The Taylor cone builds the basis of a jet that forms as a result of continuous stretching of the polymer thread. The rapid acceleration favors atomization of the charged jets and significantly increases the surface area. Thus, instantaneous solvent evaporation and solidification of the non-volatile components occur. The extremely fast nanoprecipitation prevents interfacial diffusion and also reorientation of the solidifying materials. As a result, the particles are trapped in the initial, flow-determined compartmentalized geometry. Not surprisingly, however, the preparation of bicompartmental polymer particles *via* electrohydrodynamic co-jetting requires delicate control of several competing factors.

Initial work was focused on water-soluble polymers because environmentally friendly water-based jetting systems typically exhibit lower toxicity, which is important for biomedical applications. The preparation of bicompartmental particles using the EHD co-jetting approach was first achieved using aqueous solutions of poly(acrylic acid) (PAA) and poly(ethylene glycol) (PEG).[29] The relatively high free surface energy of water led to a nearly spherical droplet, where the biphasic polymer/polymer interface is maintained and the Taylor cone is formed − literally as a virtual nozzle (Figure 3.2). Instantaneous breakup and solidification of the electrified jets into smaller flying objects under electric fields led to bicompartmental nanocolloids with well-preserved anisotropic characteristics. The preferential compartmentalization can be attributed to the manipulation of the liquid phase and the interface is maintained throughout EHD jetting and solidification. On the other hand, the use of water-soluble polymers for the fabrication of particles also requires subsequent chemical reactions to ensure stability under physiological condi-

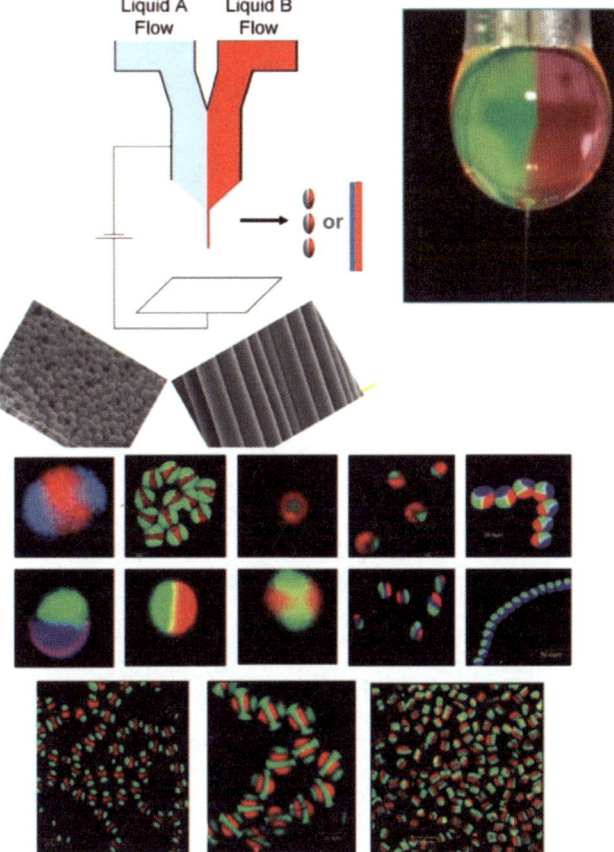

Figure 3.2   Top left: schematic description of the electrohydrodynamic (EHD) co-jetting process, which can lead to either particles (electrospraying) or fibers (electrospinning). Top right: a photograph of a typical Taylor cone and jet ejection during the EHD co-jetting. The interface between the two aligned polymer solutions is clearly observed. Bottom: examples of particles and fibers obtained from EHD co-jetting. Reproduced with permission from reference 10.

tions. In the case of poly(acrylamide-*co*-acrylic acid) [P(AAm-*co*-AA)] particles, thermal imidization of carboxylic acid and amide groups provided sufficient particle stability and resulted in controllable swelling in water.[30] The cross-linking reaction was monitored *via* FTIR spectroscopy (Figure 3.3). The incubation of bicompartmental P(AAm-*co*-AA) particles at 175 °C led to a decrease of the characteristic C=O stretching band (amide I around 1673 cm^{-1}) and the C–NH band of the amide group (amide II around 1613 cm^{-1}). Furthermore, imide bands appeared near 1720 and 1221 cm^{-1} indicating considerable cross-linking of the biphasic nanocolloids. Importantly, the initial particle shape and bicompartmental geometry were maintained throughout the

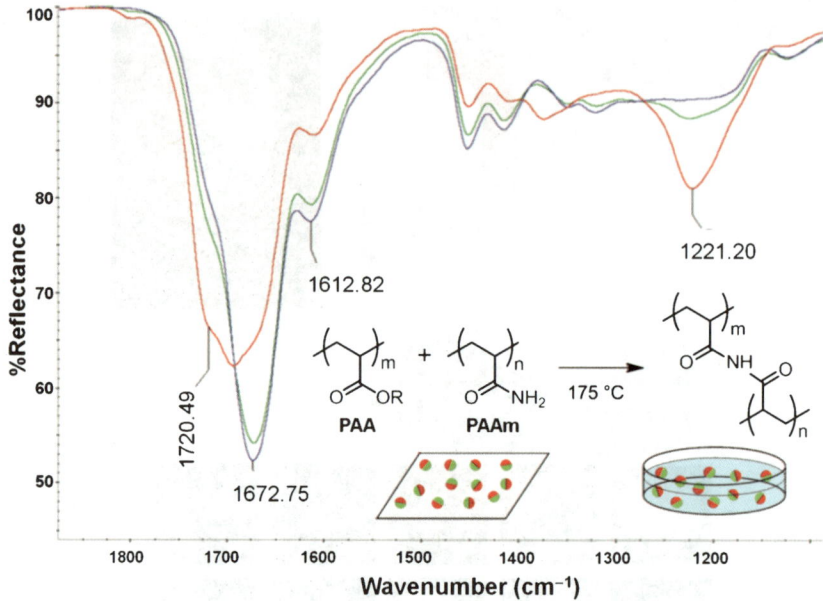

Figure 3.3   FTIR reflectance spectra of thermal cross-linking of P(AAm-*co*-AA)
bicompartmental particles for different reaction times. The blue
spectrum was recorded for nanocolloids prepared by jetting on top of
a gold substrate before the reaction, whereas the green and red spectra
were measured for the same colloids after a 1 h and a 12 h cross-linking
reaction at 175 °C, respectively. After the reaction, biphasic nanocolloids
were stable in water for extended periods of time. Reproduced with
permission from reference 30. Copyright 2007 American Chemical
Society.

cross-linking process. However, this thermal cross-linking requires exposure of
particles to elevated temperatures and this approach is only suitable when
materials or payloads within the polymer are insensitive to the elevated
temperatures.

Recently, EHD co-jetting of organic-soluble polymers have been used to
fabricate anisotropic microstructures. In particular, the EHD co-jetting of
organic solutions of poly(lactic-*co*-glycolic acid) (PLGA) has been successfully
used to prepare biodegradable bicompartmental particles and fibers
(Figure 3.4).[31] Because of their biodegradability, these particles may constitute
promising building blocks for biohybrid materials or find potential applications in drug delivery and tissue engineering.[32]

The systematic analysis of EHD co-jetting of PLGA reveals that the jetting
solution and process parameters can control compartmentalization, shape and
size.[33] The jetting parameters are directly associated with solvent choice,
concentration and vapor pressure of the jetting solution, which determine
viscosity, surface tension, dielectric constant, electrical conductivity and
density of the jetting fluids. On the other hand, process-related parameters

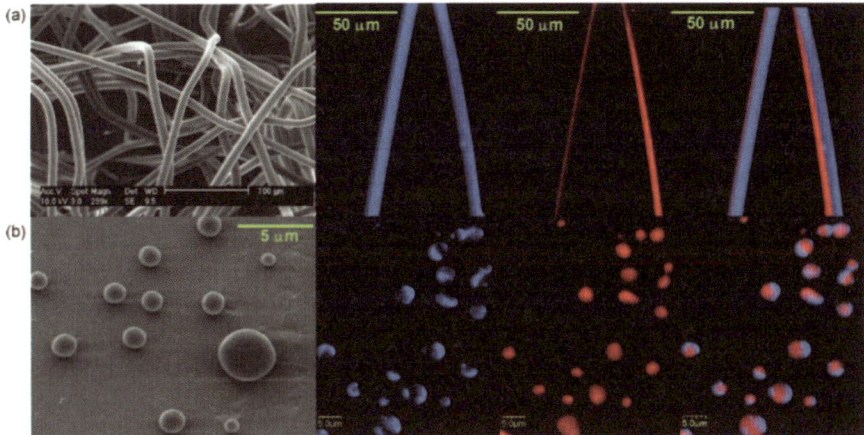

Figure 3.4 Scanning electron microscopy (left) and confocal laser scanning microscopy images (right) of (a) biphasic microfibers (10–20 μm in diameter) and (b) biphasic microparticles (2–5 μm in diameter) resulting from the EHD co-jetting of solutions containing PLGA 85:15 (red) and PLGA 50:50 (blue), respectively. Blue fluorescence represents ADS406PT and red fluorescence represents ADS406PT. Reproduced with permission from reference 31.

include solution flow rates, applied electrical potential, capillary diameter and separation between tip and biased collector, all of which contribute to the formation of a stable Taylor cone. In addition, EHD co-jetting can be affected by a number of environmental factors, such as temperature, pressure and humidity. Subtle alternations of these variables may lead to significant variations in particle size, shape, surface, morphology and porosity. We have demonstrated the transition of PLGA bicompartmental particle characteristics by focusing particularly on polymer concentration and flow rate of PLGA fluids (Figure 3.5).[33] The viscosity of a polymer solution, a function of polymer concentration and molecular weight, significantly influences jet atomization and evaporation rates. Below a critical concentration, the formation of bicompartmental disks can be observed as a result of anisotropic solvent evaporation due to high surface-to-volume ratios affected by small droplet sizes, which facilitates particle collapse due to the small amount of overall non-volatile materials within the electrified jets. Disk shapes are observed even at higher flow rates, where particle shape is retained, but the diameter of the disks increases from 3–4 to 7–10 μm due to the larger droplet sizes. As the concentration is increased, spherical particles are observed due to enhanced structural stability (3–4 μm in diameter). Increasing the polymer concentration or flow rate further leads to a size increase to ∼20 μm in diameter. At higher concentrations and low flow rates, however, fibrous architectures preferentially appear, because highly viscous solutions can resist the breakup of the jet. Therefore, individual or simultaneous control of concentrations and flow rates of the jetting solutions allows for tuning of the size and shape of anisotropic particles. In addition to changes in the solution

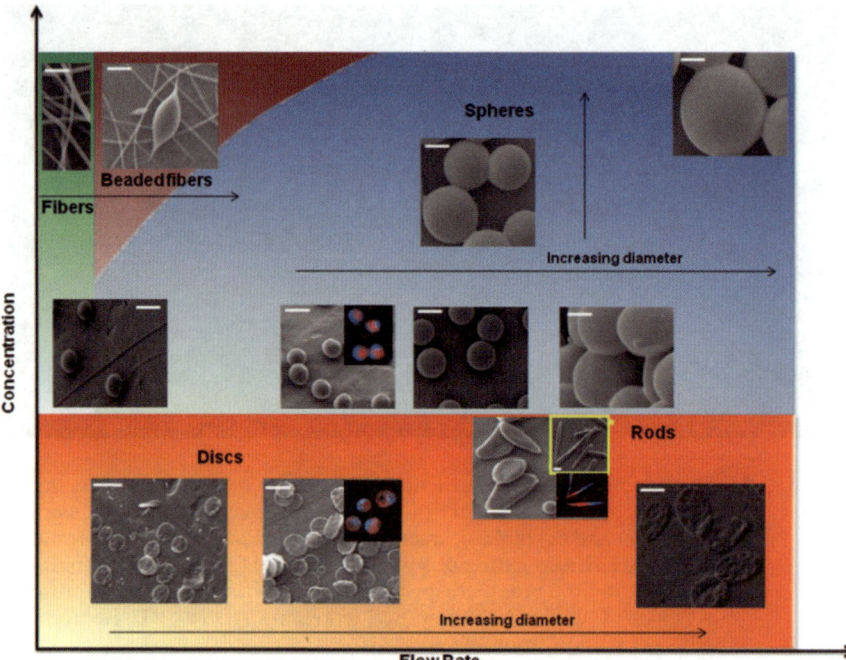

Figure 3.5   Effect of concentration and flow rate on the size and shape of
bicompartmental microparticles. At lower concentrations, bicompartmental discs are observed. Rods are observed at a critical concentration
(3.4 wt%) and flow rate (0.45 mL h^{-1}) but only upon the addition of
triethylamine. As the concentration is increased, spheres are observed.
Increasing the polymer concentration or flow rate leads to increased
sphere diameters. Fibers are observed at higher concentrations but at
low flow rates. As the flow rate increases at higher concentrations, a
transition from fibers to beaded fibers and then to particles is observed.
All scale bars represent 5 μm. Reproduced with permission from
reference 33.

viscosity, the addition of small surface-active molecules may alter other jetting
solution properties and influence the compartmentalization, size and shape of
particles.

   Because the capillary needle configuration primarily determines the
compartment geometry of particles and fibers fabricated by the EHD co-
jetting process, the increase in the number of capillaries can provide access to
particles with more than two compartments.[34,35] The EHD co-jetting of
aqueous P(AAm-*co*-AA) solutions with a three-needle setup resulted in
tricompartmental particles with diameters between hundreds of nanometers
and a few micrometers.[34] On the other hand, multicompartmental fibers with
as many as seven distinct compartments can be prepared by the EHD co-
jetting of PLGA solution (Figure 3.6).[35] Combined with low flow rates and
high solvent volatility, highly viscous PLGA solutions encountered in this case

of co-electrospinning contribute to the formation of a continuous jet, which features multiple compartments only minimally affected by the perturbations during EHD co-jetting.[36] Moreover, both the sequence and the flow geometry of the polymer solutions can also control the orientation of compartments. For example, three parallel-aligned needles in a side-by-side configuration yield sandwiched compartments, which can show all possible 'isomers' of tricompartmental fibers (Figure 3.6a–c). On the other hand, a concentric side-by-side three-needle configuration realizes three pie-shaped compartments (Figure 3.6d). Furthermore, the feasibility of the EHD co-jetting process with four to seven outlet streams presents a simple route to access novel scaffolds with a larger number and unique arrangement of compartments. When the jetting solutions were arranged in an alternating sequence, scaffolds consisting of

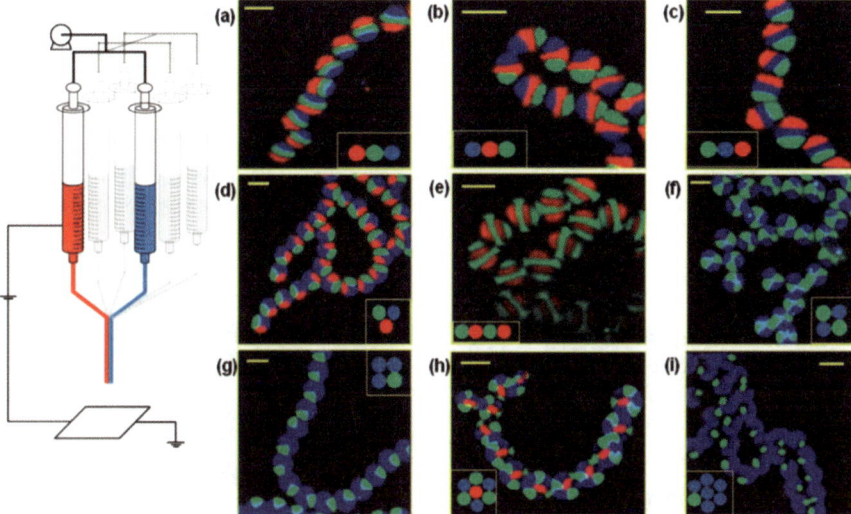

Figure 3.6   Left: schematic representation of the EHD co-jetting process using a dual capillary assembly (red and blue). This approach can be extended to the fabrication of multicompartmental microfibers by incorporating additional outlet streams (gray). Right: cross-sectional CLSM images of multicompartmental microfibers with up to seven distinct compartments. Insets indicate the number, spatial configuration and nature of the outlet streams used during electrohydrodynamic co-jetting. (a–c) Tricompartmental microfiber isomers: (a) {sRGB}; (b) {sBRG}; (c) {sRBG}. (d) Multicompartmental microfiber scaffold with a pie-shaped anisotropy, {pRBG}. (e) Tetracompartmental microfiber scaffold showing alternating red and green compartments. (f) Tetracompartmental microfiber scaffold with blue and green compartments. (g) Tetracompartmental microfiber scaffold with a green 'quarter' compartment and a threefold larger blue compartment. (h) Heptacompartmental microfiber scaffold resembling a flower. (i) Heptacompartmental fibers with one green compartment that is sixfold smaller than the other. All scale bars are 20 μm. Reproduced with permission from reference 35. Copyright 2009 American Chemical Society.

striped microfibers with four distinguishable compartments in series were prepared (Figure 3.6e). However, when a square arrangement of the outlet flows was used, the compartments of the microfibers were arranged as rosettes of alternating compartments (Figure 3.6f). Similarly, more complicated rosettes consisting of seven compartments were prepared (Figure 3.6h). Precise control of the internal fiber architectures allows for the relative size of compartments by coordinating compartments (Figure 3.6g). More importantly, these multi-compartmental microfibers can be used as platforms to transfer the well-defined anisotropic structures to cylindrical colloids by microsectioning.[37]

The distinct reactivity of individual compartments within a compositionally anisotropic particle is able to provide access to interesting particle architectures that would otherwise be impossible, or at least extremely difficult, to achieve. PLGA microdisks were prepared, which were comprised of a quarter compartment containing a mixture of poly-N-isopropylacrylamide (PNIPAM) and PLGA. Selective removal of the PNIPAM compartment by exposure to water led to 'Pacman'-like particles (Figure 3.7a).[37] Spatially controlled photo-cross-linking reactions of poly(vinyl cinnamate) (PVCi) containing compartments of bicompartmental PLGA microparticles led to novel types of particles with respect to degradation rates, solvent resistance and mechanical properties. The treatment with chloroform yielded semicapsules manifesting small pores on their surface (Figure 3.7b).[38]

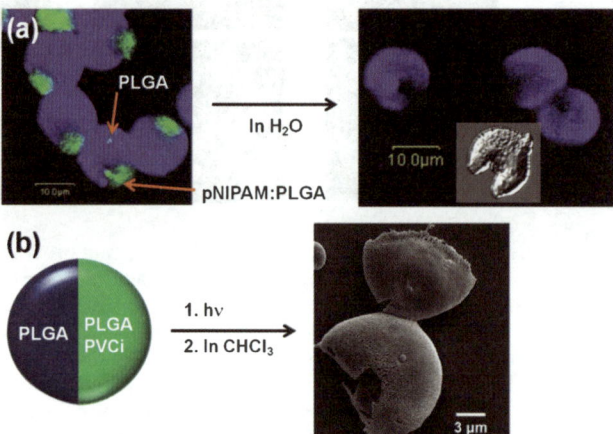

Figure 3.7   (a) A confocal laser scanning microscopy image of bicompartmental PLGA microdisks prepared by the EHD co-jetting and subsequent microsectiong. The smaller one-quarter compartment (green) comprises a water-soluble polymer (PNIPAM), which was selectively removed to yield 'Pacman'-like particles (scale bar: 10 μm). Reproduced with permission from reference 37. (b) Porous hemicapsules obtained from spatially controlled photo-cross-linking and subsequent solvent treatment of bicompartmental PLGA particles containing PVCi in one compartment. Reproduced with permission from reference 38.

3.3   Microsectioning of Compartmentalized Fibers

Non-spherical-shaped particles with reduced symmetry have shown unprecedented properties with respect to spherical colloids. In particular, recent studies have demonstrated that the anisotropic architecture of particles greatly influences the interaction between cells and artificial materials, highlighting the importance of shape for drug carriers in biomedical applications.[6,39,40] While the electrospraying process generally gives rise to spherical particles, the variation of jetting parameters may present non-spherical architectures.[33]

In addition, highly concentrated polymer solutions can be electrospun and enhanced stability of the jetting process due to higher viscosity permits compartmentalized fibers using concentric nested core–shell needles[41–43] or side-by-side needles.[44,45] However, a whipping motion of the electrospun fibers generally results in non-woven meshes or membrane-like structures. On the other hand, cylinder-shaped microparticles can be simply fabricated by microsectioning of fiber bundles if the fibers are perfectly aligned. It has been demonstrated recently that the microsectioning of aligned scaffolds is a very simple but effective method for fabricating monodisperse nano- and

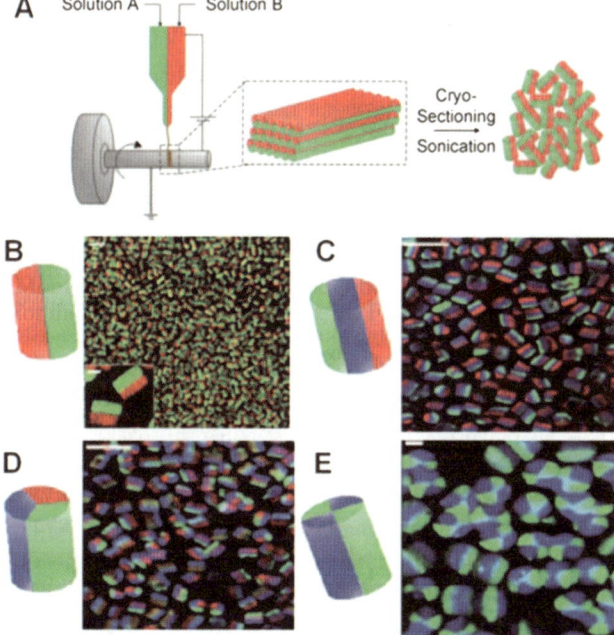

Figure 3.8   (a) Schematic diagram for producing microcylinders *via* cryosectioning of aligned fibers, which are accessible through EHD co-jetting. Confocal laser scanning microscopy images of PLGA microcylinders (mean diameter ∼ 14 μm and length 20 μm) with two (b), three (c, d) and four (e) compartments. Scale bars are 50 μm in (b)–(d) and 10 μm in (e). Reproduced with permission from reference 37.

microcolloidal structures in a well-controlled manner.[37,46,47] Beyond control of size and shape, multifunctional colloids also require well-defined internal particle architectures, as manifested by the presence of multiple substructures and compartments. Bhaskar and Lahann demonstrated the EHD co-electrospinning of concentrated PLGA solutions at low flow rates yielded whipping-free continuous multicompartmental microfibers[35] such that the electrospun fibers could be deposited as ordered bundles. The extremely stable jetted thread during the spinning of PLGA yielded aligned fibers with very uniform diameters. Subsequent cryosectioning of the microfiber bundles using an automated microsectioning device created microcylinders with different compartments ranging from bi- to heptacompartmental architectures (Figure 3.8).[37] The original compartmentalization within the fibers was maintained in the microcylinders after cryosectioning and sonication. In addition, the precise microsectioning of perfectly aligned bundles afforded monodisperse microcylinders and well-controlled aspect ratios. With further work, the resulting biodegradable multicompartmental microcylinders or microdisks may become carriers with precisely designed physical and/or chemical properties for a range of biomedical applications.

3.4   Hybrid Janus Particles

Inorganic or organic–inorganic hybrid materials have tremendous potential in a variety of applications, including catalysis, optoelectronics and energy. Anisotropic entrapment of functional nanocrystals within a single particle may provide access to new types of materials and also unprecedented applications.[25,48–50] In addition to the straightforward control of particle characteristics, an additional advantage of the EHD co-jetting process is the ability to introduce a wide range of materials into the jetting solutions. The EHD co-jetting of aqueous polyvinylpyrrolidone (PVP) containing inorganic semiconductor precursors, $Ti(OBu)_4$ and $Sn[CH_3(CH_2)_3CH(C_2H_5)CO_2]_2$, and subsequent calcination fabricated biphasic fully inorganic nanofibers made up of TiO_2 and SnO_2.[51] Recent studies demonstrated that the EHD co-jetting of nanoparticle suspensions (Figure 3.9a) allowed for the introduction of organic–inorganic nanocomposite into a selective compartment under well-controlled conditions.[52,53] Gold nanocrystals (AuNCs) were anisotropically incorporated into only one subunit of bicompartmental P(AAm-co-AA) microparticles.[52] After subsequent polymer cross-linking, TEM analysis confirmed that AuNCs were uniformly dispersed in only one compartment and transfer of AuNCs across the compartment interface was minimal (Figure 3.9b). The density of AuNCs in the bicompartmental polymer particles was controlled by variation of the AuNC concentration in the jetting solution.

Likewise, two different NCs, iron oxide and titanium dioxide NCs, were compartmentalized in bicompartmental P(AAm-co-AA) nanocolloids.[53] Electron microscopic studies revealed a clear interface and also different compartment morphologies. Chemical analysis confirmed the anisotropic

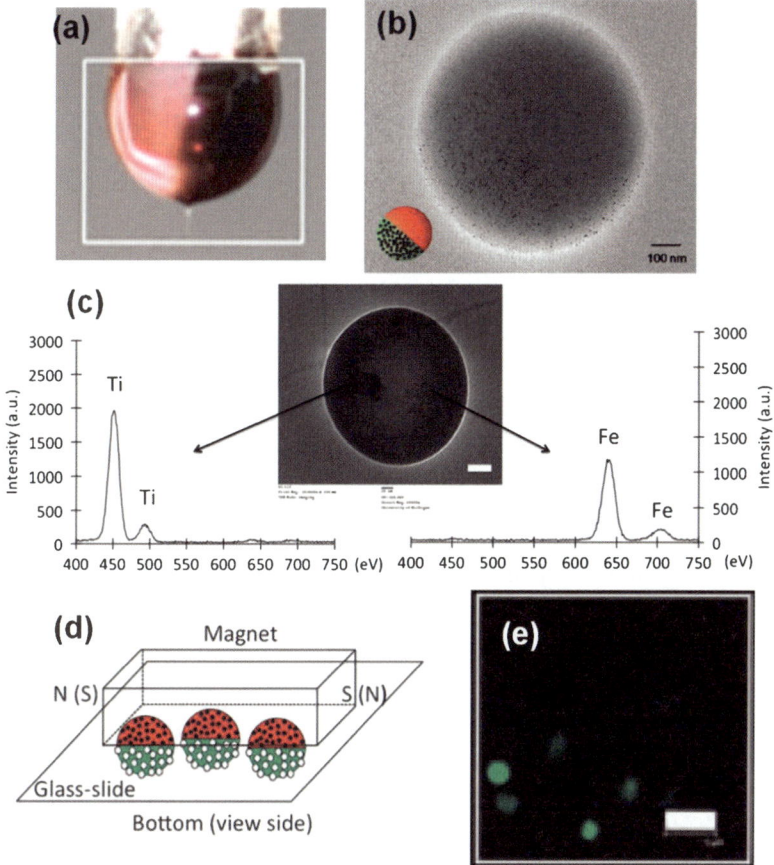

Figure 3.9 (a) Photographic image of a Taylor cone of the EHD co-jetting of polymer solutions. A dark half solution contains AuNCs. (b) TEM image of the bicompartmental particles with AuNCs confined in one compartment. (c) TEM image of bicompartmental particle prepared by EHD co-jetting of two nanoparticle suspensions, magnetite and titanium dioxide (scale bar: 40 nm). The inorganic nanoparticles are compartmentalized as shown by energy-dispersive X-ray (EDX) spectroscopy. (d) Experimental configuration for controlling the particles using a magnet and (e) its CLSM image obtained by overlay of FITC and rhodamine channels. Scale bar: 5 μm. (a, b) Reproduced with permission from reference 52 and (c–e) reproduced with permission from reference 53.

entrapment of the inorganic payloads (Figure 3.9c). Because the ability to magnetize only certain parts of a particle affect its response to magnetic fields, the application of magnetic fields to the Janus magnetic balls results in selective orientation of the bicompartmental particles (Figure 3.9d). Such control under magnetic fields may be a key feature for future magnetophoretic displays and switchable catalysts. In addition, magnetic bicompartmental particles were prepared from PLGA solutions containing iron oxide NCs, presenting controllable platforms made of biodegradable polymers in biomedical applications.[35]

3.5   Selective Surface Modification and Directional Self-assembly

Anisotropic surface modification is one of the simplest methods for preparing Janus particles with different surface patches. The surface properties of these particles can be controlled effectively. Therefore, the selective surface functionalization of compartmentalized particles may offer novel properties, because distinctive functions can be given to both external and internal design aspects in an independently controllable manner. Adding functionalized polymers to different jetting solutions used in the EHD co-jetting process allows for selective modification of certain parts of the particle. Ultimately, it may even be possible to present simultaneously multiple chemical moieties on parts of the particle surface.[29] In one specific case, the EHD co-jetting of P(AAm-*co*-AA) solutions was used to incorporate biotin- and acetylene-modified P(AAm-*co*-AA) in bicompartmental particles. In addition, surface modification of the entire surface area with PEG was necessary to minimize non-specific binding. Selectively functionalized particles had significantly higher binding compared with PEG-containing particles (Figure 3.10).[16]

In another case, selectively acetylene-functionalized bicompartmental particles were prepared by the EHD co-jetting of PLGA solutions, where one solution further contained poly[lactide-*co*-propargyl glycolide]. Importantly, the addition of the acetylene-modified PLGA did not disturb the co-jetting process.[31] The mapping of the vibrational signature *via* confocal Raman spectromicroscopy confirmed the anisotropic distribution of acetylene groups. The latter are characterized by a strong Raman band at 2100–2020 cm^{-1}.[54] Finally, the click reaction with azide–PEG–amine confirmed the selective modification of one hemisphere. In spite of the fact that the acetylene–PLGA is incorporated into the bulk, the high uniformity of the fluorescent surface band implies a high number of free acetylene groups on the surface (Figure 3.11).[31] Similarly, spatioselectively modified PLGA particles showed directional self-assembly of bicompartmental particles on to the substrates (Figure 3.12).[54] Because these particles featured biotin groups on the surface of one hemisphere (green) only, the red hemisphere is more apparent to the eye of the observer (Figure 3.12c).

Beyond spatioselective bioconjugation, the spatioselective growth of polymer brushes on bicompartmental colloidal structures can be used to create new polymer layers or lead to additional compositional anisotropy. Figure 3.13a depicts the selective polymerization on the surface of bicompartmental PLGA microcylinders.[55] Anisotropic, acetylene-modified PLGA microcylinders were prepared by EHD co-jetting and subsequent micro-sectioning. The incorporation of acetylene–PLGA into one compartment allowed for spatioselective surface modification *via* copper-catalyzed Huisgen

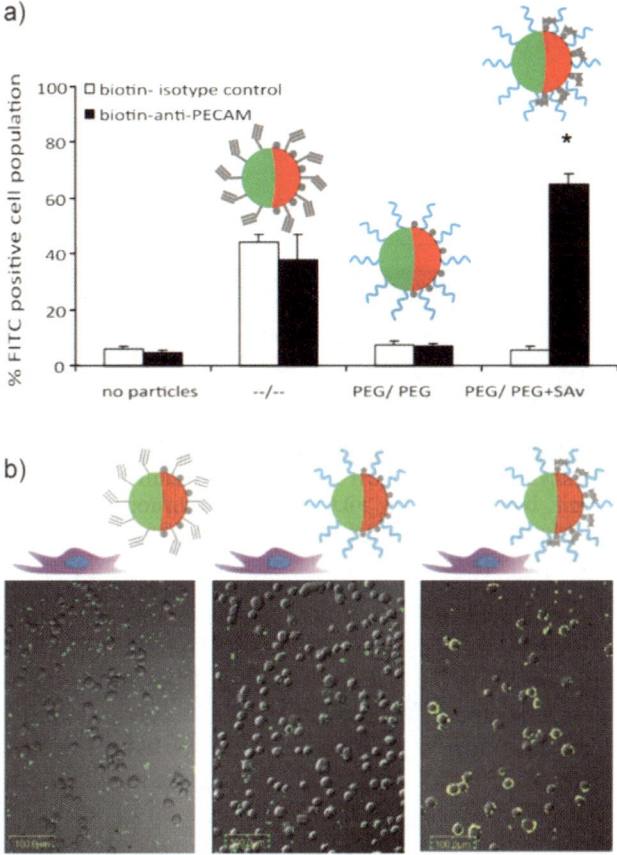

Figure 3.10   Specific cell binding of bicompartmental particles after selective surface modification of one hemisphere. Bicompartmental particles presenting PEG on the entire particle surface and biotin-functionalized polymer in only one compartment can bind selectively to human endothelial cells. Here, interactions are based on the biotin–streptavidin interaction, where biotin is displayed on the particle hemisphere and streptavidin has been introduced on the cell surface *via* a platelet/endothelial cell-adhesion molecule (PECAM, CD31) antibodies. Cell–particle binding was quantified by flow cytometry (a) and studied by fluorescence microscopy (b). Reproduced with permission from reference 16.

heterocycloaddition with an azide-functionalized atom-transfer radical polymerization (ATRP) macroinitiator. The selective growth of hydrogel brushes of poly[oligo(ethylene glycol) methyl ether methacrylate] (OEGMA) was achieved by surface-initiated ATRP reactions on the hemisphere of the PLGA microcylinders in the presence of Cu catalyst and OEGMA in water. Confocal micrographs of bicompartmental microcylinders after polymerization had well-defined surface layers comprised of swollen PEGMA. Moreover, the hydrogel compartment causes the microcylinders to bend outwards compared

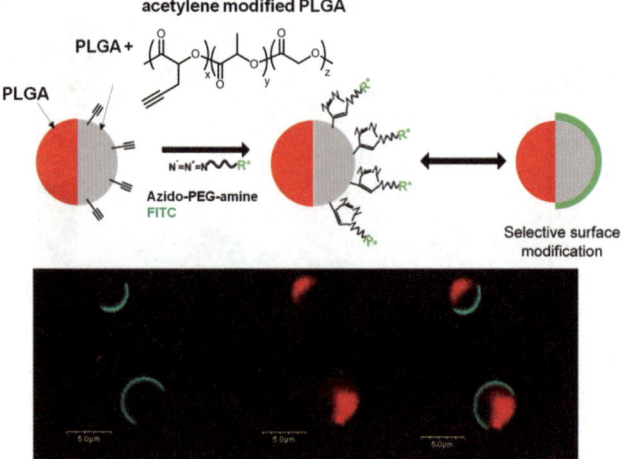

Figure 3.11   Bicompartmental PLGA microparticles labeled with a red fluorescent macromolecule (ADS306PT) in one compartment and containing free acetylene groups in the other compartment were prepared by adding poly[lactide-*co*-(propargyl glycolide)] to one jetting solution. The acetylene groups were reacted with azido–PEG–amine and the free amine groups were then reacted with FITC (PEG–amine–FITC denoted R*). Green peripheral fluorescence on one side of the particles due to FITC indicates uniform surface modification of the acetylene-containing phase. Reproduced with permission from reference 31.

with microcylinders prior to ATRP. It was demonstrated that the degree of bending depends on the length of the microcylinders; the curvature generally decreases with increase in cylinder length (Figure 3.13).

3.6   Summary and Outlook

Whereas electrospraying and electrospinning can utilize a wide range of polymers, EHD co-jetting has so far been accomplished with only a limited set of polymers that have similar solution properties. The formation of a stable Taylor cone during EHD co-jetting is typically easier to achieve when key solution parameters of the jetting solutions, such as conductivity, viscosity and density, are comparable. Beyond the expansion of the EHD co-jetting process to increasingly different sets of polymers, such as water−organic swelling polymers and thermoresponsive/photoresponsive polymers, further modification of jetting process including the use of novel needle configurations, different combinations of solvents, sacrificial templating, *etc.*, may allow for exotic types of multifunctional compartmental particles with dissimilar properties. As changes of process parameters can result in transitions between particle architectures,[56] new polymer combinations may lead to previously unseen particle shapes and novel nozzle systems may allow a plethora of different compartmentalizations.

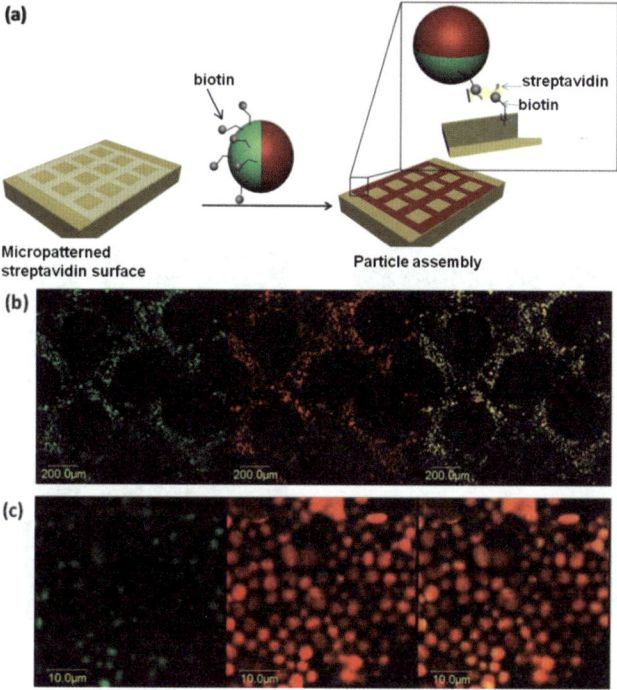

Figure 3.12 (a) Schematic description of the assembly of bicompartmental particles on to streptavidin patterned surfaces. A reactive chemical vapor deposition (CVD)-based coating containing ketone functional groups is deposited on silica surfaces. The microcontact printing of biotin–hydrazide followed by immobilization of streptavidin yields patterned substrates. Bicompartmental particles containing surface biotin groups on the green compartment then spontaneously assemble on these substrates and selective binding of the green half leads to particle orientation. (b) Confocal laser scanning microscopy image of particles as seen on the streptavidin substrates (scale bar: 200 µm). (c) High-magnification image showing particle orientation (scale bar: 10 µm). Reproduced with permission from reference 54.

The practicality of the EHD co-spraying technique might be limited by the reproducibility of particle size and architecture. The sectioning of aligned multicompartmental fibers is an excellent tool to produce very uniform multicomponent microparticles. In contrast to the whipping motion generally encountered in electrospinning, very stable PLGA microfibers prepared from the EHD co-spinning process permit a nearly perfect alignment of multi-compartmental fibers and also the formation of an apparent interface and well-defined architecture of compartments that is set by the needle geometry of the jetting solutions. Subsequent cryosectioning of fiber bundles offers monodisperse cylindrical microparticles with precisely controlled aspect ratios and compartment compositions. Increased throughput and size reduction could be improved by patterning fibers into cylindrical particles using two-

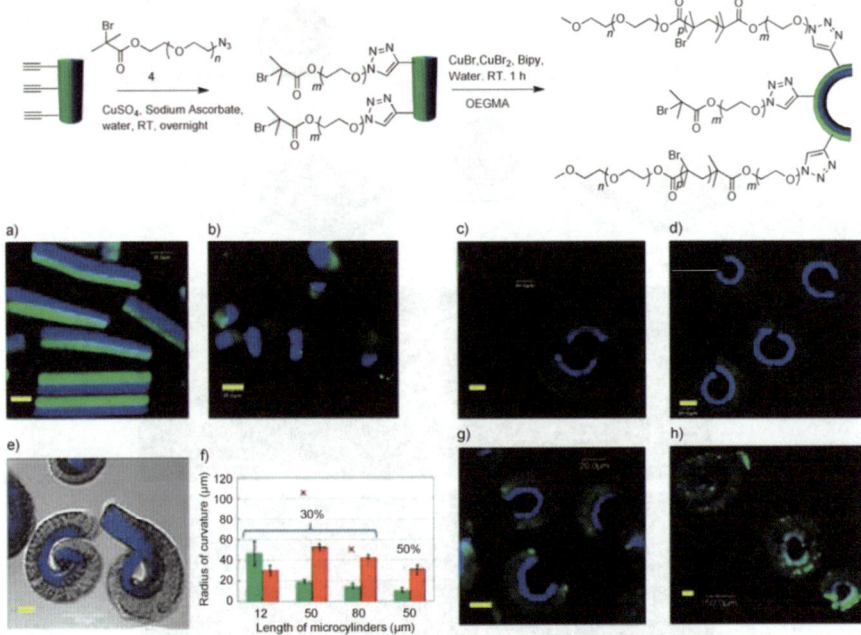

Figure 3.13   Top: scheme of selective surface modification of bicompartmental microcylinders with a PEGMA surface layer. Bottom: representative images of (a) brush-free bicompartmental microcylinders of 120 μm length (before polymerization) and after spatioselective surface-initiated polymerization of OEGMA on microcylinders of different lengths: (b) 12, (c) 50, (d) 80 and (e) 120 μm. All multiphasic cylinders contain around 30% of acetylene–PLGA in one compartment. (f) Measured inner (green) and red (outer) radii of curvature versus length of biphasic microcylinders for brush bilayers at two different acetylene concentrations (labeled above the data points). Data points × are from finite element simulations. Effect of grafting density on bending for (g) 50 and (h) 80 μm biphasic cylinders (after polymerization) which contain 50% of acetylene–PLGA in one compartment. Yellow scale bars are 20 μm. Reproduced with permission from reference 55.

beam interference lithography, although this requires that the fibers be composed of polymers compatible with photolithography.[57]

The multicompartmental particles include multiple materials in a single architecture and combine orthogonal sets of properties in unusual ways. The selective surface modification of compartments substantially expands the design space and offers synergetic effects for a diversity of potential applications. Microstructured particles with dissimilar surface patterns may exhibit the ability to self-assemble into hierarchically organized novel superstructures[58] that can serve as building blocks for future devices. Selective surface functionalization with cell-targeting materials may allow directional interactions of compartmentalized particles carrying multiple drugs. Although a much better understanding of design space and its limitations is still needed, it is our prediction that a wide range of novel particles with previously unseen particle architectures will be fabricated by EHD co-jetting in the future.

Nano- and micrometer-scale control of biodegradable materials is essential for many biomedical applications, including controlled drug delivery, regenerative medicine and simultaneous imaging and diagnosis applications. The anisotropic surface functionalization of multicompartmental PLGA particles has the potential to satisfy such important criteria.

Acknowledgements

We thank the American Cancer Society (RSG-08-284-01-CDD) for financial support and acknowledge funding from the Multidisciplinary University Research Initiative of the Department of Defense and the Army Research Office (W911NF-10-1-0518).

References

1. M. Grzelczak, J. Vermant, E. M. Furst and L. M. Liz-Marzán, *ACS Nano*, 2010, **4**, 3591.
2. S. C. Glotzer, M. J. Solomon and N. A. Kotov, *AIChE J.*, 2004, **50**, 2978.
3. S. C. Glotzer and M. J. Solomon, *Nat. Mater.*, 2007, **6**, 557.
4. S. M. Yang, S. H. Kim, J. M. Lim and G. R. Yi, *J. Mater. Chem.*, 2008, **18**, 2177.
5. A. Walther and A. H. E. Muller, *Soft Matter*, 2008, **4**, 663.
6. S. Mitragotri and J. Lahann, *Nat. Mater.*, 2009, **8**, 15.
7. D. Dendukuri and P. S. Doyle, *Adv. Mater.*, 2009, **21**, 4071.
8. J. Yoon, K. J. Lee and J. Lahann, *J. Mater. Chem.*, 2011, **21**, 8502.
9. K. J. Lee, J. Yoon and J. Lahann, *Curr. Opin. Colloid Interface Sci.*, 2011, **16**, 195.
10. J. Lahann, *Small*, 2011, **7**, 1149.
11. S. Sacanna, W. T. M. Irvine, P. M. Chaikin and D. J. Pine, *Nature*, 2010, **464**, 575.
12. Q. Chen, J. K. Whitmer, S. Jiang, S. C. Bae, E. Luijten and S. Granick, *Science*, 2011, **331**, 199.
13. Q. Chen, S. C. Bae and S. Granick, *Nature*, 2011, **469**, 381.
14. D. C. Pregibon, M. Toner and P. S. Doyle, *Science*, 2007, **315**, 1393.
15. S. Sengupta, D. Eavarone, I. Capila, G. Zhao, N. Watson, T. Kiziltepe and R. Sasisekharan, *Nature*, 2005, **436**, 568.
16. M. Yoshida, K.-H. Roh, S. Mandal, S. Bhaskar, D. W. Lim, H. Nandivada, X. P. Deng and J. Lahann, *Adv. Mater.*, 2009, **21**, 4920.
17. J. Kim, S. E. Chung, S.-E. Choi, H. Lee, J. Kim and S. Kwon, *Nat. Mater.*, 2011, **10**, 747.
18. T. Higuchi, A. Tajima, K. Motoyoshi, H. Yabu and M. Shimomura, *Angew. Chem. Int. Ed.*, 2008, **47**, 8044.

19. A. Walther, X. André, M. Drechsler, V. Abetz and A. H. E. Müller, *J. Am. Chem. Soc.*, 2007, **129**, 6187.

20. J.-W. Kim, R. J. Larsen and D. A. Weitz, *J. Am. Chem. Soc.*, 2006, **128**, 14374.

21. D. J. Kraft, J. Hilhorst, M. A. P. Heinen, M. J. Hoogenraad, B. Luigjes and W. K. Kegel, *J. Phys. Chem. B*, 2011, **115**, 7175.

22. J. P. Rolland, B. W. Maynor, L. E. Euliss, A. E. Exner, G. M. Denison and J. M. DeSimone, *J. Am. Chem. Soc.*, 2005, **127**, 10096.

23. S. Badaire, C. Cottin-Bizonne, J. W. Woody, A. Yang and A. D. Stroock, *J. Am. Chem. Soc.*, 2006, **129**, 40.

24. T. J. Merkel, K. P. Herlihy, J. Nunes, R. M. Orgel, J. P. Rolland and J. M. DeSimone, *Langmuir*, 2009, **26**, 13086.

25. T. Nisisako, T. Torii, T. Takahashi and Y. Takizawa, *Adv. Mater.*, 2006, **18**, 1152.

26. C.-H. Chen, A. R. Abate, D. Lee, E. M. Terentjev and D. A. Weitz, *Adv. Mater.*, 2009, **21**, 3201.

27. K. W. Bong, K. T. Bong, D. C. Pregibon and P. S. Doyle, *Angew. Chem. Int. Ed.*, 2010, **49**, 87.

28. A. Greiner and J. H. Wendorff, *Angew. Chem. Int. Ed.*, 2007, **46**, 5670.

29. K.-H. Roh, D. C. Martin and J. Lahann, *Nat. Mater.*, 2005, **4**, 759.

30. K.-H. Roh, M. Yoshida and J. Lahann, *Langmuir*, 2007, **23**, 5683.

31. S. Bhaskar, K.-H. Roh, X. W. Jiang, G. L. Baker and J. Lahann, *Macromol. Rapid Commun.*, 2008, **29**, 1655.

32. S. Mandal, S. Bhaskar and J. Lahann, *Macromol. Rapid Commun.*, 2009, **30**, 1638.

33. S. Bhaskar, K. M. Pollock, M. Yoshida and J. Lahann, *Small*, 2010, **6**, 404.

34. K.-H. Roh, D. C. Martin and J. Lahann, *J. Am. Chem. Soc.*, 2006, **128**, 6796.

35. S. Bhaskar and J. Lahann, *J. Am. Chem. Soc.*, 2009, **131**, 6650.

36. P. Gupta and G. L. Wilkes, *Polymer*, 2003, **44**, 6353.

37. S. Bhaskar, J. Hitt, S. W. L. Chang and J. Lahann, *Angew. Chem. Int. Ed.*, 2009, **48**, 4589.

38. K. J. Lee, S. Hwang, J. Yoon, S. Bhaskar, T.-H. Park and J. Lahann, *Macromol. Rapid Commun.*, 2011, **32**, 431.

39. J. A. Champion, Y. K. Katare and S. Mitragotri, *Proc. Natl. Acad. Sci. U. S. A.*, 2007, **104**, 11901.

40. S. E. A. Gratton, P. A. Ropp, P. D. Pohlhaus, J. C. Luft, V. J. Madden, M. E. Napier and J. M. DeSimone, *Proc. Natl. Acad. Sci. U. S. A.*, 2008, **105**, 11613.

41. J. T. McCann, D. Li and Y. Xia, *J. Mater. Chem.*, 2005, **15**, 735.

42. I. C. Liao, S. Chen, J. B. Liu and K. W. Leong, *J. Control. Release*, 2009, **139**, 48.

43. C. Wang, K.-W. Yan, Y.-D. Lin and P. C. H. Hsieh, *Macromolecules*, 2010, **43**, 6389.

44. T. Lin, H. Wang and X. Wang, *Adv. Mater.*, 2005, **17**, 2699.

45. S. Chen, H. Hou, P. Hu, J. H. Wendorff, A. Greiner and S. Agarwal, *Macromol. Mater. Eng.*, 2009, **294**, 781.
46. C. Huang, B. Lucas, C. Vervaet, K. Braeckmans, S. Van Calenbergh, I. Karalic, M. Vandewoestyne, D. Deforce, J. Demeester and S. C. De Smedt, *Adv. Mater.*, 2010, **22**, 2657.
47. D. J. Lipomi, M. A. Kats, P. Kim, S. H. Kang, J. Aizenberg, F. Capasso and G. M. Whitesides, *ACS Nano*, 2010, **4**, 4017.
48. S. Crossley, J. Faria, M. Shen and D. E. Resasco, *Science*, 2010, **327**, 68.
49. J. R. Howse, R. A. L. Jones, A. J. Ryan, T. Gough, R. Vafabakhsh and R. Golestanian, *Phys. Rev. Lett.*, 2007, **99**, 048102.
50. S.-H. Kim, S. Y. Lee and S.-M. Yang, *Angew. Chem. Int. Ed.*, **49**, 2535.
51. Z. Liu, D. D. Sun, P. Guo and J. O. Leckie, *Nano Lett.*, 2006, **7**, 1081.
52. D. W. Lim, S. Hwang, O. Uzun, F. Stellacci and J. Lahann, *Macromol. Rapid Commun.*, 2010, **31**, 176.
53. S. Hwang, K.-H. Roh, D. W. Lim, G. Y. Wang, C. Uher and J. Lahann, *Phys. Chem. Chem. Phys.*, 2010, **12**, 11894.
54. S. Bhaskar, C. T. Gibson, M. Yoshida, H. Nandivada, X. P. Deng, N. H. Voelcker and J. Lahann, *Small*, 2011, **7**, 812.
55. S. Saha, D. Copic, S. Bhaskar, N. Clay, A. Donini, A. J. Hart and J. Lahann, *Angew. Chem. Int. Ed.*, 2012, **51**, 660.
56. A. Kazemi and J. Lahann, *Small*, 2008, **4**, 1756.
57. M. C. George and P. V. Braun, *Angew. Chem. Int. Ed.*, 2009, **48**, 8606.
58. T. D. Nguyen, E. Jankowski and S. C. Glotzer, *ACS Nano*, 2011, **5**, 8892.

CHAPTER 4

Synthesis of Janus Particles by Emulsion-based Methods

CHENGLIANG ZHANG, WEI WEI, FUXIN LIANG AND ZHENZHONG YANG*

State Key Laboratory of Polymer Physics and Chemistry, Institute of Chemistry, Chinese Academy of Sciences, Zhongguancun North First Street 2, Beijing 100190, China
*E-mail: yangzz@iccas.ac.cn

4.1 Introduction

The efficiency of the preparation of Janus structures was extremely low in the early approaches and practical utilization of the materials was greatly hampered, although various promising performances have been demonstrated. It is urgently required to develop new approaches to produce Janus materials on a large scale. Among the various approaches, emulsion-based synthesis is most effective. In emulsions, one phase is dispersed in another immiscible phase mediated with the interface to compartmentalize the dispersed phase and continuous phase. There are several different ways of synthesizing Janus particles using emulsions. In the first method, particles after proper modification can be immobilized at the interface due to the Pickering effect. The particles are naturally separated into two distinct parts by two emulsion phases, which can be selectively modified to render different chemistry resulting in Janus geometry. In the second method, within the traditional liquid–liquid emulsions stabilized with surfactants, liquid droplets are dispersed. Inside the droplets, many species can be incorporated, including

RSC Smart Materials No. 1
Janus Particle Synthesis, Self-Assembly and Applications
Edited by Shan Jiang and Steve Granick

monomers, polymers and nanoparticles. Alteration of variables, for example by solvent evaporation and application of external fields, can induce phase separation inside the droplets to achieve Janus particles. In the third method, many preformed solid particles are capable of dispersion in emulsions. The particles can be engineered in many ways. Swelling of polymer particles followed by a sequential polymerization can induce phase separation to achieve Janus particles. Polymerization-induced de-wetting on to the particle surface can also give anisotropic particles. Interestingly, by dynamic protrusion of the core material from a core–shell particle forming a bulb on the shell can result in the rapid formation of non-spherical Janus particles with distinct compartmentalization of two different components. The following sections provide more detailed descriptions of the three types of emulsion-based synthesis of Janus materials.

4.2   Synthesis at a Pickering Emulsion Interface

Partial protection of particles has been a basic rational approach to the synthesis of Janus particles. The most straightforward way is partially to embed particles with proper inert materials in two-dimensional flat substrates. The exposed part is available for post-modification. Various modification methods, including electron beam sputtering and micro-contact printing, can be employed to modify the exposed part to derive different chemistry. However, the yield of Janus particles by this two-dimensional planner protection is rather low, and a typical batch can only generate milligram quantities of Janus particles. The interfacial area of emulsion droplets is much larger for the synthesis of Janus particles and the yield can be significantly increased. Thanks to the discovery by Ramsden[1] and Pickering,[2] particles with the correct hydrophobicity can stabilize the interface between two immiscible liquids and form so-called Pickering emulsions. Particles incorporated at the interface are clearly divided into two parts by the interface. This process offers an opportunity to modify selectively the two parts to achieve different chemistry. Synthesis of Janus particles at a Pickering emulsion interface has attracted growing interest, since the method is valid for a variety of particles with different sizes and compositions.[3–6] Success in the synthesis of Janus particles is determined by how to restrict motion, especially rotation of particles at the interface. Granick and co-workers first proposed an elaborate approach to restrict the motion of silica particles at the interface using frozen wax as an oil phase at lower temperature, the exposed part of the particles being selectively modified with a silane forming a thin layer of a different chemical moiety, while the part embedded within the wax remains intact.[7] A huge family of silanes with various functional groups can be introduced to modify the exposed part, and in this way the chemistry of the Janus particles can be widely varied. The notion of 'Janus balance' was proposed to define the dimensionless ratio of work to transfer a Janus particle from the oil/water interface into the oil phase, normalized by the work required to move it into the water phase. The Janus balance value can be simply calculated from the interfacial contact angle and the geometry of Janus particles, without the need to know the interfacial energy. This concept may help to predict how a Janus particle behaves as solid surfactant with respect to efficiency and function, as the Janus balance of particles is analogous to the classical hydrophile–lipophile balance (HLB) of molecular surfactants.[8] They further developed a simple method to tune the Janus balance of particles.[9] For charged particles, introduction of oppositely charged surfactants on to the particle surface can control the degree of penetration of the particles absorbed at the liquid/liquid interface. The surface area exposed to the water phase is chemically modified in a subsequent reaction.

Using the approach proposed by Granick and co-workers, stimuli-responsive Janus particles were synthesized.[10] The amine-functionalized silica particles were partially covered with frozen wax generated by emulsions, and atom transfer radical polymerization (ATRP) initiators were selectively introduced to the exposed area. For an amine-functionalized particle, 20–25% of the surface is immersed in wax and thus protected. In comparison, only 10% of the non-functionalized particle is immersed. On to the ATRP initiator grafted area, a polymer brush is sequentially grafted from the surface, forming Janus particles. This grafting approach is fairly universal, and responsive polymers can be conjugated on to one part of the particle to achieve the corresponding stimuli-responsive performance.

Excess growth of another material on to the exposed part of particles at the frozen Pickering emulsion interface usually leads to coalescence among the particles, and it may become difficult to achieve individual dispersed Janus particles. Further, it is not easy to control the shape of Janus particles. We proposed an alternative approach to synthesize Janus non-spherical silica particles by asymmetric wet etching, as shown in Figure 4.1.[11] The silica particles are first modified with a silane to introduce a very thin desired functional corona on to the silica surface. At the Pickering frozen wax/water emulsion interface, the particles are partially embedded and thus protected. By selectively etching the exposed corona with aqueous hydrofluoric acid in the outer aqueous phase, a fresh silica surface with Si–OH group emerges after removal of the corona moiety. Janus spherical particles are derived after separation from the interface. With further etching of the fresh silica surface, the exposed spherical part evolves progressively to be non-spherical, for example, mushroom-like. On to both the fresh silica side and the protected side with the corona, other materials can be selectively grown to derive other different compositions. On to the vinyl corona side of the silica after being modified with 3-acryloxypropyltrimethoxysilane, a polystyrene (PS) cap is grown by one-stage emulsion polymerization (Figure 4.1a). The fresh silica side is larger than the PS cap and remains coarse. Silica can be gradually etched from the Janus particles to control the size ratio of silica to PS sides. Eventually, PS nano-caps are derived after complete removal of silica (Figure 4.1b). No PS is grown on to the coarse side, indicating that the vinyl

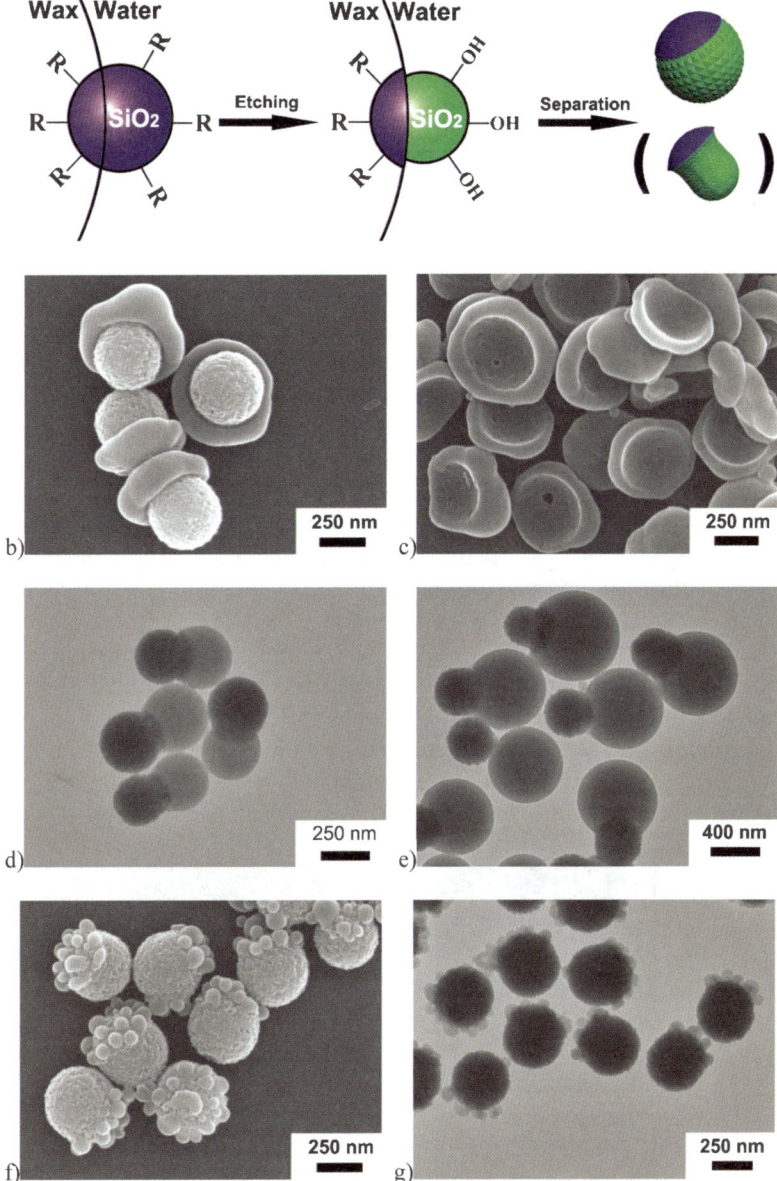

Figure 4.1 Illustrative synthesis of Janus non-spherical particles by asymmetric etching of the exposed side of the silica particles at a Pickering frozen wax/water emulsion interface (a). Morphologies of the Janus SiO_2–PS composite particles with a PS cap grown on the corona side (b); the PS nano-caps after removing SiO_2 (c); the Janus SiO_2–PS dimers with increased PS volume fraction (the dark area corresponding to silica and the gray area to PS) (d, e); the Janus SiO_2–PS composite colloids with PS nano-flowers grown on the corona side (f, g).[11]

corona exists exclusively on the smooth side. By increasing the amount of initiator, the PS cap becomes larger and gradually evolves into a spherical shape, forming a silica–PS dimer (Figure 4.1c). By further increasing the amount of monomer, the PS side becomes dominant (Figure 4.1d). The Janus balance of the Janus PS–silica particles is thus continuously tuned by varying the amount of PS. When the monomer–initiator mixture is added dropwise to alter the polymerization dynamics, some nano-flowers are grown on the corona side instead of a spherical contour (Figure 4.1e, f).

Structural fluctuation of the Pickering emulsion interface upon freezing would possibly cause some uncertainties for particle absorption. Using a liquid/liquid interface is always the most straightforward approach although rotation of particles at the liquid interface will be a problem. Kawaguchi and co-workers attempted to prepare Janus particles at a liquid/liquid Pickering emulsion interface from responsive poly[*N*-isopropylacrylamide-*co*-(acrylic acid)] microgels.[12] Acrylic acid was chosen as a co-monomer to introduce functional groups into the microgel hemisphere by a simple carbodiimide coupling reaction. As shown in Figure 4.2, a Pickering oil-in-water emulsion was formed by stirring hexadecane and water with the microgel as stabilizer. After removal of the unattached microgel from the continuous phase, amine groups were introduced by a carbodiimide coupling with water-soluble ethylenediamine and 1-ethyl-3-(3-dimethylaminopropyl)carbodiimide hydrochloride. The composition of such Janus colloids is rather restricted. It remains unclear whether the particle rotation could influence the formation of the Janus particles. Calculations predicted that a Janus particle with amphiphilic characteristics will be absorbed more strongly at an emulsion interface.[13] It is conjectured that once the two parts of a uniform particle absorbed at a

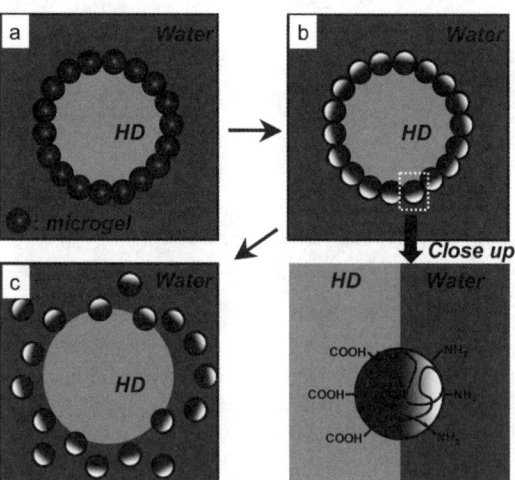

Figure 4.2 The use of an oil/water interface of Pickering emulsions allows the introduction of a distributed functional group into the microgels if the functionalization occurs on only one side of the microgels.[12]

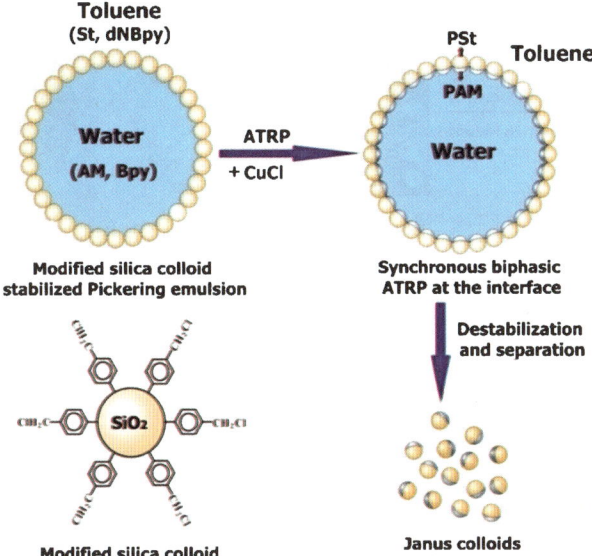

Figure 4.3 Illustrative synthesis of Janus colloids by biphasic grafting at a Pickering liquid/liquid emulsion interface and favorable growth of other materials on to the desired side of the Janus particles to control their composition and microstructure.[14]

Pickering emulsion interface have been simultaneously grafted with polymers with opposite wettability, the particle rotation will be significantly confined to facilitate further polymer growth, forming Janus particles.

We performed Janus particle synthesis by simultaneous biphasic grafting of different polymer brushes on to the two parts of a silica particle at a Pickering liquid/liquid emulsion interface as shown in Figure 4.3.[14] The two parts of a particle at the Pickering emulsion interface are divided into the two immiscible phases, which can be synchronously modified by ATRP in the two phases. The imposed hydrophilic–lipophilic interaction on the original Pickering effect of the particles will further restrict their rotation at the interface. These Janus particles can be further modified to derive various other new Janus composite colloids with different compositions. For example, desired corona capped nanoparticles were grown on to the desired side of the Janus particles by strong interaction, and the miscrostructure of the particles was tuned.

4.3 Synthesis in a Liquid Droplet

Starting from a liquid droplet dispersed in a liquid medium, many variables, including polymerization, solvent evaporation, temperature and external fields, can induce phase separation, forming asymmetric particles. As shown in Figure 4.4, Janus composite magnetic particles are synthesized during solvent evaporation from PS-*b*-PAA modified apolar magnetite nanocrystal

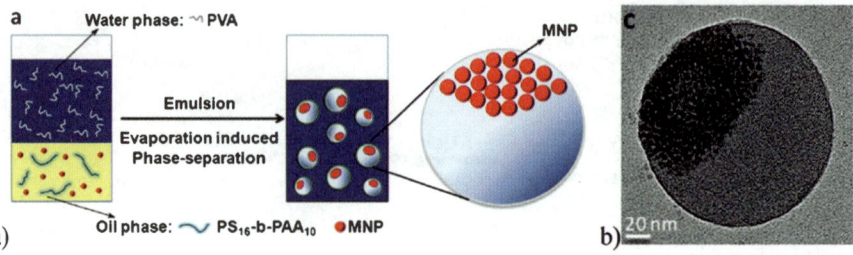

a)

b)

Figure 4.4   Schematic diagram of the synthesis of multifunctional nanocomposites with spatially separated functionalities (a) and TEM of the typical magneto-optical nanocomposite Janus particles (b).[15]

droplets in an aqueous phase stabilized with PVA.[15] During the subsequent solvent evaporation, magnetic particles segregate in one hemisphere, achieving spatial separation of imaging and magnetic therapy functionalities.

Solvent evaporation from droplets fabricated by microfluidics can greatly enhance the monodispersity of particles.[16] Similarly, an emulsion stabilized with silica particles was generated with a microfluidic device, the dispersed phase being photo-curable resin which contained iron oxide nanoparticles.[17] When an external magnetic field was applied right after the formation of the monodispersed droplets, magnetic phase separation generated the Janus particles. However, the yield of particles by microfluidic fabrication was rather low. In emulsions, Dyab *et al.* reported the synthesis of a Janus structure by drive directional movement of magnetite nanoparticles inside polymerizable droplets.[18] After the successive arrangement of inner compartments in the presence of a magnetic field, polymerization of the oil phase generated bicompartmental particles as shown in Figure 4.5.

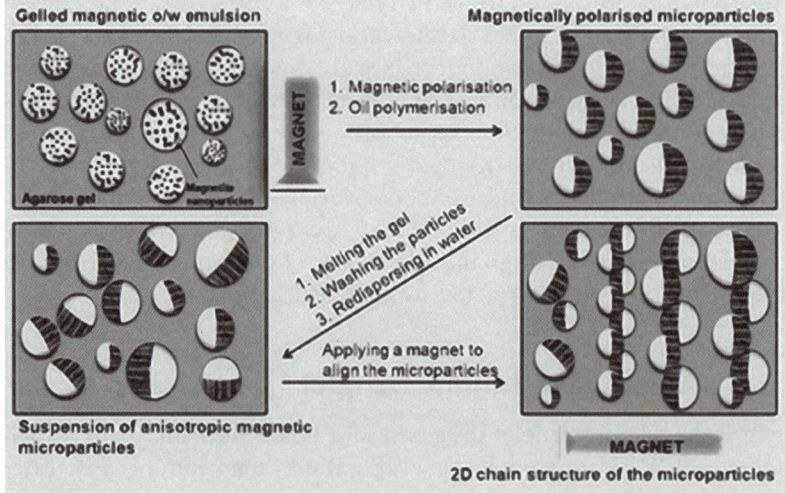

Figure 4.5   Schematic diagram of the synthesis of anisotropic magnetic microparticles from a polymerizable oil-in-water magnetic emulsion.[18]

We proposed an emulsion approach to synthesize non-spherical Janus particles from *in situ* polymerizing deformable polymers.[19] It is significant that in the dispersed oil phase polymerization induces phase separation, which forms nascent deformable polymer particles. They move towards the interface and are immobilized there due to the Pickering effect as shown in Figure 4.6. Mismatch among the three interfacial tensions at the triple phase contact line is responsible for the deformation of the particles at the interface. Meanwhile, hydrophilic monomers in the aqueous phase are initiated by the residual free radicals on the exposed side, forming hydrophilic polymer brushes. The non-spherical particles are strictly Janus in both shape and chemistry. Controlling the surfactant concentration and extent of cross-linking can further control the macroscopic shape of the Janus materials from a disk to hemisphere and ball-on-disk structures.

Janus sheets deserve more attention owing to their highly anisotropic shape and asymmetric chemistry. Compared with Janus spheres, rotation of Janus sheets at an interface is greatly restricted, thus making the emulsion more stable as the planar configuration facilitates their favorable orientation. Nanometer-sized polymeric Janus disks have been synthesized by cross-linking supramolecular structures from block copolymers.[20,21] The polymeric disks are swellable in the presence of organic solvents, and thus easily deformed to lose their original shape. Robust inorganic Janus platelets can tolerate organic solvents. Janus silica platelets are produced on a large scale by crushing modified commercial glass hollow spheres after selective modification of their exterior and interior surfaces. The Janus platelets are several micrometers in thickness and their composition is poorly controlled.[22] We extended this simple crushing approach to prepare Janus silica nanosheets from their parent Janus hollow spheres (Figure 4.7a, b).[23] The silica Janus hollow spheres are

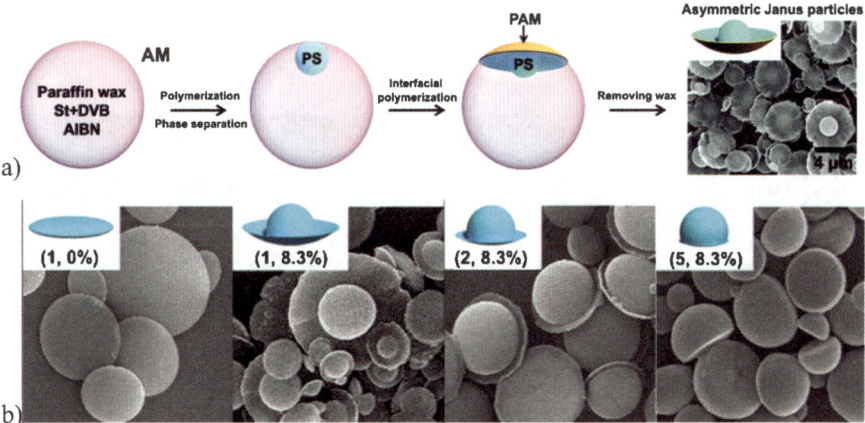

Figure 4.6 Schematic diagram of the synthesis of the asymmetric Janus polymer particles at the interface of emulsion droplets and the as-synthesized Janus particles (a) and other morphologies (b).[19]

first synthesized to form a Janus interface by suitable chemistry, for example, a self-organized sol–gel process at an emulsion interface. The key issue is that oil-soluble silanes with different organic moieties can self-organize at the interface, probably induced by the surfactant already existing at the interface. Further sol–gel processing makes the cross-linked shell more robust. The thickness of the shell is tunable within nanometers and the pendant groups on both surfaces facing polar and apolar phases are broadly tunable from a diversity of silane candidates. By favorable growth of other materials on the desired side, the composition and microstructure of the Janus nanosheets can be greatly extended and additional new properties can be introduced. The Janus nanosheets can readily stabilize immiscible liquids to form emulsions. When magnetic nanoparticles are preferentially absorbed on the hydrophilic side, the dispersed oil droplets in the emulsions can be magnetically manipulated. An exciting prospect is that the Janus nanosheets can serve as nano-robots when they are capable of recognizing their targets, for example, malignant cells.[24] Favorable polymerization on to the vinyl group-containing side of the Janus silica nanosheets leads to Janus polymer–silica composite nanosheets (Figure 4.7c).[25] By tuning the weight ratio of polymer to silica components, the Janus balance of the nanosheets is tunable from more hydrophilic to more lipophilic. The continuous phase of the emulsions stabilized with the Janus composite nanosheets can be inverted from oil to water. The Janus composite nanosheets will become smart materials when responsive species are favorably grown on desired sides.

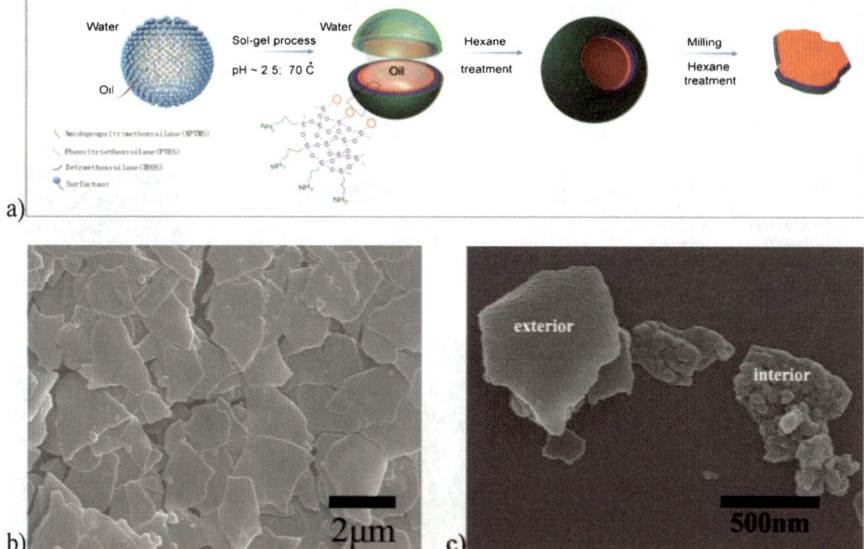

Figure 4.7   Schematic diagram of the fabrication of silica Janus nanosheets by crushing Janus hollow spheres (a) and SEM images of the silica Janus nanosheets (b) and Janus polymer–silica composite nanosheets (c).[23,25]

4.4 Synthesis upon Preformed Particles

Seeded polymerization to form Janus particles usually involves swelling a cross-linked seed particle with monomers that are subsequently polymerized, or swelling a premade multi-polymer particle by switching solvent quality. It is a prerequisite that the newly formed polymer obtained by swelling polymerization should not be immiscible with the seed polymer matrix and segregates from the matrix particle, resulting in protrusion of a polymer phase from the seed particle.[26] Since the synthesis of monodisperse polystyrene (PS) microspheres is well developed and PS is easily swellable with solvents or various monomers, this approach has been extensively employed to swell PS microspheres. Weitz *et al.* synthesized uniform anisotropic particles consisting of two immiscible polymers such as PS–poly(methyl methacrylate) (PMMA) and PS–poly (butyl methacrylate) (PBMA), as shown in Figure 4.8.[27,28] It is a key issue that the introduction of cross-linking agents is conducive to phase separation by release of the formed high internal stress.

The shape of the particles is essentially determined by the cross-linking density and phase separation kinetics.[29] Similarly, raspberry-like PS–poly-acrylonitrile (PAN) particles with multiple protrusions of PAN on the PS surface have been prepared.[30] In addition to polymeric particles, inorganic materials can also be incorporated during the process. Janus PS–silica composite particles containing magnetite (Fe_3O_4) have been prepared.[31] It is important to control the size and shape of the secondary particles attached to the seed particles. More complex anisotropic structures of PS–poly(*N*-isopropylacrylamide) (PNIPAM) particles have been prepared by controlling the surfactant concentration, swelling ratio of seed particles and polymerization temperature (Figure 4.9).[32]

Okubo and co-workers exploited alternative attempts to focus on chemically anisotropic particles with methacrylate monomers grown from PS particles including disk-shaped particles (Figure 4.10).[33,34] PS–PMMA–silica composite particles having the silica particles encapsulated in the PMMA bulb were successfully fabricated. After removal of the silica particles from the inside of the composite particles, hollow snowman-like structures were created.[35]

We synthesized anisotropic silica–polymer composite particles by radical polymerization-induced de-wetting on a particle surface as shown in

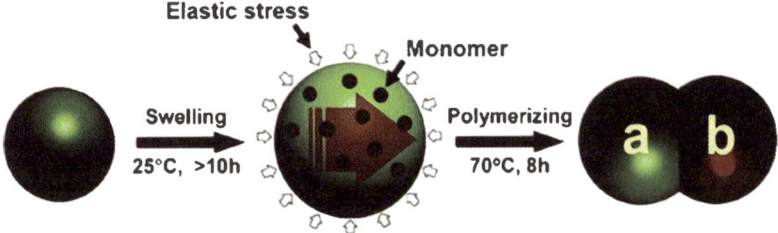

Figure 4.8 Schematic diagram of the synthesis of anisotropic non-spherical dumbbell particles by seeded polymerization.[28]

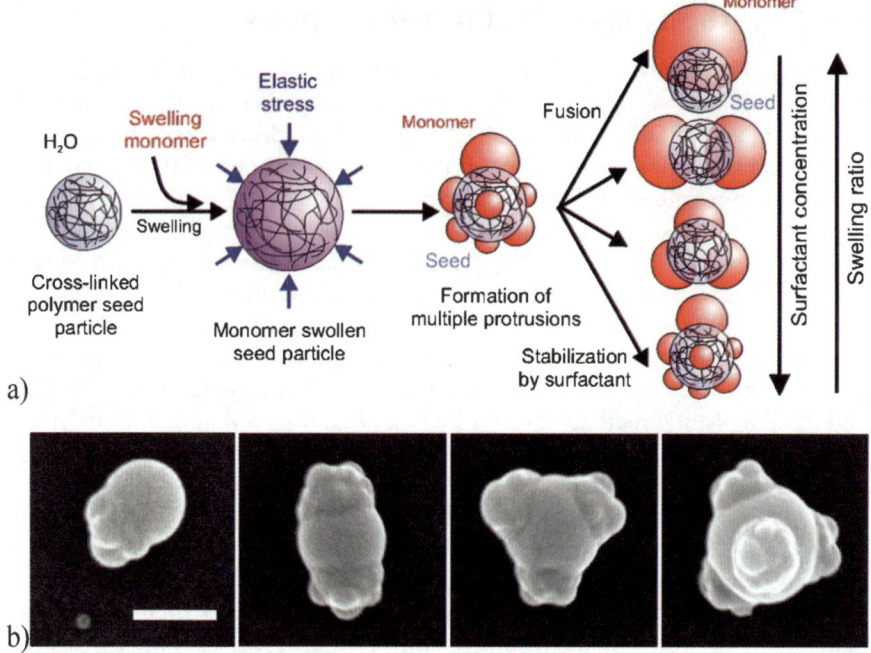

a)

b)

Figure 4.9 (a) Schematic diagram of the synthesis and several factors on patchy particles. (b) By fusion of their liquid protrusions, the highly cross-linked polystyrene seeds form colloidal molecules. Scale bar: 300 nm.[32]

Figure 4.11. The anisotropic structure can be tuned by modifying the surfactant concentration and degree of cross-linking. Janus particles were derived by selectively etching the anisotropic composite particles from the weak, thin organic layer. This method is facile and can be scaled up to produce Janus particles in large quantities.[36] Similarly to other reported anisotropic composite particles, there is a layer of PS coating on the silica surface, although it is rather thin.[37,38] Such composite particles are not true Janus in

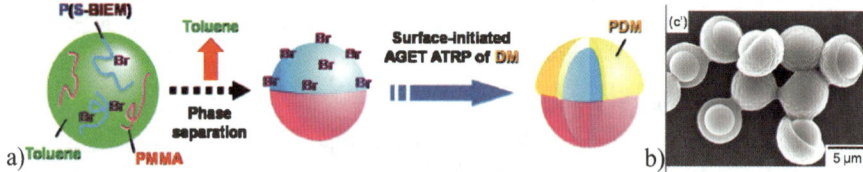

Figure 4.10 (a) Schematic diagram of the synthesis and systematic influence of several factors on the resulting patchy particles. ££(b) SEM of PMMA–poly(styrene-2-(2-bromoisobutyryloxy)ethyl methacrylate) P(S-BIEM)-g-poly(2-(dimethylamino)ethyl methacrylate) (PDM) Janus particles prepared by surface-initiated activator generated by electron transfer (AGET) ATRP of 2-(dimethylamino)ethyl methacrylate (DM) in an aqueous medium using spherical PMMA–P(S-BIEM) macroinitiator Janus particles.[33]

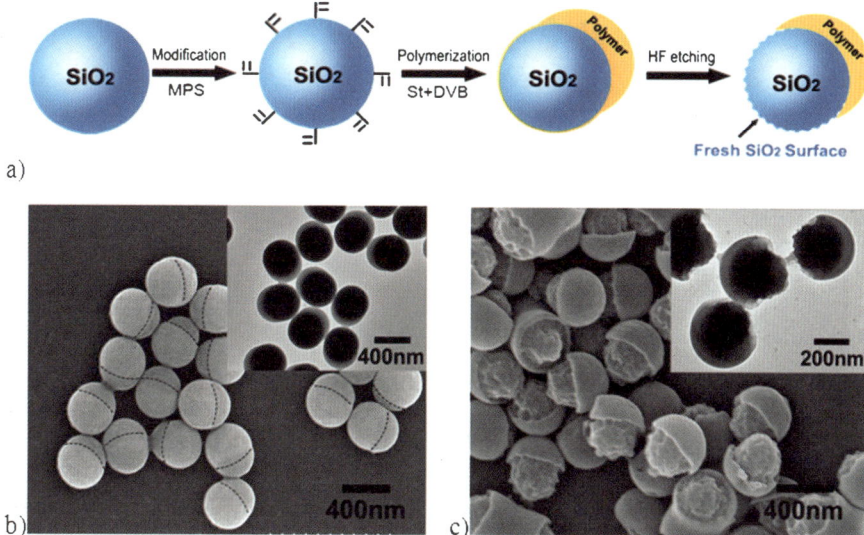

a)

b) c)

Figure 4.11   (a) Schematic diagram of the synthesis of Janus colloids by wet etching the anisotropic composite particles formed by polymerization-induced de-wetting. SEM images of (b) silica–polymer anisotropic composite particles prepared at a monomer:silica weight ratio of 0.5:1) and (c) the as-synthesized Janus particles after etching the anisotropic composite particles with aqueous HF.[36]

chemistry since they are not amphiphilic. We further selectively etched the thin layer of PS to expose the fresh SiO_2 surface beneath, while the other thick PS bulge remained. Silica can be gradually etched from the Janus particles to derive PS hollow particles after complete removal of silica. In addition to submicron particles, inorganic nanoparticles can be used as seeds to induce polymer phase separation, forming hybrid Janus structures.

We recently reported a facile approach to the large-scale production of submicron Janus polymer or hybrid particles by seeded emulsion polymerization against a cross-linked polymer hollow particle seed as shown in Figure 4.12.[39] A monomer mixture of styrene (St) and divinylbenzene (DVB) is dropped into the seed emulsion of cross-linked PAN hollow particles, accompanied by polymerization at high temperature. Since PAN and PS are typically immiscible, the PS phase separates from the PAN shell especially driven by the polymer network elastic–retractile force during cross-linking. Anisotropic Janus PAN–PS polymer particles were obtained, similar to other reported anisotropic particles.[40–44] Although the PS bulb contains no PAN, the PAN part should contain a minority of PS. In order to shield the effect of PS on the purity of the PAN part, the PAN part was selectively hydrolyzed to achieve functional groups including carboxylic acid (–COOH). The functional groups can induce further favorable growth of other materials to coat the PAN part completely. Janus composite particles were formed. In addition to the

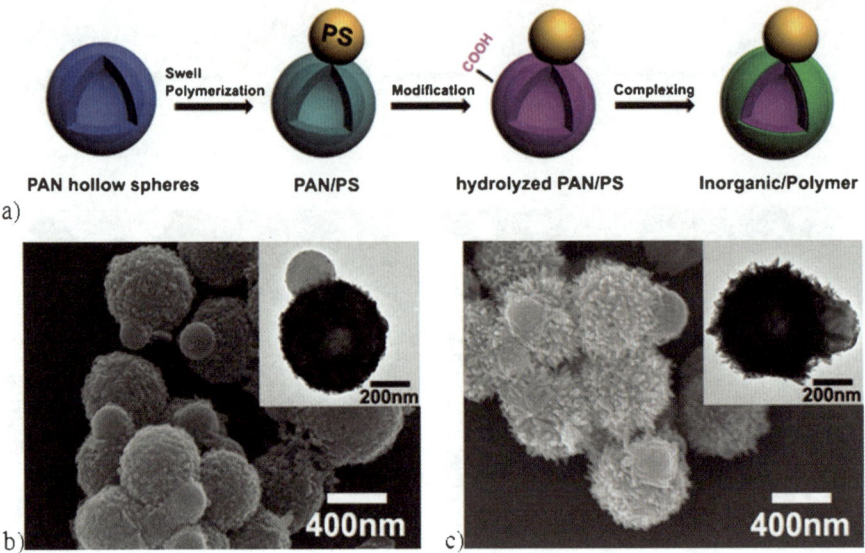

a)

PAN hollow spheres   PAN/PS   hydrolyzed PAN/PS   Inorganic/Polymer

b)   c)

Figure 4.12   (a) Schematic diagram of the synthesis of Janus submicrometer-sized particles by seeded emulsion polymerization. SEM images of (b) Janus titania–PS composite particles and (c) Janus β-FeOOH–PS composite particles.[39]

composition, the microstructure of the Janus colloids is tunable. The colloids are strictly Janus in both shape and chemistry.

Xia and co-workers simply modified the precipitation polymerization of PS by adding Au (or Ag) nanoparticles to the system after the polymerization had proceeded for a few minutes (Figure 4.13).[45] The asymmetric particles are uniform in both size and morphology and each PS bead contains only one Au or Ag nanoparticle on its surface. The asymmetric particles are considered to be PS–Au (Ag) Janus hybrid particles. The time of adding the Au colloids is crucial to achieving asymmetric hybrid particles. If Au colloids are added before polymerization, the PS bead might contain a large number of Au nanoparticles due to aggregation, which may be induced by the initiator and other ionic species. If Au colloids are added 30 min after starting polymerization, the yield of hybrid particles decreases to 50%. These results suggest that the PS oligomers formed in the early stage of polymerization could stabilize the Au nanoparticles to prevent agglomeration.

In contrast to additional growth of another material against preformed particles in emulsions, dynamic swelling of a core–shell structure to protrude the core material on to the shell forming a bulb is a fairly simple approach to derive anisotropic Janus particles both in shape and chemistry. Wang *et al.* reported the synthesis of chemically anisotropic snowman-like particles by protrusion of the PS core from the polyelectrolyte multilayer shell in THF–water mixed solvent.[46] Dissolved PS should precipitate upon contact with the continuous phase, which is determinative in forming the PS bulb on the shell.

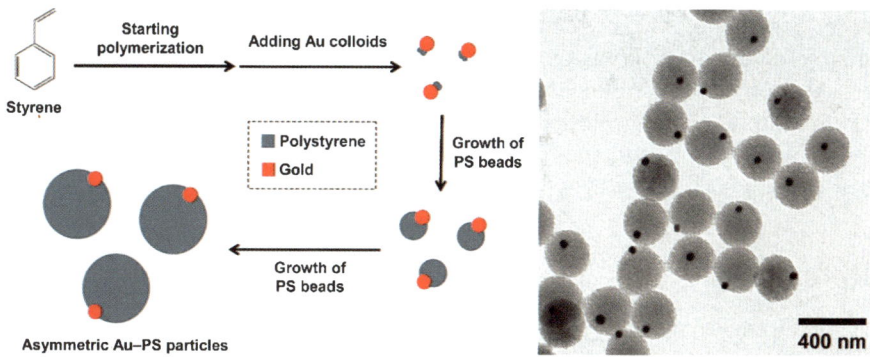

Figure 4.13 Schematic diagram of the synthesis of asymmetric Au–PS Janus nanoparticles and their morphology.[45]

The process is completed rapidly, within a few minutes. While PS–poly(St-*co*-trimethoxysilylpropylacrylate (TMSPA)) core–shell particles were further swollen with sequential addition of monomers, the hydrophobic PS core is protruded on to the hydrophilic shell, forming a bulb. Dumbbell-shaped polymer Janus particles were created.[47] However, the bulb consisted of linear polymers, and thus is rather weak and easily dissolved from the Janus particles in the presence of good solvents. The Janus particles are not robust when they meet solvents, and they cannot serve as solid surfactants to emulsify the organic agents. It is necessary to elaborate dynamic protrusion further to synthesize robust inorganic/organic Janus particles at high solid contents.

4.5 Summary and Outlook

Among the methods used to fabricate Janus particles, emulsion-based synthesis is the most promising owing to its easier scale-up, low cost and generality for diversified composition, including organic, inorganic and hybrid materials. An adequate quantity of Janus particles is crucial for the systematic characterization of the performance of Janus particles and eventually their industrial applications. In combination with the selective post-modification of one compartment to introduce functional groups and thereby sequential preferential growth of other materials, the composition and microstructure of Janus materials can be broadly tunable. The interplay between colloids and surfactants at the emulsion interface can effectively control the penetration depth of the colloid at the interface and the Janus balance after selective modification is further tunable. Particularly for those soft particles at the interface, mismatch among the three interfacial tensions along the triple phase contact line results in deformation of the particles and leads to non-spherical Janus objects. In a liquid droplet or with a preformed particle dispersed in emulsions, many variables, including solvent evaporation, polymerization and external field, can induce phase separation to form anisotropic Janus particles on a large scale. Although emulsion-based synthesis is currently focused on submicron or larger Janus particles, the synthesis of Janus nanoparticles with varied compositions will be developed rapidly in the near future. Co-assembly of amphiphilic molecules and Janus nanoparticles will provide more opportunities to gain further insight into the interaction physics and to design new materials and devices.

References

1. W. Ramsden, *Proc. R. Soc. London*, 1903, **72**, 156.
2. S. U. Pickering, *J. Chem. Soc.*, 1907, **91**, 2001.
3. A. D. Dinsmore, M. F. Hsu, M. G. Nikolaides, M. Marquez, A. R. Bausch and D. A. Weitz, *Science*, 2002, **298**(5595), 1006.
4. R. Aveyard, B. P. Binks and J. H. Clint, *Adv. Colloid Interface Sci.*, 2003, **100–102**, 503.
5. S. Sacanna, W. K. Kegel and A. P. Philipse, *Langmuir*, 2007, **23**(21), 10486.
6. M. Lattuada and T. A. Hatton, *J. Am. Chem. Soc.*, 2007, **129**(42), 12878.
7. L. Hong, S. Jiang and S. Granick, *Langmuir*, 2006, **22**(23), 9495.
8. S. Jiang and S. Granick, *J. Chem. Phys.*, 2007, **127**(16), 161102.
9. S. Jiang and S. Granick, *Langmuir*, 2008, **24**(6), 2438.
10. S. Berger, A. Synytska, L. Lonov, K. J. Eichhorn and M. Stamm, *Macromolecules*, 2008, **41**(24), 9669.
11. B. Liu, C. L. Zhang, J. G. Liu, X. Z. Qu and Z. Z. Yang, *Chem. Commun.*, 2009, (26), 3871.
12. D. Suzuki, S. Tsuji and H. Kawaguchi, *J. Am. Chem. Soc.*, 2007, **129**(26), 8088.
13. B. P. Binks and P. D. I. Fletcher, *Langmuir*, 2001, **17**(16), 4708.
14. B. Liu, W. Wei, X. Z. Qu and Z. Z. Yang, *Angew. Chem. Int. Ed.*, 2008, **47**(21), 3973.
15. X. H. Gao and S. H. Hu, *J. Am. Chem. Soc.*, 2010, **132**(21), 7234.
16. D. A. Weitz, R. K. Shah and J. W. Kim, *Adv. Mater.*, 2009, **21**(19), 1949.
17. S. H. Kim, J. Y. Sim, J. M. Lim and S. M. Yang, *Angew. Chem. Int. Ed.*, 2010, **49**(22), 3786.
18. A. K. F. Dyab, M. Ozmen and M. Ersoz and V. N. Paunov, *J. Mater. Chem.*, 2009, **19**(21), 3475.
19. Y. H. Wang, C. L. Zhang, C. Tang, J. Li, K. Shen, J. G. Liu, X. Z. Qu, J. L. Li, Q. Wang and Z. Z. Yang, *Macromolecules*, 2011, **44**(10), 3787.
20. A. Walther, X. André, M. Drechsler, V. Abetz and A. H. E. Müller, *J. Am. Chem. Soc.*, 2007, **129**(19), 6187.
21. A. Walther, M. Drechsler and A. H. E. Müller, *Soft Matter*, 2009, **5**(2), 385.
22. B. Gruning, U. Holtschmidt, G. Koerner and G. Rossmy, US Patent 4 715 986, 1987.
23. F. X. Liang, K. Shen, X. Z. Qu, C. L. Zhang, Q. Wang, J. L. Li, J. G. Liu and Z. Z. Yang, *Angew. Chem. Int. Ed.*, 2011, **50**(10), 2379.

24. Janus materials: a crushing future. *NPG Asia Mater.*, published online 16 May 2011, doi:10.1038/asiamat.2011.77;http://www.nature.com/am/journal/2011/201105/full/am201199a.html
25. Y. Chen, F. X. Liang, H. L. Yang, C. L. Zhang, Q. Wang, X. Z. Qu, J. L. Li, Y. L. Cai, D. Qiu and Z. Z. Yang, *Macromolecules*, 2012, **45**(3), 1460.
26. H. R. Sheu, M. S. Elaasser and J. W. Vanderhoff, *J. Polym. Sci., Part A: Polym. Chem.*, 1990, **28**(3), 629.
27. L. T. Yan, N. Popp, S. K. Ghosh and A. Boker, *ACS Nano*, 2010, **4**(2), 913.
28. D. A. Weitz, J. W. Kim and R. J. Larsen, *J. Am. Chem. Soc.*, 2006, **128**(44), 14374.
29. D. A. Weitz, J. W. Kim and R. J. Larsen, *Adv. Mater.*, 2007, **19**(15), 2005.
30. H. R. Liu and H. F. Huang, *J. Polym. Sci., Part A: Polym. Chem.*, 2010, **48**(22), 5198.
31. J. P. Ge, Y. X. Hu, T. R. Zhang and Y. D. Yin, *J. Am. Chem. Soc.*, 2007, **129**(29), 8974.
32. D. J. Kraft, J. Hilhorst, M. A. P. Heinen, M. J. Hoogenraad, B. Luigjes and W. K. Kegel, *J. Phys. Chem. B*, 2011, **115**(22), 7175.
33. T. Tanaka, M. Okayama, Y. Kitayama, Y. Kagawa and M. Okubo, *Langmuir*, 2010, **26**(11), 7843.
34. T. Tanaka, M. Okayama, H. Minami and M. Okubo, *Langmuir* 2010, **26**(14), 11732.
35. D. Nagao, M. Hashimoto, K. Hayasaka and M. Konno, *Macromol. Rapid Commun.*, 2008, **29**(17), 1484.
36. C. L. Zhang, B. Liu, C. Tang, J. G. Liu, X. Z. Qu, J. L. Li and Z. Z. Yang, *Chem. Commun.*, 2010, **46**(25), 4610.
37. S. Reculusa, C. Poncet-Legrand, A. Perro, E. Duguet, E. Bourgeat-Lami, C. Mingotaud and S. Ravaine, *Chem. Mater.*, 2005, **17**(13), 3338.
38. A. Perro, S. Reculusa, F. Pereira, M. H. Delville, C. Mingotaud, E. Duguet, E. Bourgeat-Lami and S. Ravaine, *Chem. Commun.*, 2005, (44), 5542.
39. C. Tang, C. L. Zhang, J. G. Liu, X. Z. Qu, J. L. Li and Z. Z. Yang, *Macromolecules*, 2010, **43**(11), 5114.
40. H. Sheu, M. Ei-Aasser and J. Vanderhoff, *J. Polym. Sci., Part A: Polym. Chem.* 1990, **28**, 653.
41. J. W. Kim, R. J. Larsen and D. A. Weitz, *J. Am. Chem. Soc.*, 2006, **128**(44), 14374.
42. J. W. Kim, R. J. Larsen and D. A.Weitz, *Adv. Mater.*, 2007, **19**, 2005.
43. J. W. Kim, D. Lee, H. C. Shum and D. A.Weitz, *Adv. Mater.*, 2008, **20**, 3239.
44. H. Ahmad, N. Saito, Y. Kagawa and M. Okubo, *Langmuir*, 2008, **24**(3), 688.
45. A. Ohnuma, E. C. Cho, P. H. C. Camargo, L. Au, B. Ohtani and Y. N. Xia, *J. Am. Chem. Soc.*, 2009, **131**(4), 1352.
46. D. Y. Wang,, H. K. Yu and Z. W. Mao, *J. Am. Chem. Soc.*, 2009, **131**(18), 6366.
47. J. G. Park, J. D. Forster and E. R. Dufresne, *J. Am. Chem. Soc.*, 2010, **132**(17), 5960.

CHAPTER 5

Particle Replication in Non-wetting Templates: a Platform for Engineering Shape- and Size-specific Janus Particles

JOSEPH M. DESIMONE*[a,b], JIE-YU WANG[a] AND
YAPEI WANG[c]

[a] Department of Chemistry, University of North Carolina at Chapel Hill,
Chapel Hill, NC 27599, USA; [b] Department of Chemical and Biomolecular
Engineering, North Carolina State University, Raleigh, NC 27695, USA;
[c] Department of Chemistry, Renmin University of China, Beijing 100872,
China
*E-mail: desimone@email.unc.edu

5.1 Introduction

Because of their unique anisotropic characteristics, Janus particles have been assessed as attractive building blocks for self-assembly with desirable supramolecular architectures, leading to the creation of novel functional materials.[1-5] Janus particles with tunable chemistry and physical properties have attracted much attention in the fields of colloidal physics and chemistry for various applications ranging from optoelectronics, e-ink, drug delivery and bioimaging.[6]

RSC Smart Materials No. 1
Janus Particle Synthesis, Self-Assembly and Applications
Edited by Shan Jiang and Steve Granick
© The Royal Society of Chemistry 2012
Published by the Royal Society of Chemistry, www.rsc.org

Early synthetic strategies for generating Janus particles were focused on self-assembly and surface modification.[7–9] So far, they are still popular as self-assembly can begin with a wide range of molecular moieties, including DNA and block copolymers, and surface modification is being extended to more and more mild reactions which hardly damage the particle interiors. Although the number of chemical compositions and surface chemistries utilized in these particle systems is increasing, particle shapes have been largely limited to a small number of simple geometries. Several recently developed top-down strategies have overcome this limitation and the creation of high-level structures has become possible.[10–15] This chapter highlights the strategy of particle replication in non-wetting templates (PRINT) recently developed in our laboratories to fabricate monodisperse Janus particles with precise control over the size, shape, chemistry and distribution of the chemically distinct regions.

5.2 PRINT Technique

The history of the PRINT technique began with the synthesis of a new highly fluorinated perfluoropolyether (PFPE) elastomer by Rolland *et al.* in 2004.[16,17] This photo-curable resin was demonstrated to have remarkably low surface energy (8–10 dyn cm^{-1}) with respect to other materials. In addition, possessing the properties of high gas permeability, high elastic recovery, good mechanical strength and high chemical and solvent resistance sets PFPE apart from the more commonly used polydimethylsiloxane (PDMS) as a template for micromolding. The PRINT technique is a soft lithography technique based on the use of PFPE molds that is able to mold most hydrophobic and hydrophilic liquids to generate useful materials in the form of arrays of patterned features, arrays of particles and isolated particles.

The PRINT process for fabricating isolated particles starts with an etched silicon master created using standard lithographic techniques in which the features are raised (Figure 5.1). The photo-curable liquid PFPE resin is then poured on to the silicon master and allowed to distribute evenly across and wet the surface of the master template. The resin is then photochemically cross-linked to form a robust elastomeric PFPE mold that is subsequently peeled away from the master to reveal micro- or nanoscale cavities on its surface. Next, a pre-particle solution is cast on a high surface energy sheet [typically poly(ethylene terephthalate) (PET)] using a Myer rod to form a uniform film of specified thickness. The sheet is then laminated to the empty PFPE mold using heat and pressure. When the high surface energy sheet is peeled away, capillary forces keep the liquid trapped in the mold cavities whereas excess solution is wicked away by the high surface energy sheet. The pre-particle solution in the filled mold cavities is then solidified using an appropriate means (UV light, thermal heating, lyophilization, *etc.*). Lastly, the solidified particles in the mold are removed by laminating the filled mold to a sacrificial harvesting film such as polyvinylpyrrolidinone, poly(vinyl alcohol) or

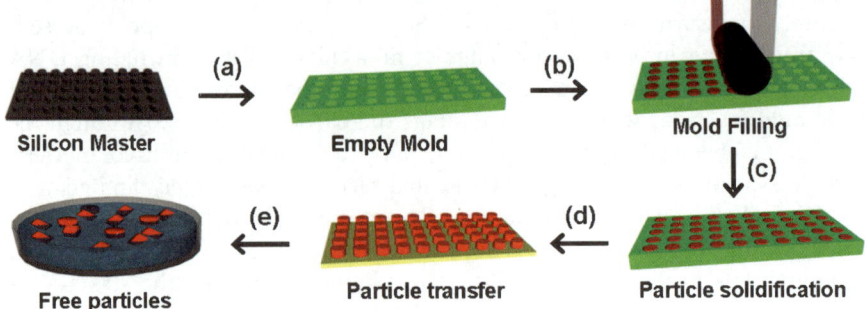

Figure 5.1   Schematic illustration of the PRINT process for particle fabrication. (a) A
PFPE mold is made from a silicon master template; (b) the PFPE mold
cavities are then filled with a liquid precursor *via* capillary flow in
connection with a high surface energy countersheet to remove the excess
material from the land areas; (c) various means are used to solidify the
precursor material contained in the PFPE mold cavities; (d) the particles
are transferred from mold cavities using a sacrificial harvesting sheet; (e)
free particles are obtained by dissolving away the sacrificial adhesive film.

cyanoacrylate. Peeling the harvesting film away from the mold results in an
array of particles on the harvesting sheet. Free particles can then be collected
by dissolving away the sacrificial adhesive film with an appropriate solvent for
the adhesive and a non-solvent for the particles.

The PRINT technique has several attractive features that make it ideally
suited for particle fabrication owing to the unique properties of the PFPE
mold: (i) the low surface adhesion and elastic deformation of the PFPE mold
facilitate the removal of the connecting flash layer, whereas other lithographic
methods usually need an etching step to remove this scum layer; (ii) the highly
fluorinated nature of the PFPE mold also facilitates the removal of particles
from the mold; and (iii) PFPE has been shown to be compatible with a number
of organic solvents that swell the traditionally used PDMS mold material, thus
allowing PRINT to generate particles using a range of materials with very high
fidelity to the original silicon master.

Another advantage of the PRINT technique is its scalability, allowing for
the fabrication of monodisperse particles with precise and independent control
over the particle size, shape and composition in relatively large quantities using
roll-to-roll processing.[18] In addition, the PRINT process is delicate and mild
enough to be compatible with a variety of important cancer therapeutic agents,
detection and imaging agents, various cargos (*e.g.* DNA, siRNA, protein,
chemotherapy drugs, *etc.*), targeting ligands (*e.g.* antibodies, cell targeting
peptides) and functional matrix materials (bioabsorbable polymers, stimuli-
responsive materials, *etc.*).[19] As a result, this new technique is really a platform
technology that makes it possible to study the physical and chemical effects of
particles in drug delivery, electronic devices, colloid science and other
application areas.

   Until recently, only particles composed of a homogeneous matrix had been fabricated. However, through some adaptations to the PRINT process, the fabrication of chemically distinct multicomponents segregated in a single particle has now been successfully achieved using the PRINT technology. As a result, the formation of two-component Janus particles and even particles with more than two chemically distinct regions with precisely controlled size and shape is now possible.

5.3   Janus Particles Fabricated by the PRINT Technique

Janus particles are fabricated through adaptation of the PRINT process using a multi-step mold filling operation. Through this adaptation, PRINT particles can be fabricated consisting of multiple compositions if two or more distinct materials are stepwise trapped in the same mold cavity at each filling step. Current studies of pre-particle materials for multi-step mold filling have been focused on photo-curable monomers. Depending on the dimensions of the mold features, the stepwise mold filling can occur in either vertical or horizontal directions. Ideally, one could anticipate partially filling the cavities in the PRINT mold with one component then in a subsequent step filling the rest of the cavity with a different chemistry to generate a Janus PRINT particle. In reality, though, it is difficult to partially fill the cavities uniformly. On the other hand, it is easy to fill the cavities completely with a composition that includes a diluent with a high vapor pressure that can be volatilized, leaving behind other components having a lower vapor pressure that could be partially cured followed by a subsequent step to 'top-off' the mold cavities with a second component to create a Janus particle.

5.1.1   Stepwise Vertical Mold Filling

A typical process for vertically filling mold cavities to fabricate Janus particles is illustrated in Figure 5.2. In the first step, the mold cavities are filled with a photo-curable monomer solution in which the monomer has been dissolved in a volatile solvent. The solvent is then removed by evaporation and the resulting monomer is concentrated in the bottom of the mold cavities, thus leaving the mold partially filled. The monomer is then exposed to low-intensity UV irradiation and partially cured. In the partially cured state, the monomer still retains some photo-curable reactive groups but is solid enough to allow for a second filling of the mold cavity by another monomer without phase mixing. Both monomer phases are then completely photo-cured using intense UV irradiation, which results in the two different monomers being covalently bound together, inducing the formation of a Janus particle. The array of Janus particles can then be transferred from the mold cavities on to a sacrificial harvesting film as is done in the standard PRINT process. Free Janus particles with two distinct phases are finally obtained by dissolving away the sacrificial adhesive.

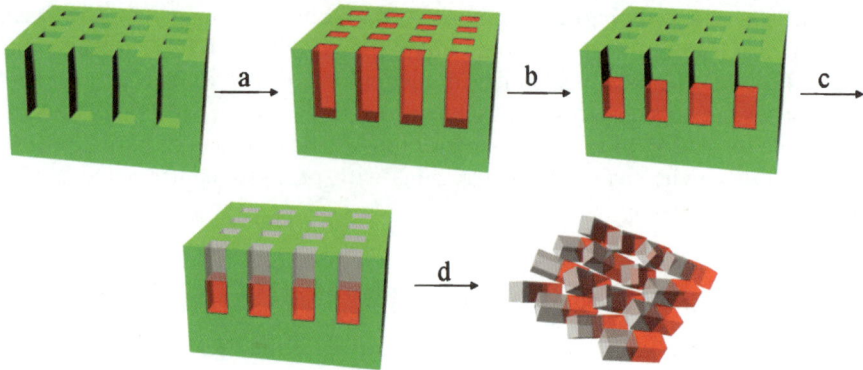

Figure 5.2   Fabrication of Janus particles using PRINT technology. (a) The initial
monomer solution that has been diluted with a volatile solvent is filled
into the mold; (b) after evaporation of the solvent, the remaining
monomer is partially photo-cured; (c) another monomer is then filled into
the partially filled mold cavities and both monomer phases are
completely cured; (d) Janus particles are obtained after harvesting and
purification. Reproduced from reference 20 with permission from the
Institute of Physics.

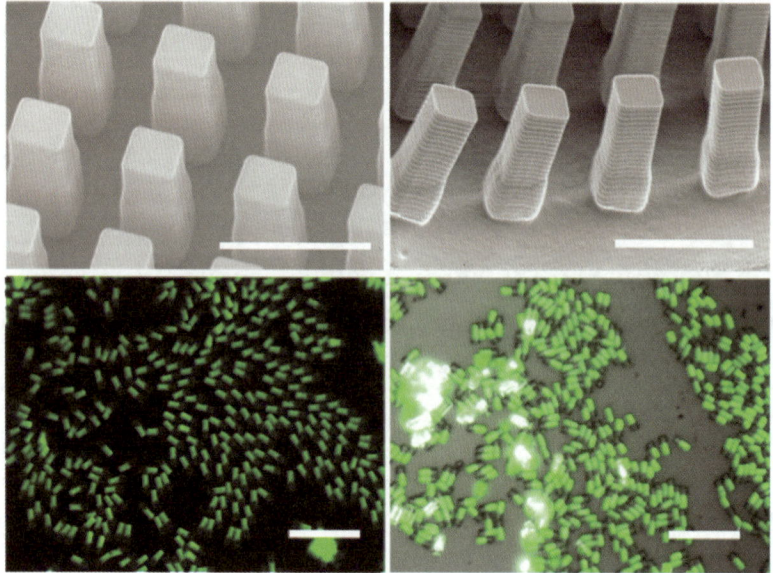

Figure 5.3   SEM (top) and microscope (bottom) images of amphiphilic (left) and dipolar
(right) Janus particles. Particles in the SEM images were harvested on a
polycyanoacrylate harvesting layer. The microscope image of the amphiphilic
Janus particles was obtained from the microscrope fluorescent channel
whereas that of the dipolar Janus particles was obtained from the overlaid
image of the fluorescent and bright field channels. The scale bars in the SEM
images represent 5 μm and in the microscope images 20 μm. Reproduced
from reference 20 with permission from the Institute of Physics.

Early examples were constructed using a mold with a feature size of a $2 \times 2 \times 6$ µm rectangular prism. Two-phase Janus particles were fabricated according to the previously described vertical two-step mold filling strategy.[20] Amphiphilic Janus particles were obtained by combining a hydrophilic monomer [poly(ethylene glycol) monomethyl ether monomethacrylate, $M_n = 1000$ g mol^{-1}] with a hydrophobic monomer (lauryl acrylate). In addition, dipolar Janus particles were fabricated by combining a positively charged monomer (2-aminoethyl methacrylate hydrochloride dissolved in trimethylol-propane ethoxylate triacrylate) with a negatively charged monomer (acrylic acid mixed with trimethylolpropane ethoxylate triacrylate). The scanning electron microscopy (SEM) and optical microscopy images in Figure 5.3 clearly show the biphasic architecture of the resulting Janus particles. For better identification of the disparate regions in the Janus particles, one of the blocks was made fluorescent by doping with fluorescein *o*-acrylate while the other block without the fluorophore remained dark. By varying the concentration of the monomer in the monomer solution of the first block, the volume fraction ratio of the two different regions could be easily controlled.

5.1.2 Horizontal Stepwise Mold Filling

The stepwise mold filling fabrication of particles in the vertical direction is limited owing to two issues: (1) higher aspect ratio particles require a stronger force to pull the particles out of the mold cavities and (2) the etching of silicon master templates with a high aspect ratio pillar is very challenging. Because of these limitations, the multistep mold filling PRINT process has been extended to the horizontal direction as high aspect ratio particles that are lying down can be much more easily etched in silicon templates than vertically oriented high aspect ratio features.

Starting with a $20 \times 20 \times 240$ µm rectangular rod as the starting template, a two-step mold filling process was developed to fabricate triphasic particles.[21] First, a dilute solution of a photo-curable hydrophilic monomer in DMF was used to completely fill the PFPE mold. The solvent was then evaporated and the remaining monomer was drawn by capillary forces to the ends of the rectangular cavity. The monomer was then partially cured using a low-intensity UV light source to convert the monomer into a soft gel while leaving enough reactive groups to covalently bind subsequent blocks. Subsequently, the middle, empty cavity of the mold was completely filled with a second hydrophobic monomer. As shown in Figure 5.4, after the final monomer composition had been fully cured by intense UV irradiation, a triphasic architecture with the regions covalently connected together was obtained. The hydrophilic:hydrophobic ratio in the triphasic structure could be precisely tuned by simply changing the concentration of the original monomer solution. As shown in Figure 5.5, the hydrophilic heads of ABA amphiphilic triphasic rods could be controlled by varying the concentration of the first monomer

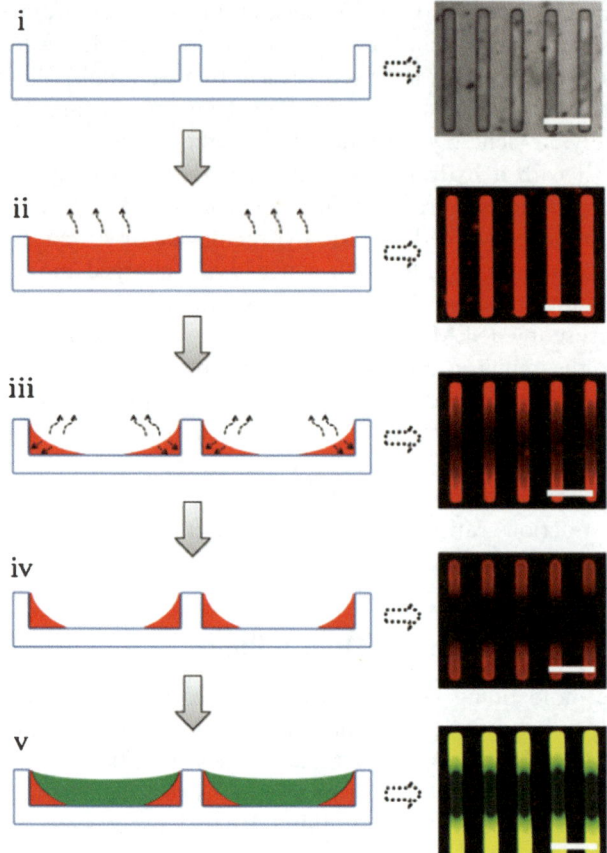

Figure 5.4 Schematic illustration of the formation of triphasic rods and the corresponding microscope images of the molds in each step (all scale bars 100 μm). To distinguish the middle filled hydrophobic regions from the hydrophilic end regions, a photo-curable red dye and a green dye were premixed into the hydrophilic and hydrophobic monomers, respectively, prior to photo-curing. Reproduced from reference 21 with permission from the American Chemical Society.

mixture. Solely hydrophobic (Figure 5.5a) and hydrophilic (Figure 5.5b) particles were fabricated as reference samples.

The principle of fabricating triphasic rods has been extended to the generation of multi-region ABABA rods. In this technique, the mold is partially filled as before by first diluting a hydrophobic monomer solution containing a green dye. A second region is then generated using a dilute hydrophilic monomer solution containing a red dye (Figure 6a), while retaining an open space in the middle of the mold cavity which is able to be filled by a third hydrophobic monomer (undiluted) containing a green dye. As

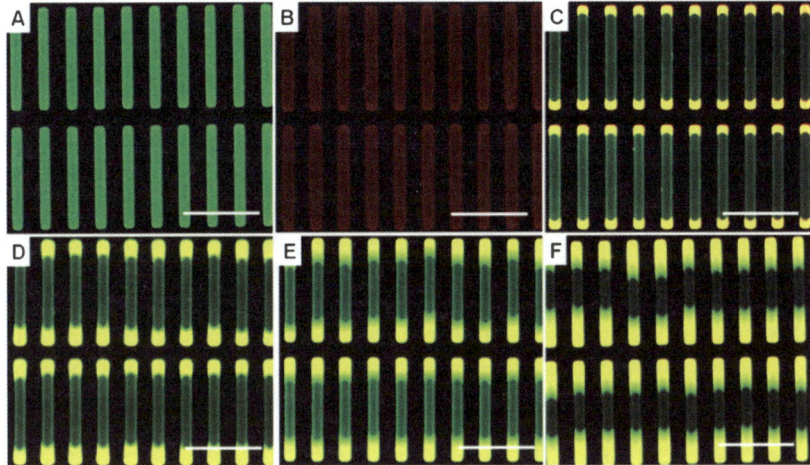

Figure 5.5 Array of ABA 20 × 20 × 240 μm triphasic rod particles on harvesting film with tunable block dimensions. (a) One-component hydrophobic particles. (b) One-component hydrophilic particles. (c)–(f) ABA triphasic particles with different hydrophilic:hydrophobic:hydrophilic ratios corresponding to the four hydrophilic monomer concentrations of (c) 10, (d) 20, (e) 30 and (f) 50 wt%. Images (c)–(f) were captured by overlaying the images under red and green channels. Scale bar: 200 μm. Reproduced from reference 21 with permission from the American Chemical Society.

shown in Figure 6b and c, five regions arranged as **ABABA** were observed to coexist in a particle, demonstrating this to be a powerful tool for building a library of anisotropic rods possessing tunable multiphases of different compositions.

Multiphase particles having either tri- or pentaphasic architectures are centrosymmetric. In order to fabricate asymmetric two-phase rods, centrifugal

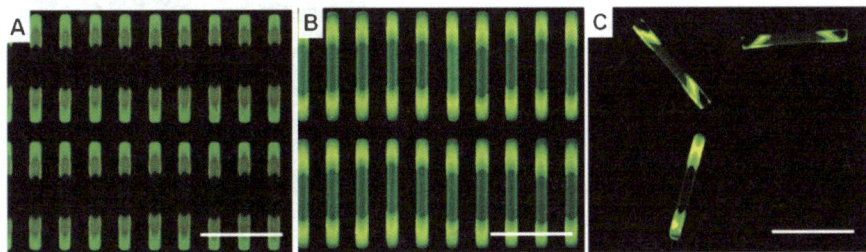

Figure 5.6 Array of anisotropic rods having multiphases. (a) The diphasic particles transferred from the partially filled mold at two ends on to a cyanoacrylate film. (b) Multiphase particles harvested on a cyanoacrylate film. (c) Free multiphase particles separated from cyanoacrylate film. The images were captured by overlaying the images under red and green channels. Scale bar: 200 μm. Reproduced from reference 21 with permission from the American Chemical Society.

force was used to pull the monomer from one end of the cavity to the other end after partially filling the mold with the first monomer. Typically, a partially filled mold containing two hydrophilic end regions was subjected to rotation around an axis perpendicular to the mold plane. As shown in Figure 5.7, the resulting centrifugal force drew the first monomer composition to the outer end of the mold cavity, leaving the rest of the cavity open. The open space was then subsequently filled with a second hydrophobic monomer, which after solidification *via* complete photo-curing yielded amphiphilic diphasic rods, as shown in Figure 5.8. In contrast to previous examples where the length of the initial region could be controlled by varying the concentration of the monomer solution, in this method it was observed that the length of the resulting initial region, on varying the initial monomer concentration from 30 to 80 wt%, was independent of the monomer concentration. Instead, the length of the initial region was found to be dependent on the rotational velocity as increasing the velocity decreased the region dimensions. Moreover, the length of the hydrophilic region remained unchanged as the rotation time was increased at a given angular velocity. This new technique could thus provide a strategy for constructing asymmetric particles of varying hydrophilic/hydrophobic ratios.

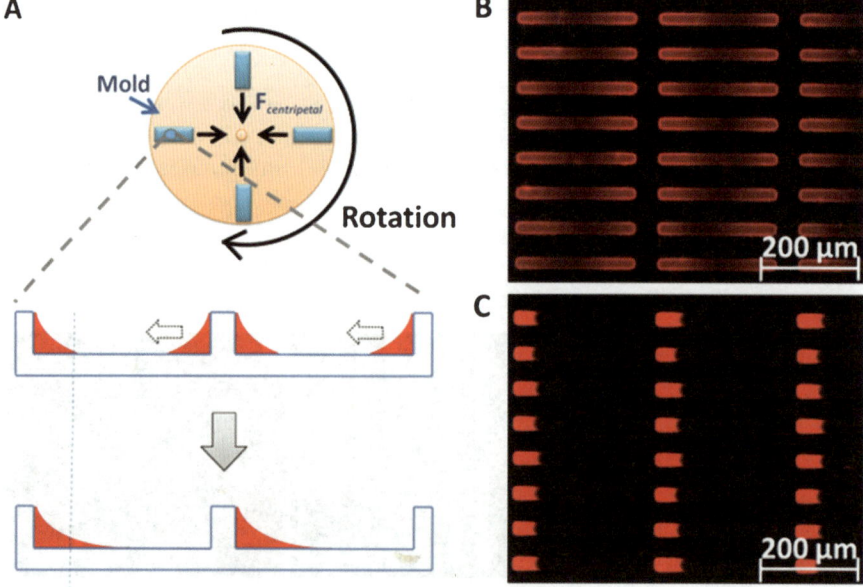

Figure 5.7   (a) Schematic illustration of the cross-section of the mold before and after rotation. Partially filled molds with a hydrophilic monomer containing a red dye dissolved in DMF solution at concentrations of 80 wt% before (b) and after rotation (c) at 14 000 rpm. Prior to rotation, the solvent was completely removed on a hot-plate at 70 °C for 30 min. Reproduced from reference 21 with permission from the American Chemical Society.

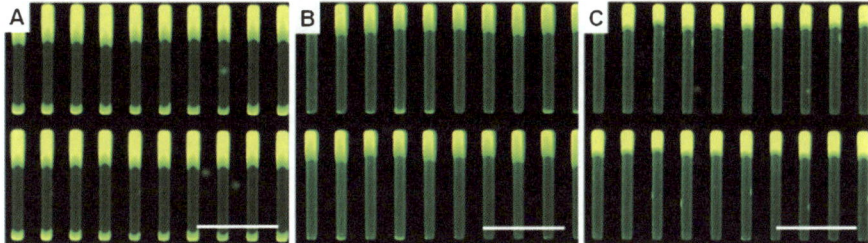

Figure 5.8 Amphiphilic diphasic particles harvested on a harvesting film of cyanoacrylate, which were fabricated by filling the mold with 80 wt% of hydrophilic monomer solution and then rotating at various angular velocities, (a) 8000, (b) 10 000 and (c) 14 000 rpm, before being filling with a second hydrophobic monomer and cured. Images were captured by overlaying the images under red and green channels. Scale bar: 200 μm.

Stepwise mold filling in the horizontal direction can also be tailored to fabricate Janus particles with a diversity of particle shapes. Having mold cavities of non-centrosymmetric curvature concentrates the diluted monomer solution at the region having the strongest capillary effect. Several non-centrosymmetric features were attempted using the same synthetic route as previously described. As shown in Figure 5.9, Janus particles consisting of two separated phases with asymmetric dumbbell and tear-drop features were successfully fabricated. The first region containing a green dye is drawn into the region with a relatively smaller curvature, where stronger capillary force exists. This two-step filling process could certainly be applied to making Janus particles of other shapes.

As before, the ratio of the two regions in these asymmetric Janus particles could be tuned by adjusting the concentration of the initial monomer solution. The optical microscope images outlined in Figure 5.10 demonstrate a range of asymmetric Janus particles having tunable composition ratios.

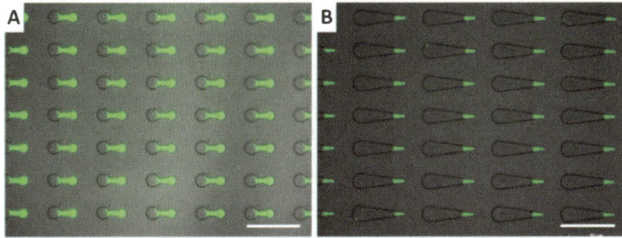

Figure 5.9 Fluorescent microscopy images of Janus particle arrays with asymmetric (a) dumbbell and (b) tear-drop shapes harvesting on a sacrificial film. Scale bar: 50 μm.

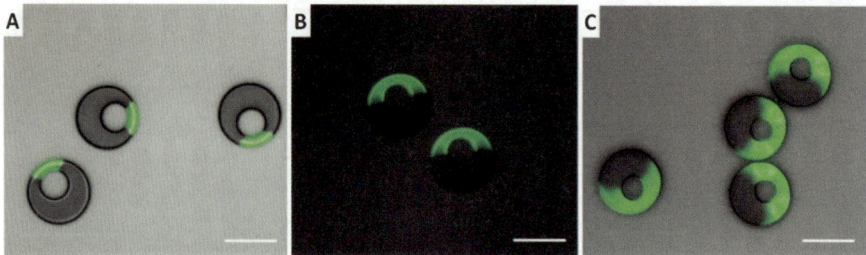

Figure 5.10   Fluorescent microscope images of asymmetric particles demonstrating different domain ratios resulting from varied concentrations of the initial monomer solution. Scale bar: 20 μm.

5.4   Patchy PRINT Particles

Many examples on selectively creating patches on particle surfaces by means of partially covering the particles have been reported. For example, patchy particle surfaces can be achieved either by trapping the particles at an oil/water interface[8] or by arraying particles on a two-dimensional substrate.[4] The PRINT technique is unique in its ability to fabricate more complex shapes rather than regular spheres. Using PRINT, patchy particles can be fabricated by surface modification of the solidified particle while either in the mold or on the harvesting layer. When particles are in the mold, one open face could be modified to form patchy particles. Once the particles have been transferred on to a harvesting layer, the other faces could also be modified. Surface modifications of the PRINT particles using chemical grafting and metal deposition have been attempted.[20]

5.4.1   Surface-modified Particles by Chemical Grafting

Two particle shapes made using the PRINT process were selectively modified *via* chemical functionalization on one face. The particle shapes used were a 2 × 2 × 6 μm rectangular-prism and the other a 3 μm diameter hexnut. The particles were comprised of 67 wt% of the cross-linker PEG_{428}-triacrylate, 20 wt% of PEG_{1000}-monomethyl ether monomethacrylate, 10 wt% of the primary amine monomer aminoethyl methacrylate (AEM), 2 wt% of the fluorescent monomer fluorescein *o*-acrylate and 1 wt% of the photoinitiator 2,2-diethoxyacetophenone (DEAP). After photo-curing the monomer mixtures in the mold, the full mold with particles was inverted into a solution of buffered N-Hydroxysuccinimide (NHS)-rhodamine, which reacted with the primary amine group in the particles to introduce the rhodamine dye selectively on one exposed particle surface. The process is illustrated in Figure 5.11. After reaction, the molds were thoroughly rinsed with water and dried. The particles were then transferred out of the mold using cyanoacrylate and further washed with acetone for several times to remove the adhesive completely.

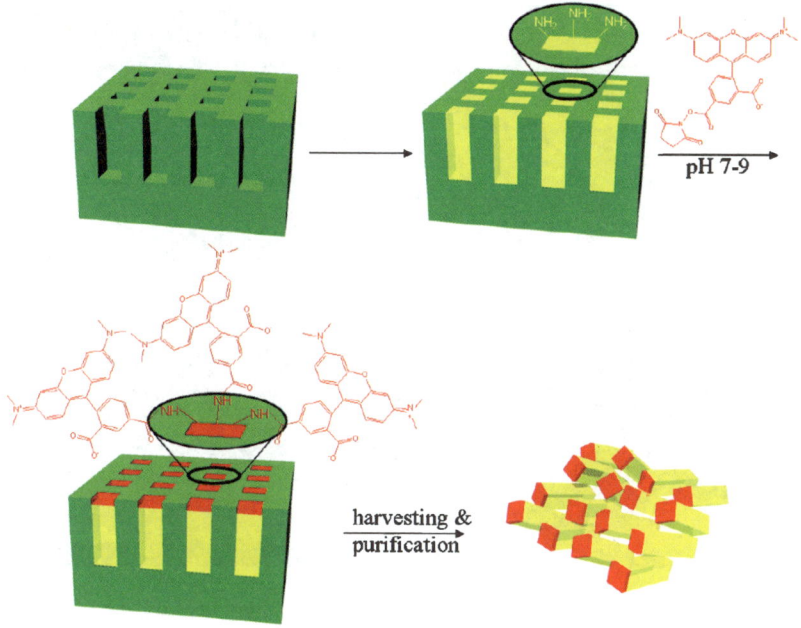

Figure 5.11   Reaction scheme for the anisotropic surface functionalization of primary amine-containing particles in the mold with NHS-rhodamine. Reproduced from reference 20 with permission from the Institute of Physics.

The fluorescence microscopy images of the resulting particles with a selectively functionalized surface are shown in Figure 5.12. The particles clearly exhibit both green and red fluorescence. The green fluorescence is the result of the fluorescein *o*-acrylate incorporated into the bulk of the particles and the red fluorescence is from the surface-bound rhodamine dye.

5.4.2   Surface-functionalized Particles by Metal Deposition

In addition to chemical grafting to the particle surface while in the mold, selective vapor deposition of a metal can also be used to achieve chemical anisotropy on a particle surface. To prepare metal end-capped PRINT particles, a 10 nm layer of a Pd–Au alloy was sputtered on to the surface of a filled $2 \times 2 \times 6$ μm rectangular prism mold. After coating, the particles were harvested by cyanoacrylate and purified. Scanning electron microscopy (SEM) and energy-dispersive X-ray spectroscopy (EDS) were used to confirm the presence of the metal alloy selectively on only one face of the particle surface (Figure 5.13). The metal-coated surface was brighter in the SEM image compared with the particle body. In addition, EDS analysis indicated the presence of the elements carbon and oxygen along the body of the particle whereas gold and palladium were identified only on the coated end.

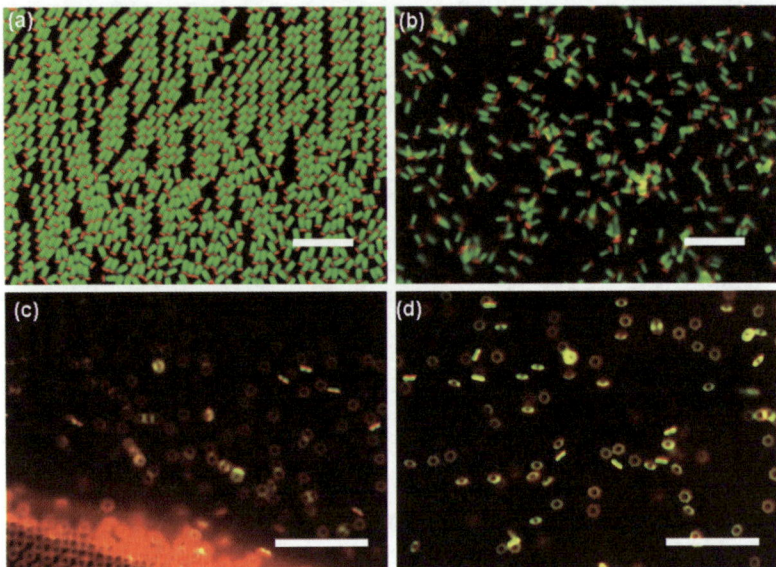

Figure 5.12   Fluorescence images of regiospecifically functionalized PRINT particles where (a) shows 2 × 2 × 6 μm regiospecifically functionalized rectangular prisms collapsed on polycyanoacrylate, (b) the purified and dried regiospecifically functionalized prisms and (c) and (d) regiospecifically functionalized 3 μm hexnuts. Scale bars: 20 μm. Reproduced from reference 20 with permission from the Institute of Physics.

5.5   Self-assembly of Janus PRINT Particles

Being able to fabricate particles comprised of two or more phases allows the opportunity to direct the self-assembly behaviors of these particles in solution or at an interface. The PRINT technique in particular allows for the fabrication of Janus particles with more complex geometries than the typically used simple spheres. A directed self-assembly study using 20 × 20 × 240 μm multiphase PRINT rods was conducted at a water/perfluorodecalin (PFD) interface to demonstrate the dependence of the self-assembly behavior on the particle architecture.[21] The self-assembly of micron-sized particles at a water/oil interface is mainly attributed to the lateral capillary forces acting between the particles due to deformation of the oil/water interface.[22–24] As controls, single-phase hydrophilic particles were shown to aggregate side-to-side into bundles (Figure 5.14a). The morphology of the assembled particles at the interface was studied using a gel trapping method which allowed replication of the oil/water interface on a cured PFPE film.[25] The single-phase hydrophilic particles were found to reside on the oil/water interface with the particles projecting into the water phase. In this case, the particles could not be trapped in the oil phase as they were mostly confined by the water gel phase. As shown in Figure 5.14c, the interfacial distortion indicative of a negative meniscus on the oil surface was replicated. As a result, the hydrophilic particles tended to

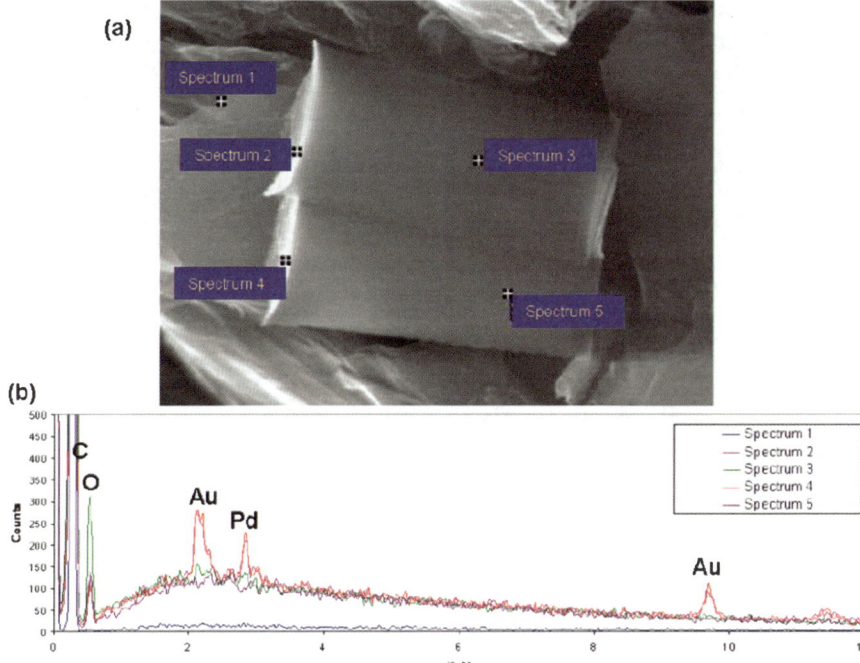

Figure 5.13 (a) SEM image of two 2 × 2 × 6 μm metal-capped rectangular prisms on a graphite pin stub, where the indicated regions were scanned with an EDS detector, and (b) EDS spectra from the different scanned regions. Scan 1 is of the graphite pin background, scans 2 and 4 are of the particle functionalized end and clearly show the presence of gold and palladium, and scans 3 and 4 of the particle body do not show any evidence of metals being present. Reproduced from reference 20 with permission from the Institute of Physics.

interact over more of their perimeter in a side-to-side fashion, leading to a more pronounced reduction of the total energy at the interface. In contrast, the single-phase hydrophobic particles adopted a tip-to-tip aggregation at the water/PFD interface, forming a branched network (Figure 5.14b). Unlike the hydrophilic particles, the hydrophobic particles preferred wetting by the oil. As shown in Figure 5.14d, the hydrophobic particles contacted the oil phase by a rectangular long edge corner with nearly half of the perimeter projecting into the oil phase. Owing to this particular surface wetting phenomenon, the long sides of the particles have less excess area than the ends, hence the capillary attraction resulting from the elimination of excess area is stronger at the particle tips. As a result of the relatively weak attraction between the long sides, conventional steering breaks the metastable side-to-side aggregates, allowing the formation of branched structures with minimum energy in a tip-to-tip manner.

The self-assembly of amphiphilic tri- and diphasic rod particles was also investigated at the water/PFD interface. The triphasic particles, behaving like

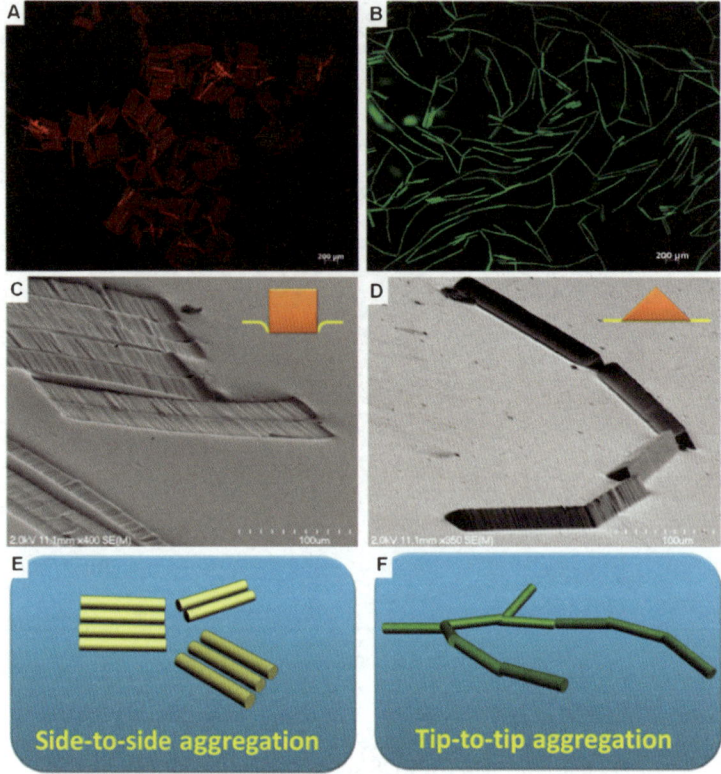

Figure 5.14   Fluorescence microscopic images of different particles assembled at the water/PFD interface for the (a) hydrophilic and (b) hydrophobic particles. Scanning electron micrographs are shown for the (c) PFPE mold obtained for the particle trapping experiment for the single-phase hydrophilic particles on the oil surface and (d) PFPE mold for the single-phase hydrophobic particles trapped at the oil phase. Self-assembly models are illustrated for the (e) hydrophilic and (f) hydrophobic particles at the water/oil interface. Reproduced from reference 21 with permission from the American Chemical Society.

bolaamphiphiles, preferred a side-to-side assembly, forming ordered ribbon structures at the interface, as shown in Figure 5.15a. With regard to the particle orientation at the interface, a very similar lateral capillary force to single-phase hydrophilic particles is assumed to provide a driving force in this directed self-assembly process. However, the triphasic architecture allows each particle to match the others in a better side-to-side fashion than does the single-phase hydrophilic particle. It is the difference in interfacial distortion around the hydrophilic and hydrophobic regions that limits each region of the triphasic particles to direct the assembly only with others having similar composition. It was observed that the triphasic particles adopted a bent conformation at the interface. This is presumably caused by the curved interface between the hydrophilic and hydrophobic regions and the swelling of

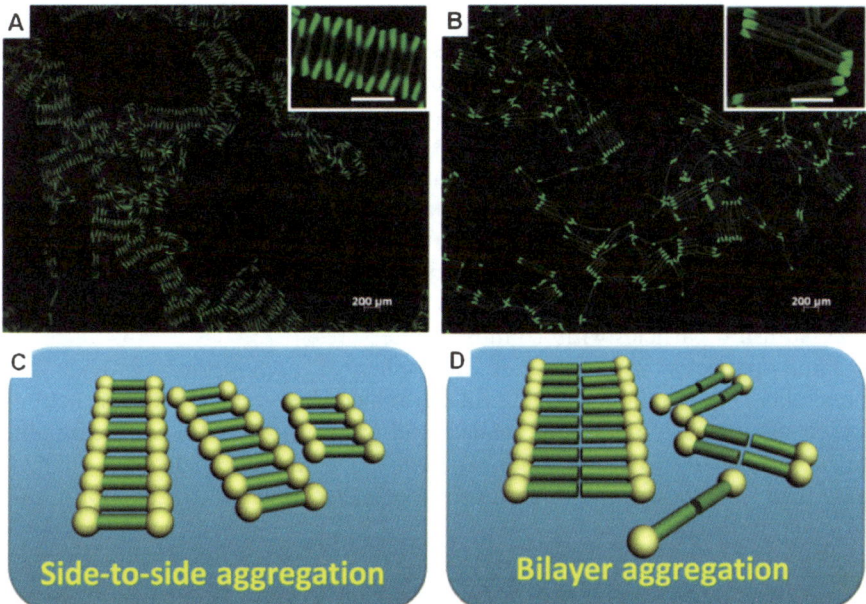

Figure 5.15 Fluorescent microscopy images of di- and triphasic particles assembled at the water/PFD interface. (a) ABA hydrophilic–hydrophobic–hydrophilic triphasic particles with 50 wt% hydrophilic monomer content as shown previously in Figure 5.7f. (b) AB diphasic particles comprised of 80 wt% hydrophilic monomer as previously described in Figure 5.10c. Scale bar of insets: 200 μm. Self-assembly models are illustrated for the particles at a water/oil interface for the (c) tri- and (d) diphasic particles. Reproduced from reference 21 with permission from the American Chemical Society.

the hydrophilic regions in water. Interestingly, the diphasic particles, having an asymmetric structure similar to a molecular surfactant, self-assembled into a 2D bilayer structure at the water/PFD interface (Figure 5.15b). In a bilayer structure, the side-to-side lateral interactions pull the particles together forming bundles while the heads of the hydrophobic regions tended to aggregate by a tip-to-tip lateral interaction. The cooperation of both assembly fashions thus induces the formation of a bilayer structure, being like the shell structure of vesicles assembled by traditional surfactants.

5.6 Conclusion and Future Perspectives

The PRINT technique is a leading technology for the fabrication of particles with precise control over the geometry and composition. The synthesized particles represent a new paradigm for assembling nanoscale materials into microscale structures, creating opportunities in optoelectronics, reinforcement and encapsulation/release strategies. Two-phase Janus particles and also multiphase particles have been shown to be able to be easily fabricated *via* a multi-step mold filling adaptation of the PRINT technique, thus allowing precise control over the particle phases. Traditionally, numerous questions in the field of colloid physics and chemistry are encountered because of the limitations in particle geometry. Multiphase PRINT particles are now able to provide a platform to study the effects of composition and particle geometry on particle–particle interactions in colloid science. The ability to fabricate multiple components within a single particle is anticipated to open up avenues to new drug carriers, allowing the encapsulation of several drugs within hydrophilic or hydrophobic phases in one carrier. In addition, it is envisioned that the internalization of particles into the various types of cells could be realized using particles with a Janus architecture.

Janus particles with complex non-spherical architectures provide the possibility of obtaining novel self-assembled hierarchical superstructures which show great potential in medical and optoelectronic devices. Being able to fabricate nanoscale non-spherical Janus particles, although more challenging than the fabrication of micron-sized particles, is very attractive for a host of research fields.

Acknowledgements

The authors thank the National Science Foundation under Grants DMR-0923604 and DMR-0906985 for their support of this research, and also the STC program of the National Science Foundation for shared facilities. The authors acknowledge the careful editing of Dr Douglas E. Betts of Reichhold Inc.

References

1. L. Hong, A. Cacciuto, E. Luijten and S. Granick, *Langmuir*, 2008, **24**, 621–625.
2. Q. Chen, J. K. Whitmer, S. Jiang, S. C. Bae, E. Luijten and S. Granick, *Science*, 2011, **331**, 199–202.
3. Q. Chen, S. C. Bae and S. Granick, *Nature*, 2011, **469**, 381–385.
4. Q. Chen, E. Diesel, J. K. Whitmer, S. C. Bae, E. Luijten and S. Granick, *J. Am. Chem. Soc.*, 2011, **133**, 7725–7727.
5. N. Glaser, D. J. Adams, A. Böker and G. Krausch, *Langmuir*, 2006, **22**, 5227–5229.
6. A. Walther and A. H. E. Müller, *Soft Matter*, 2008, **4**, 663–668.
7. T. Tanaka, M. Okayama, H. Minami and M. Okubo, *Langmuir*, 2010, **26**, 11732–11736.
8. L. Hong, S. Jiang and S. Granick, *Langmuir*, 2006, **22**, 9495–9499.
9. Y. K. Takahara, S. Ikeda, S. Ishino, K. Tachi, K. Ikeue, T. Sakata, T. Hasegawa, H. Mori, M. Matsumura and B. Ohtani, *J. Am. Chem. Soc.*, 2005, **127**, 6271–6275.
10. S. Park, S.-W. Chung and C. A. Mirkin, *J. Am. Chem. Soc.*, 2004, **126**, 11772–11773.

11. K. Liu, Z. Nie, N. Zhao, W. Li, M. Rubinstein and E. Kumacheva, *Science*, 2010, **329**, 197–200.
12. J.-W. Kim, R. J. Larsen and D. A. Weitz, *J. Am. Chem. Soc.*, 2006, **128**, 14374–14377.
13. D. Dendukuri, T. A. Hatton and P. S. Doyle, *Langmuir*, 2007, **23**, 4669–4674.
14. C.-H. Choi, J. Lee, K. Yoon, A. Tripathi, H. A. Stone, D. A. Weitz and C.-S. Lee, *Angew. Chem. Int. Ed.*, 2010, **49**, 7748–7752.
15. Y. Wang, T. J. Merkel, K. Chen, C. A. Fromen, D. E. Betts and J. M. DeSimone, *Langmiur*, 2011, **27**, 524–528.
16. J. P. Rolland, E. C. Hagberg, G. M. Denison, K. R. Carter and J. M. DeSimone, *Angew. Chem. Int. Ed.*, 2004, **43**, 5796–5799.
17. Y. P. Wang, D. E. Betts, J. A. Finlay, L. Brewer, M. E. Callow, J. A. Callow, D. E. Wendt and J. M. DeSimone, *Macromolecules*, 2011, **44**, 878–885.
18. S. E. A. Gratton, S. S. William, M. E. Napier, P. D. Pohlhaus, Z. Zhou, K. B. Wiles, B. W. Maynor, C. Shen, T. Olafsen, E. T. Samulski and J. M. DeSimone, *Acc. Chem. Res.*, 2008, **41**, 1685–1695.
19. T. J. Merkel, K. P. Herlihy, J. Nunes, R. M. Orgel, J. P. Rolland and J. M. DeSimone, *Langmuir*, 2010, **26**, 13086–13096.
20. H. Zhang, J. K. Nunes, S. E. A. Gratton, K. P. Herlihy, P. D. Pohlhaus and J. M. DeSimone, *New J. Phys.*, 2009, **11**, 075018.
21. J. Y. Wang, Y. Wang, S. S. Sheiko, D. E. Betts and J. M. Desimone, *J. Am. Chem. Soc.*, 2012, **134**, 5801–5806.
22. P. A. Kralchevsky and K. Nagayama, *Adv. Colloid Interface Sci.*, 2000, **85**, 145–192.
23. H.-J. Butt and M. Kappl, *Adv. Colloid Interface Sci.*, 2009, **146**, 48–60.
24. A. Böker, J. He, T. Emrick and T. P. Russell, *Soft Matter*, 2007, **3**, 1231–1248.
25. N. Bowden, F. Arias, T. Deng and G. M. Whitesides, *Langmuir*, 2001, **17**, 1757–1765.

CHAPTER 6

Theoretical Calculations of Phase Diagrams and Self-assembly in Patchy Colloids

ACHILLE GIACOMETTI[a], FLAVIO ROMANO[b,c] AND
FRANCESCO SCIORTINO*[c]

[a] Dipartimento di Scienze Molecolari e Nanosistemi, Università di Venezia,
S. Marta DD2137, 30123 Venice, Italy; [b] Dipartimento di Fisica, Sapienza
Università di Roma, Piazzale A. Moro 2, 00185 Rome, Italy; [c] Physical and
Theoretical Chemistry Laboratory, Department of Chemistry, University of
Oxford, South Parks Road, Oxford OX1 3QZ, UK
*E-mail: francesco.sciortino@uniroma.it

6.1 Introduction

Self-assembly processes in patchy colloids represent one of the most striking
examples where experimental methodologies and theoretical tools have
progressed in parallel within a relatively short time scale.[1–3] While the former
have been addressed elsewhere in this volume and in recent reviews,[4,5] in this
contribution we address the latter and, more specifically, the main
methodologies that have been envisaged over the years to tackle the
computation of the phase diagrams and phase transitions from one phase to
another in dispersions of new-generation colloids, *i.e.* particles interacting *via*
non-spherical potentials. Indeed, chemists and material scientists are starting
to gain control of the shapes[6] and local properties of colloids. Hard cubes,
tetrahedra, cones, rods and also composed shapes of nano- or microscopic size

RSC Smart Materials No. 1
Janus Particle Synthesis, Self-Assembly and Applications
Edited by Shan Jiang and Steve Granick
© The Royal Society of Chemistry 2012
Published by the Royal Society of Chemistry, www.rsc.org

have made their appearance in the laboratory and are becoming available in bulk quantities. Patterning of the surface properties of these particles[7-9] provides additional degrees of freedom to be exploited by scientists to engineer materials with peculiar properties. Patches on the particle surface can be functionalized with specific molecules[10,11] (including DNA single strands[12,13]) to create hydrophobic or hydrophilic areas, providing specificity to the particle–particle interaction.[5,14]

Statistical physics provides a very rich and flexible toolbox to study the thermophysical and structural properties of complex fluids,[15,16] especially when coupled with the most recent and powerful computing techniques devised to deal with systems with a large number of degrees of freedom.[17,18] Whereas theoretical studies of simple liquids and conventional colloidal systems have a long and venerable tradition,[19] numerical and analytic investigation on patchy colloids are relatively new, as in the past it was always tacitly assumed that even the unavoidable inhomogeneities in their surface composition could be neglected at a sufficiently coarse-grained scale. This is not the case, however, for patchy colloids that have surface patterns, chemical compositions and functionalities that are explicitly meant to be inhomogeneous.[4,5,14] Hence the corresponding pair potentials describing inter-particle interactions depend on their relative orientations in addition to distances, and the analysis clearly becomes more complex. This is, however, by no means an insurmountable difficulty, as several analytical and computational techniques have been devised in statistical physics to cope with the orientational dependence of the potentials.[15,16]

In this chapter, we discuss some of them in the framework of a particular pair potential that can be reckoned as a reasonable compromise between the complexity of the real interactions and the necessary simplicity required to keep the analysis practicable. The basic idea of the model is built on the hard-sphere model, by providing a fraction of the surface sphere with a square-well character. This attractive region can be either condensed into a single large patch or distributed over two (or more) patches symmetrically placed over the surface. Different spheres then interact *via* square-well or hard-sphere potentials depending on their relative orientations and distances.

This model was proposed in 2003 by Kern and Frenkel[20] and since then has attracted considerable attention. There are two main reasons for this. On the one hand, the model is very flexible, as both the size and the number of the patches can be independently tuned, and this allows mimicking of several different physical situations ranging from nanocolloids with more isolated attractive spots[21] to globular proteins with large regions of solvophobic exposed surfaces.[22,23] On the other hand, the phase diagram obtained from the model can be directly compared with those obtained from experiments, as recently shown in several cases.[24-27] In addition, the model displays some unusual features that can be paradigmatic for more complex systems.[28-31]

The aim of this chapter is to introduce the main theoretical techniques for the evaluation of the phase diagram. This includes various Monte Carlo techniques (Section 6.4), integral equations (Section 6.5) and perturbation theory (Section 6.6). The level is intended to be pedagogic, with the main ideas behind each method outlined for non-experts in the field. Emphasis is placed on the calculations of thermodynamic quantities necessary for the phase diagram analysis and hence a number of additional important results related to structural properties and other thermodynamic probes have been omitted.

6.2   The Kern–Frenkel Model

Consider a set of N identical hard spheres of diameter σ in a volume V at temperature T suspended in a microscopic fluid. When the surfaces of the spheres are uniform with no interactions other than their steric hindrance, the model has often been employed as a paradigm of sterically stabilized colloidal suspensions in the limit of high temperature or good solvent.

As discussed elsewhere in this volume, the colloids that are envisaged as elementary building blocks for the self-assembly process are patchy colloids with different philicities in different parts of the surface.[4,5] This means, for instance, that one fraction of the surface may be solvophilic and the other solvophobic. In solution, the solvophobic part will tend to avoid contact with the solvent and hence will act as an effective attractive force in the presence of another solvophobic patch lying on a different sphere.

One can then consider the following model that was introduced in 2003 by Kern and Frenkel in the present form,[20] but it is worth remarking that the idea of considering hard spheres and decorating them with patches of various forms and patterns dates back much earlier and several earlier versions in different fields can be considered as its ancestors.[32–35]

A circular patch is attached to the surface of each sphere, as depicted in Figure 6.1, with the central position of the patch identified by the unit vector $\hat{\mathbf{n}}$ and its amplitude measured by the angle θ_0. Unlike the case of uniform spheres, the interactions among spheres are anisotropic as they depend on the relative orientation of the unit vectors on each sphere with the direction connecting their centers. Then the idea is that two spheres attract each other if they are within the range of the attractive potential, with the corresponding attractive patches on each sphere properly aligned.

If $\hat{\mathbf{n}}_1$ and $\hat{\mathbf{n}}_2$ are the unit vectors associated with each patch on spheres 1 and 2 and $\hat{\mathbf{r}}_{12}$ is the direction connecting the centers of the two spheres, then the interparticle potential reads

$$\Phi(12) = \phi_0(r_{12}) + \Phi_1(12) \tag{6.1}$$

where the first term is the hard-sphere contribution:

$$\phi_0(r) = \begin{cases} \infty, & 0 < r < \sigma \\ 0, & \sigma < r \end{cases} \tag{6.2}$$

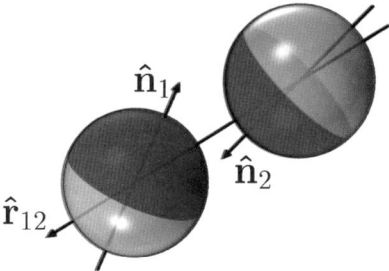

Figure 6.1   Sketch of one-patch patchy particles as modeled by the Kern–Frenkel potential. The surface of each sphere is partitioned into an attractive part (color code: blue) and a repulsive part (color code: white). The unit vectors $\hat{n}_1$ and $\hat{n}_2$ define the orientations of the patches, whereas the vector $\hat{r}_{12}$ joins the centers of the two spheres, from sphere 1 to sphere 2. The particular case shown corresponds to a 50% fraction of attractive surface (coverage $\chi = 0.5$).

and the second term

$$\Phi_1(\hat{n}_1,\hat{n}_2,r_{12}) = \phi_{SW}(r_{12})\Psi(\hat{n}_1,\hat{n}_2,\hat{r}_{12}) \qquad (6.3)$$

is the orientation-dependent attractive part, which can be factorized into an isotropic square-well tail:

$$\phi_{SW}(r) = \begin{cases} -\varepsilon, & \sigma < r < \lambda\sigma \\ 0, & \lambda\sigma < r \end{cases} \qquad (6.4)$$

multiplied by an angular dependent factor:

$$\Psi(\hat{n}_1,\hat{n}_2,\hat{r}_{12}) = \begin{cases} 1, & \text{if} \quad \hat{n}_1 \cdot \hat{r}_{12} \geq cos\theta_0 \quad \text{and} \quad -\hat{n}_2 \cdot \hat{r}_{12} \geq cos\theta_0 \\ 0, & \text{otherwise} \end{cases} \qquad (6.5)$$

Here σ is the sphere diameter, $(\lambda - 1)\sigma$ is the width of the square-well interaction and ε its depth. $2\theta_0$ defines the angular amplitude of the patch. The unit vectors $\hat{n}_i(\omega_i),(i=1,2)$, are defined by the spherical angles $\omega_i = (\theta_i,\varphi_i)$ in an arbitrarily oriented coordinate frame and $\hat{r}_{12}(\Omega)$ is identified by the spherical angle Ω in the same frame. Reduced units, for temperature $T^* = k_B T/\varepsilon$, pressure $P^* = \beta P/\rho$ and density $\rho^* = \rho\sigma^3$, will be used throughout, with k_B being the Boltzmann constant. For future reference, we also introduce the packing fraction $\eta = \pi\rho^*/6$. The coverage χ is the fraction of attractive surface on the particle; χ can be related to the patch half-width θ_0

$$\chi = \langle \Psi(\hat{n}_1,\hat{n}_2,\hat{r}_{12})\rangle_{\omega_1\omega_2}^{\frac{1}{2}} = \frac{1-cos\theta_0}{2} \qquad (6.6)$$

where we have introduced the angular average

$$\langle\ldots\rangle_\omega = \frac{1}{4\pi}\int d\omega\ldots \qquad (6.7)$$

6.3   The Tools of Statistical Physics

Statistical physics has developed a number of different theoretical approaches to compute the thermophysical properties of a fluid.[15–17] In order to compare with experiments, we are most interested in the computation of the fluid–fluid and fluid–solid phase diagram on the one hand and on the specific mechanism driving aggregation, and hence self-assembly, on the other. Among this arsenal of different available techniques, here we review three different methodologies that were recently exploited in the framework of the Kern–Frenkel model: Monte Carlo simulations,[17,18] integral equation theories[15] and thermodynamic perturbation theories.[16]

Monte Carlo simulations are undoubtedly one the most efficient ways to compute accurately the properties of a model fluid. As discussed in more detail below, the main limitations of simulations are that they can be very demanding from a computational point of view, especially for sufficiently realistic potentials, and that they are unable to distinguish metastable from stable equilibrium states. On the other hand, they provide virtually exact estimates of all static quantities of interest. Several improvements have been proposed over the years, some of them triggered by the problems discussed in this chapter, so that the methodology is very well established and by now extensively reviewed and described in detail in several books (see references 17 and 18 and references cited therein). The case of patchy colloids, however, is relatively recent, although it builds upon previous established procedures on other complex fluids.

Integral equation and thermodynamics perturbation theories are two of the main methodologies from the toolbox of statistical physics that are at the basis of our current understanding of simple and molecular fluids.[15,16] In spite of their known drawbacks and shortcomings, they are known to provide reliable predictions for both structural and thermodynamic properties each in their own domain of applications. Their applications to patchy colloids is a natural, albeit not trivial, extension of formalisms already developed in the last two decades for molecular fluids.[16] As will become clear, they both become particularly attractive in view of the large computational effort involved in Monte Carlo simulations. In addition, they are able to access to some details and nuances that are not easily accessible by other methods.

6.4   Monte Carlo Simulations

The aim of Monte Carlo simulations is the computation of thermodynamic quantities by performing an average over a suitable ensemble of microstates.

The choice of the ensemble is dictated both by the quantities to be computed and by the specific system under investigation, for which one ensemble can be more convenient than the others. Below we review the most interesting techniques that have been used to calculate phase diagrams of patchy colloid models.

6.4.1 Canonical NVT and NPT Methods

Simulations in the NVT (isothermal–isochoric) and NPT (isothermal–isobaric) ensembles are probably the most common example. In these ensembles, the number of particles N, the temperature T and the volume V (in the NVT ensemble) or the pressure P (in the NPT ensemble) are held constant. The Markov chain in configuration space is constructed *via* a sequence of translational and rotational moves, accepted with an appropriate probability that depends upon the change in potential energy and T. In the NPT case, the volume is also varied. With a proper choice of the acceptance probability, the system first evolves towards equilibrium and then starts to sample equilibrium configurations with the Boltzmann statistical weight. The equilibration process can be rather long, especially in cases where kinetic traps are present (as in the vicinity of gel or glass transitions) or when an activation barrier needs to be overcome. This last case arises when the system is metastable with respect to a lower energy phase or when it organizes into mesoscopic structures and specific self-assembly processes involving large numbers of particles are requested. The approach to equilibrium can be monitored by focusing on the time evolution of collective properties (*e.g.* the potential energy, the density, the pressure). Since equilibration can be rather slow, it is highly recommended to make use of a logarithmic time scale when searching for a drift in the time dependence of the investigated property.

When a sufficiently large number of statistically independent equilibrium configurations have been generated and stored, all possible structural (static) information can be calculated. Typical quantities that are computed are the total energy U, the radial distribution function $g(r)$ and the structure factor. In the case of anisotropic systems, such as patchy particles, the orientational dependence of the structural properties also needs to be evaluated. The center–center $g(r)$ is not sufficient to evaluate the average potential energy or the pressure, in contrast to the isotropic case, where calculation of U or P from $g(r)$ usually simply requires a one-dimensional integration.

In the case of hard bodies or in the presence of stepwise potentials (*e.g.* the square-well potential), direct evaluation of P in NVT simulations is in principle possible[36,37] but not straightforward. To evaluate the equation of state, *i.e.* the relation between density and P at constant T, the NPT ensemble is often preferred.

Various additional improvements can be (and are) used to improve the convergence of the scheme in a way that will be described in each specific example.

6.4.2 Gibbs Ensemble Method

A convenient scheme was devised by Panagiotopulos[38] to address specifically the problem of the direct evaluation of the gas–liquid phase coexistence by Monte Carlo simulations. This is known as the Gibbs Ensemble Monte Carlo (GEMC) method. N particles are partitioned into two distinct simulation boxes. In addition to intra-box translational and rotational moves, particle and volume swap moves (keeping both the total number of particles and the total volume fixed) are proposed and accepted with the appropriate probability.[17] In this way, the two coexisting phases are simulated without the intervention of an interface between them. When convergence is reached, the densities in the two boxes provide the value of the coexisting densities of the liquid and gas phases. It should be pointed out that the GEMC method becomes inefficient when the density of the liquid phase becomes large, since the probability of inserting a particle in a favorable state, *i.e.* not overlapping with any other, becomes extremely small.

For the specific case of Kern–Frenkel (KF) particles discussed later in this chapter, GEMC simulations have been performed for a system of 1200 particles in a total volume of $(16\sigma)^3$. On average, the code attempts one volume change every five particle swap moves and 500 displacement moves. Each displacement move is composed of a simultaneous random translation of the particle center (uniformly distributed between $\pm 0.05\sigma$) and a rotation (with an angle uniformly distributed between ± 0.1 rad) around a random axis.

6.4.3 Grand-canonical Ensemble μVT

In the neighborhood of the gas–liquid critical point, the free-energy barrier separating the two phases becomes comparable to the amplitude of the thermal fluctuations of the relatively small systems that can be accessed in simulation. In this case, the GEMC method cannot be used to investigate the system, since it is subject to size effects and spontaneous swapping of the average densities between the two boxes.

A precise evaluation of the critical parameters (density and temperature) can be obtained by performing simulations in the grand-canonical μVT ensemble,[17] where density fluctuations are accounted for at fixed volumes and temperature, coupled with the finite-size scaling analysis as envisaged by Wilding.[39] Monte Carlo simulations in the grand-canonical ensemble are implemented by performing trial insertions and deletions of particles, in addition to trial displacements and rotations. The critical parameters of the system can be extracted by matching the calculated distribution of density fluctuations to the expected distribution at the critical point, a feature which is largely system independent.[39]

In the implementation of the grand-canonical simulations to the KF model reported later, one insertion/deletion attempt was performed, on average, every 500 trial translational/rotational displacements.

6.4.4   Fluid–Solid Coexistence: Thermodynamic Integration

To compute numerically the free energies of the fluid and the crystals and their coexistence lines, it is possible to resort to thermodynamic integration methods. Details of this procedure are for example provided in the detailed review by Vega *et al.*[40]

The starting point of the procedure requires the identification of a state point in the pressure–temperature plane where two phases, I and II, share the same chemical potential, *i.e.* $\mu_I(P,T) = \mu_{II}(P,T)$. The chemical potential of the fluid can be computed by thermodynamic integration using the ideal gas as a reference state and by integrating the equation of state, $P(\rho)$, at fixed temperature:

$$\frac{\beta F(T,\rho)}{N} = log\left(\rho\sigma^3\right) - 1 + \int_0^\rho \frac{\beta P/\rho' - 1}{\rho'} d\rho' \tag{6.8}$$

where F/N is the Helmholtz energy per particle. The first term on the right-hand-side is the ideal gas part and depends upon the system dimensionality. The chemical potential can then be recovered as

$$\beta\mu[P(\rho),T] = \frac{\beta F[P(\rho),T]}{N} + \beta P(\rho)/\rho \tag{6.9}$$

To compute the chemical potential of a crystal, one can perform thermodynamic integration at fixed density and temperature using an ideal Einstein crystal as the reference system. This method, known as Frenkel–Ladd procedure,[17,40] is very efficient and by now standard. Integration of the crystal equation of state provides a way to evaluate the chemical potential at different T and P. The pressure at which the chemical potential of the fluid and of the crystal are identical along an isotherm provides the coexisting pressure at the selected T.

Starting from a coexistence point, coexistence lines can finally be inferred by using Gibbs–Duhem integration, as described by Kofke,[41] numerically integrating over the Clausius–Clapeyron equation.

6.5   Integral Equation Theories

6.5.1   General Scheme

First, let us consider simple fluids where the particles can be regarded as spherically symmetrical. All thermodynamic properties of such fluids can be straightforwardly computed from the radial distribution function $g(r)$. In integral equation theory,[15] the strategy to infer the thermophysical properties of a fluid hinges on the calculation of the total correlation function $h(r) \equiv g(r) - 1$, which, in turn, is related to the direct correlation function $c(r)$ by the Ornstein–Zernike (OZ) equation:

$$h(r) = c(r) + \rho \int d\mathbf{r}' \, c\left(r'\right) h\left(|\mathbf{r} - \mathbf{r}'|\right) \tag{6.10}$$

Once $h(r)$ is known, all statistical and thermodynamic properties can in principle be computed. However, as $h(r)$ depends upon the unknown quantity $c(r)$, an additional equation involving both quantities is required for the solution. Unlike eqn (6.10), which is exact, the second equation always involves some approximation. This gives rise to some well-known thermodynamic inconsistencies, which are the main shortcomings of this method and may severely limit its applicability.

The second relation between $h(r)$ and $c(r)$ also involves the two-body potential $\varphi(r)$ and can be cast in the general form

$$c(r) = exp[-\beta\varphi(r) + \gamma(r) + B(r)] - 1 - \gamma(r) \tag{6.11}$$

where $\gamma(r) = h(r) - c(r)$ is an auxiliary function. Although this equation is again exact in principle, it involves the bridge function $B(r)$ that in general depends upon higher body correlation functions, so in practice an approximation (closure) is always necessary. The quality of the results obtained will depend crucially on the reliability of the approximations involved; several closures have been proposed over the years with their pros and cons well classified and under control. Among them, the reference hyper-netted chain (RHNC) stands out as an optimal trade-off between simplicity and precision of its predictions and this is the one that will be used in the present chapter.

Having closed the systems of two equations in two unknowns, $h(r)$ and $c(r)$, the system may then be solved iteratively with the convolution appearing in the OZ equation [eqn (6.10)] simplified in Fourier space as

$$\hat{h}(k) = \frac{\hat{c}(k)}{1 - \rho\hat{c}(k)} \tag{6.12}$$

$\hat{h}(k)$ and $\hat{c}(k)$ being the Fourier transforms of $h(r)$ and $c(r)$ respectively.

The RHNC closure was introduced by Lado[42] for spherical potentials and later extended to molecular fluids.[43,44] In the RHNC closure, one replaces the exact $B(r)$ appearing in eqn (6.11) by its hard-sphere counterpart $B_0(r)$, which is the only system for which a reliable expression (the Verlet–Weiss expression[45]) is available. Rosenfeld and Ashcroft[46] demonstrated that the effectiveness of the reference system could be magnified by treating its parameters as variables to be optimized in some fashion. It is in fact possible to determine them *via* a variational free energy principle[47] that enhances internal thermodynamic consistency. With the effective hard sphere diameter σ_0 suitably chosen in this way, the RHNC has been shown to provide fairly precise estimates of the chemical potential and pressure, which are the two crucial thermodynamic quantities needed for the calculation of phase diagrams.

The case of anisotropic potentials is significantly more complex from the algorithmic point of view, but the philosophy behind the methodology is identical. The procedure hinges on a remarkable piece of work carried out by Lado and reported in a series of papers[43,44,47] in the framework of molecular fluids and more recently adapted to the case of patchy colloids. Here we just sketch the idea; more details can be found elsewhere.[48–50]

The angular-dependent counterparts of eqns (6.10) and (6.11) are given in terms of $\gamma(12) = h(12) - c(12)$ and are

$$\gamma(12) = \frac{\rho}{4\pi} \int d\mathbf{r}_3 d\omega_3 [\gamma(13) + c(13)]c(32) \tag{6.13}$$

for the OZ equation and

$$c(12) = exp[-\beta\Phi(12) + \gamma(12) + B(12)] - 1 - \gamma(12) \tag{6.14}$$

for the closure equation. Again, the RHNC approximation amounts to assuming $B(12) = B_0(r_{12})$, but clearly this is a much more drastic approximation in the present case, as the real $B(12)$ depends on the relative orientations of the particles whereas the reference $B_0(r_{12})$ does not. As a result, one might expect the results to be less precise in this case compared with the isotropic fluid counterpart.

6.5.2 Iterative Procedure

As discussed before, the iterative procedure in the case of the spherical isotropic potential is very simple and is outlined in Figure 6.2. It requires a series of transformations to and from Fourier space, where the solution of the OZ equation is more conveniently carried out in view of eqn (6.12). The iterative solution of the angular-dependent Ornstein–Zernike (OZ) equation [eqn (6.13)] along with the approximate closure equation [eqn (6.14)] again requires a series of direct and inverse Fourier transforms between real and momentum space involving the bridge function $B(12)$, the direct correlation function $c(12)$, the pair distribution function $g(12)$, the total correlation function $h(12)$ and the auxiliary function $\gamma(12) = h(12) - c(12)$.

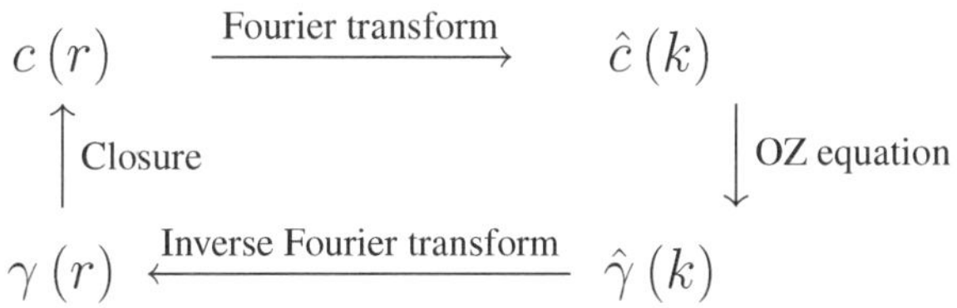

Figure 6.2 Schematic flow chart for the solution of the OZ equation for isotropic potentials.

In addition, however, the orientational degrees of freedom introduce additional direct and inverse Clebsch–Gordan (CG) transformations between the coefficients of the angular expansions in different frames.[16] Two important examples are the so-called 'axial' or 'molecular' frames, with $\hat{\mathbf{z}}=\hat{\mathbf{r}}_{ij}$ in real space and the $\{k\}$ representation with $\hat{\mathbf{z}}=\hat{\mathbf{k}}$ in momentum space, because in those representations some of the calculations become particularly simple. This set of transformations also allows the definition of the correlation functions [in particular g(12)] in an arbitrarily oriented axes. We further note that, in the presence of an anisotropic potential, the Fourier transform is implemented through a Hankel transform, and that the cylindrical symmetry of the angular dependence included in the Kern–Frenkel potential in Section 6.2 (when the number of patches present on each sphere is one or two) allows us to use the simpler version of the procedure for linear molecules.

All necessary equations can be found in reference 16, which is the standard reference for this topic, and only the most important ones will be reported in the following. The resulting scheme is illustrated in Figure 6.3 and is the extension of the isotropic case given in Figure 6.2.[49] Consider the expansion in spherical harmonics of the auxiliary function $\gamma(12)$ in the axial frame:

$$\gamma(12)=\gamma(r,\omega_1,\omega_2)=4\pi\sum_{l_1,l_2,m}\gamma_{l_1l_2m}(r)\,Y_{l_1m}(\omega_1)\,Y_{l_2\bar{m}}(\omega_2) \qquad (6.15)$$

where $\bar{m}\equiv-m$, and its inverse reads

$$\gamma_{l_1l_2m}(r)=\frac{1}{4\pi}\int d\omega_1 d\omega_2\gamma(r,\omega_1,\omega_2)\,Y_{l_1m}^*(\omega_1)\,Y_{l_2\bar{m}}^*(\omega_2) \qquad (6.16)$$

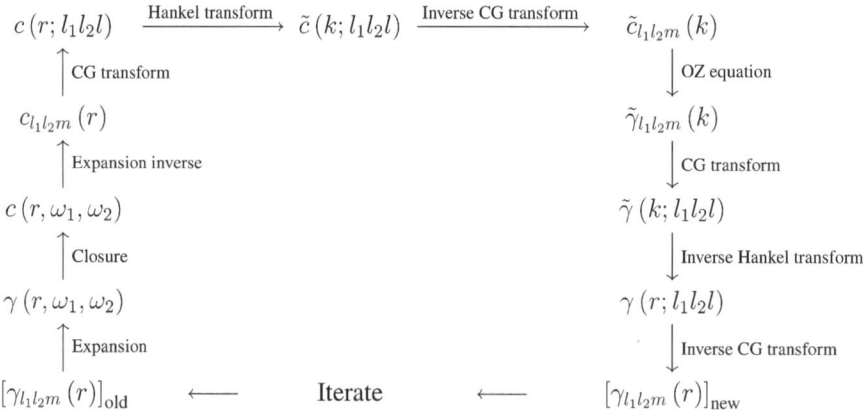

Figure 6.3   Schematic flow chart for the solution of the OZ equation for the Kern–Frenkel angle-dependent potential. See Section 6.5 for a description of the scheme.

Equation (6.16) provides the initial set $\left[\gamma_{l_1 l_2 m}(r)\right]_{\text{old}}$ described as the initial point in Figure 6.3, whereas eqn (6.15) yields the next term in the iteration map $\gamma(r12, \omega1, \omega2)$.

Using the aforementioned RHNC closure approximation $B(12) = B_0(r_{12})$, the bridge function is constructed and then inserted in eqn (6.14) to obtain $c(r, \omega_1, \omega_2)$. Equation (6.16) with the replacement $\gamma \rightarrow c$ is then exploited to infer the corresponding axial coefficients $c_{l_1 l_2 m}(r_{12})$.

The next step is to implement a Clebsch–Gordan transform in direct space, in order to transform from axial coefficients $c_{l_1 l_2 m}(r)$ where $\hat{\mathbf{z}} = \hat{\mathbf{r}}$ to space coefficients $c(r; l_1 l_2 l)$ in an arbitrarily oriented frame. The necessary expressions are the equation pairs

$$c(r; l_1 l_2 l) = \left(\frac{4\pi}{2l+1}\right)^{\frac{1}{2}} \sum_m C(l_1 l_2 l; m\bar{m}0) c_{l_1 l_2 m}(r) \tag{6.17}$$

where the $C(l_1 l_2 l; m\bar{m}0)$ are Clebsch–Gordan coefficients, with the inverse transform given by

$$c_{l_1 l_2 m}(r) = \sum_l C(l_1 l_2 l; m\bar{m}0) \left(\frac{2l+1}{4\pi}\right)^{\frac{1}{2}} c(r; l_1 l_2 l) \tag{6.18}$$

and the coefficients $c(r; l_1 l_2 l)$ are then given by eqn (6.18).

The last required tool is the Fourier transform of the radial parts, that is, a Henkel transform as given by the pairs

$$\tilde{c}(k; l_1 l_2 l) = 4\pi i^l \int_0^\infty \mathrm{d}r r^2 c(r; l_1 l_2 l) j_l(kr) \tag{6.19}$$

with the inverse transform reading

$$c(r; l_1 l_2 l) = \frac{1}{2\pi^2 i^l} \int_0^\infty \mathrm{d}k k^2 \tilde{c}(k; l_1 l_2 l) j_l(kr) \tag{6.20}$$

Having obtained $\tilde{c}(k; l_1 l_2 l)$ from eqn (6.19), we have then reached the turning point in Figure 6.3, from which the returning part can then be started with a parallel sequence of operations in Fourier space. These include a backward Clebsch–Gordan transformation to return to an axial frame in k space and obtain $\tilde{c}_{l_1 l_2 m}(k)$, and an OZ equation in k-space to obtain $\tilde{\gamma}_{l_1 l_2 m}(k)$, followed by a forward Clebsch–Gordan transformation and an inverse Hankel transform, to find $\gamma(r; l_1 l_2 l)$. A final backward Clebsch–Gordan transformation brings a new estimate of the original coefficients $\left[\gamma_{l_1 l_2 m}(r)\right]_{\text{new}}$. This cycle is repeated until self-consistency between input and output $\gamma_{l_1 l_2 m}(r)$ is achieved as before.

Note that the OZ equation required in {k} representation, as expressed in terms of the axial expansion coefficients of the transformed pair functions, is given by the following matrix form:

$$\tilde{\gamma}_{l_1 l_2 m}(k) = (-1)^m \rho \sum_{l_3=m}^{\infty} \left[\tilde{\gamma}_{l_1 l_3 m}(k) + \tilde{c}_{l_1 l_3 m}(k) \right] \tilde{c}_{l_3 l_2 m}(k) \qquad (6.21)$$

6.5.3   Thermodynamics

Once the correlation function $h(12)$ [and hence the distribution function $g(12)$ $= h(12) + 1$] is known, the excess free energy can be computed as[44,49]

$$\frac{\beta F_{ex}}{N} = \frac{\beta F_1}{N} + \frac{\beta F_2}{N} + \frac{\beta F_3}{N} \qquad (6.22)$$

where

$$\frac{\beta F_1}{N} = -\frac{1}{2}\rho \int d\mathbf{r}_{12} \langle \frac{1}{2}h^2(12) + h(12) - g(12)\ln\left[g(12)e^{\beta\Phi(12)}\right]\rangle_{\omega_1\omega_2} \qquad (6.23)$$

$$\frac{\beta F_2}{N} = -\frac{1}{2}\rho \int \frac{dk}{(2\pi)^3} \sum_m \left\{ \ln \text{Det}\left[\mathbf{I} + (-1)^m \rho \tilde{\mathbf{h}}_m(k)\right] - (-1)^m \rho \text{Tr}\left[\tilde{\mathbf{h}}_m(k)\right] \right\} \qquad (6.24)$$

$$\frac{\beta F_3}{N} = \frac{\beta F_3^0}{N} - \frac{1}{2}\rho \int d\mathbf{r}_{12} \langle [g(12) - g_0(12)]B_0(12)\rangle_{\omega_1\omega_2} \qquad (6.25)$$

In eqn (6.24), $\tilde{\mathbf{h}}_m(k)$ is a Hermitian matrix with elements $\tilde{h}_{l_1 l_2 m}(k)$, $l_1, l_2 \geq m$, and $\mathbf{I}$ is the unit matrix. The last equation, for F_3, directly expresses the RHNC approximation. Here F_3^0 is the reference system contribution, computed from the known free energy F_{ex}^0 of the reference system as $F_3^0 = F_{ex}^0 - F_1^0 - F_2^0$, with F_1^0 and F_2^0 calculated as above but with reference system quantities.

The bridge function $B_0(12)$ appearing in eqn (6.25) is the key approximation in the RHNC scheme, since it replaces the unknown bridge function $B(12)$ in the general closure equation [eqn (6.14)]. This is taken from the Verlet–Weiss expression of the hard-sphere model,[45] as anticipated, in view of its simplicity and of the fact that it works reasonably well for the case of the square well as we will see, but with a renormalized diameter σ_0 for the hard sphere that is selected by enforcing the variational condition[47]

$$\rho \int d\mathbf{r}[g_{000}(r) - g_{HS}(r; \sigma_0)] \frac{\partial B_{HS}(r; \sigma_0)}{\partial \sigma_0} = 0 \qquad (6.26)$$

From the free energy F, one can of course compute all thermodynamics following standard procedures. For the computation of the phase diagram, the pressure and chemical potential are required at any fixed temperature.

The virial pressure P is obtained as[16]

$$P = \rho k_B T - \frac{1}{3V} \left\langle \sum_{i=1}^{N} \sum_{j>i}^{N} r_{ij} \frac{\partial}{\partial r_{ij}} \Phi(ij) \right\rangle$$

$$= \rho k_B T - \frac{1}{6} \rho^2 \int d\mathbf{r}_{12} \left\langle g(12) r_{12} \frac{\partial}{\partial r_{12}} \Phi(12) \right\rangle_{\omega_1 \omega_2}$$

(6.27)

that in turn can be cast in the form involving the cavity function $y(12) = g(12) e^{\beta \Phi(12)}$ with the result[49]

$$\frac{\beta P}{\rho} = 1 + \frac{2}{3} \pi \rho \sigma^3 \left\{ \left\langle y(\sigma, \omega_1, \omega_2) e^{\beta \varepsilon \Psi(\omega_1, \omega_2)} \right\rangle_{\omega_1 \omega_2} \right.$$

$$\left. - \lambda^3 \left\langle y(\lambda \sigma, \omega_1, \omega_2) \left[e^{\beta \varepsilon \Psi(\omega_1, \omega_2)} - 1 \right] \right\rangle_{\omega_1 \omega_2} \right\}$$

(6.28)

As already remarked, one of the main advantages of the RHNC closure stems from the fact that the calculation of the chemical potential does not introduce any approximation in addition to that already included in the closure. It can be obtained from the thermodynamic relation

$$\beta \mu = \frac{\beta F}{N} + \frac{\beta P}{\rho}$$

(6.29)

that was already used in eqn (6.9).

Finally, we note that the ideal quantities for the free energy, the virial pressure and the chemical potential are

$$\frac{\beta F_{id}}{N} = \ln(\rho \Lambda^3) - 1 \qquad \frac{\beta P_{id}}{\rho} = 1 \qquad \beta \mu_{id} = \ln(\rho \Lambda^3)$$

(6.30)

6.6 Barker–Henderson Perturbation Theory

Another powerful method to access the thermophysical properties of a fluid is thermodynamic perturbation theory, which directly extracts the free energy of the system from the knowledge of the free energy F_0 of a reference fluid (the hard-sphere fluid in the present case). This is a well-known technique in several fields of physics, including simple[15] and complex[16] fluids, and hinges on the fact that often the system under investigation is not very different from the reference one, so that an expansion in this perturbation term is a reasonable approximation. Under these conditions, the results are expected to be fairly reliable, even when stopping the expansion at the lowest orders.

In the square-well fluid case, the analysis was carried out in details in the late 1970s,[51,52] starting from the pioneering work of Zwanzig in 1954,[53] and recently extended to the patchy case.[23,54]

Working in the grand-canonical ensemble, as this is the most convenient one,[52] we assume the total potential U to have the following form:

$$U_\gamma(1,\ldots,N) = U_0(1,\ldots,N) + \gamma U_I(1,\ldots,N)$$

$$= \sum_{i<j} \Phi_\gamma(ij) = \sum_{i<j} \Phi_0(ij) + \gamma \sum_{i<j} \Phi_I(ij) \qquad (6.31)$$

where $U_0(1,\ldots,N) = \sum_{i,j}\Phi_0(ij)$ is the unperturbed part and $U_I(1,\ldots,N) = \sum_{i,j}\Phi_I(ij)$ is the perturbation part. Here $0 \le \gamma \le 1$ is used as perturbative parameter. Note that when each coordinate i includes both the coordinate r_i and patch orientation $\hat{\mathbf{n}}_i$, so that $i \equiv (r_i, \hat{\mathbf{n}}_i)$, then the expression is valid also for the Kern–Frenkel model.[23,54] For simple fluids, $i = r_i$ only. Introducing the following short-hand notation:

$$\int_{1,\ldots,N} (\cdots) \equiv \int \left[\prod_{i=1}^{N} d\mathbf{r}_i \langle (\cdots) \rangle_{\omega_i} \right] \qquad (6.32)$$

for the integration over all particle coordinates, the grand-canonical partition function

$$Q_\gamma = \sum_{N=0}^{+\infty} \frac{e^{\beta \mu N}}{N! \Lambda_T^{3N}} \int_{1,\ldots,N} e^{-\beta U_\gamma} = e^{-\beta \Omega_\gamma} \qquad (6.33)$$

where Λ_T is the de Broglie thermal wavelength and Ω_γ is the grand-potential, can then be used to obtain an expansion of the Helmholtz free energy:[52]

$$F_\gamma = F_0 + \gamma \left(\frac{\partial F_\gamma}{\partial \gamma} \right)_{\gamma=0} + \frac{1}{2!}\gamma^2 \left(\frac{\partial^2 F_\gamma}{\partial \gamma^2} \right)_{\gamma=0} + \cdots \qquad (6.34)$$

which is valid for arbitrary γ.

Taking the derivative of $\ln Q_\gamma$ at fixed chemical potential μ, then using eqn (6.31) one has

$$\left[\frac{\partial}{\partial \gamma} \ln Q_\gamma \right]_\mu = \frac{1}{2} \int_{1,2} \frac{\partial}{\partial \gamma} \left[-\beta \Phi_\gamma(12) \right] \rho_\gamma(12) \qquad (6.35)$$

where

$$\rho_\gamma(1 \ldots h) = \frac{1}{Q_\gamma} \sum_{N=h}^{+\infty} \frac{e^{\beta \mu N}}{(N-h)! \Lambda_T^{3N}} \int_{1,\ldots,N} e^{-\beta U_\gamma} \qquad (6.36)$$

When $\gamma = 1$, this yields the free energy correct to first order in the expansion eqn (6.34).

The second-order correction is far more laborious. Indeed, the extension of this analysis involves higher order correlation functions and this forces additional approximations to come into play,[51,52] thus hampering its practical utility. In 1967, Barker and Henderson gave a much simpler recipe[55] that was found to be fairly effective in predicting the phase diagram of the square-well fluid[52] and was recently extended to the Kern–Frenkel case.[54]

The final result for arbitrary angular-dependent potential, correct to second order, reads[54]

$$\frac{\beta F}{N} = \frac{\beta F_0}{N} + \frac{\beta F_1}{N} + \frac{\beta F_2}{N} + \dots \tag{6.37}$$

where

$$\beta F_1 = \frac{1}{2}\rho N \int d\mathbf{r} g_0(r) \langle \beta \Phi_1(r,\Omega,\omega_1,\omega_2) \rangle_{\omega_1,\omega_2} \tag{6.38}$$

and

$$\beta F_2 = -\frac{1}{4} k_B T \rho N \left(\frac{\partial \rho}{\partial P}\right)_0 \int d\mathbf{r} g_0(r) \langle [\beta \Phi_1(r,\Omega,\omega_1,\omega_2)]^2 \rangle_{\omega_1,\omega_2} \tag{6.39}$$

In the particular case of the Kern–Frenkel potential, one obtains[23,54]

$$\frac{\beta F_1}{N} = \frac{12\eta}{\sigma^3} \int d\mathbf{r} g_0(r) \langle \beta \Psi(12) \rangle_{\omega_1,\omega_2} \tag{6.40}$$

and

$$\frac{\beta F_2}{N} = -\frac{6\eta}{\sigma^3} \left(\frac{\partial \eta}{\partial P_0^*}\right)_T \int d\mathbf{r} g_0(r) \phi_{SW}^2(r) \langle [\beta \Psi(12)]^2 \rangle_{\omega_1,\omega_2} \tag{6.41}$$

where $P_0^* = \beta P_0/\rho$ is the reduced pressure of the HS reference system and $g_0(r)$ the corresponding radial distribution function.

As before, the pressure and chemical potential can be derived from the reduced free energy per particle $\beta F/N$, using the exact thermodynamic identities

$$\frac{\beta P}{\rho} = \eta \frac{\partial}{\partial \eta}\left(\frac{\beta F}{N}\right) \tag{6.42}$$

and

$$\beta \mu = \frac{\partial}{\partial \eta}\left(\eta \frac{\beta F}{N}\right) \tag{6.43}$$

The phase diagram in the temperature–density plane can then be computed using a numerical procedure that will be discussed in connection with eqns (6.44) and (6.45) below, as it is identical with that used for integral equations.

6.7 Calculation of the Fluid–Fluid Coexistence Curves for the Integral Equation and Perturbation Theory

The common feature of integral equations and thermodynamic perturbation theory is that one is able to obtain approximate expressions for both the pressure P and the chemical potential μ as a function of the temperature T and the density ρ. In the presence of a fluid–fluid (gas–liquid) transition, both P and μ will have well-defined dependence on T and ρ in the gas and liquid branches, but not in the coexistence regions. Hence, in order to obtain the coexistence curves and hence the phase diagram in the temperature–density plane, two procedures are possible. The first is a graphical procedure where one plots the chemical potential versus the pressure at a given fixed temperature and seeks the intersections between the gas and the liquid branches. An example of this procedure in the case of the square-well potential can be found elsewhere.[48] This procedure, however, is neither very practical nor very precise, as it involves qualitative deduction of the crossing points. Yet it can be used as a first preliminary estimate of a more precise numerical calculations as follows.

For a fixed temperature T, one can then compute the pressure of the gas (colloidal-poor) phase P_g and of the liquid (colloidal-rich) phase P_l and the corresponding chemical potentials μ_g and μ_l. The fluid–fluid (gas–liquid) coexistence line then follows from a numerical solution of a system of non-linear equations:

$$P_g(T,\rho_g) = P_l(T,\rho_l) \tag{6.44}$$

$$\mu_g(T,\rho_g) = \mu_l(T,\rho_l) \tag{6.45}$$

the solutions of which are the gas ρ_g and liquid ρ_l densities associated with the coexistence lines. By plotting the resulting ρ_g and ρ_l as a function of T, the coexistence curve can be constructed in the region where the transition occurs. This procedure will be followed in the analysis of the Kern–Frenkel fluid–fluid phase diagram for both integral equations (see Section 6.8.1) and thermodynamic perturbation theory (see Section 6.8.4), but can also be exploited for the computation of the fluid–solid transition, as explained in Section 6.8.5.

6.8 Results

6.8.1 Fluid–Fluid Coexistence Curves from the RHNC Integral Equation

We start by reviewing the quality of the RHNC integral equation theory for the simple case of the isotropic square-well case, a model which has often be used as an effective potential (in the implicit solvent representation) to model colloidal particles interacting *via* depletion interactions.[56,57] Figure 6.4 shows

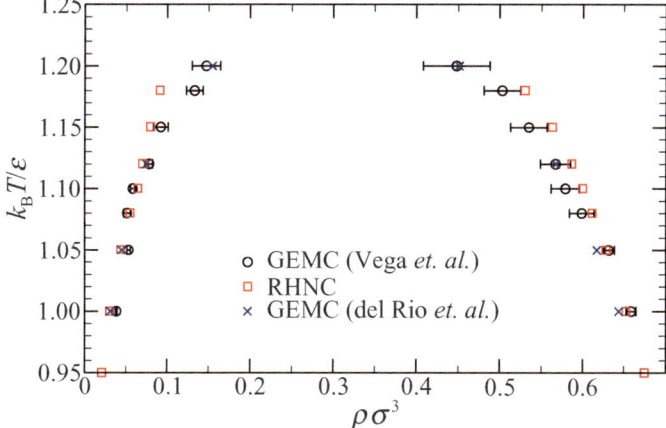

Figure 6.4 Fluid–fluid coexistence curves $k_B T/\varepsilon$ versus $\rho\sigma^3$ for a square-well fluid of range $\lambda = 1.5$. Data from the RHNC integral equation (squares) are contrasted with Monte Carlo simulations by Vega *et al.*[58] (circles) and del Río *et al.*[59] (crosses). Adapted from reference 48.

the RHNC predictions for the gas–liquid coexistence from the literature[48] contrasted with Monte Carlo simulations on the same system carried out by different groups.[58,59]

Although the integral equation results appear to reproduce reasonably well those from numerical simulations, two features are worth noting. First, thermodynamic inconsistencies associated with the approximate nature of the closure are at the origin of the incorrect evaluation of the pressure and chemical potentials and hence of the exact location of the coexistence lines. This is an unavoidable feature of all integral equations and its origin and effects are well known. For most cases, in fact, the whole critical region is inaccessible to integral equation theories.

A second additional point is related to the non-linear nature of the self-consistency procedure and gives rise to numerical instabilities that may or may not be controlled depending on the state point considered. As a general rule, lower temperatures and higher densities are more difficult to converge and hence for some points the solution of the system of eqns (6.44) and (6.45) might not even exist.

These considerations are even more compelling in the more complex case of a Kern–Frenkel fluid, where condensation takes place at lower T associated with the lower coverages involved. Hence one might expect an agreement with respect to numerical simulations not better than what has been reported for the isotropic square-well case. This is indeed the case, as shown in Figure 6.5 for the representative example of single-patch particles with coverage $\chi = 0.8$, for which 80% of the surface has a square-well character, condensed into a unique patch. The width of the square-well was selected to be $\lambda = 1.5$ as before, so that the limit of full coverage gives back the result of Figure 6.4, and the value

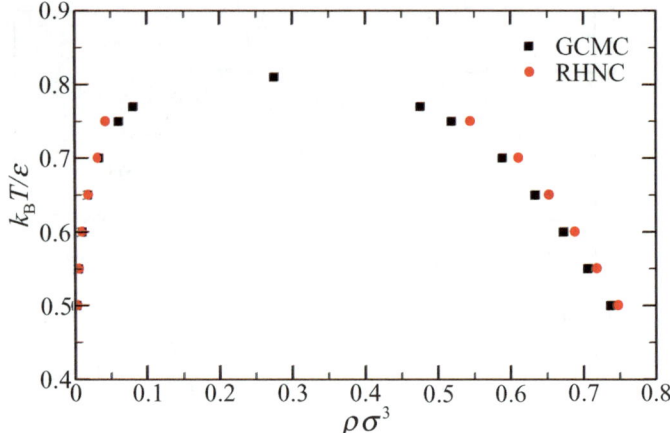

Figure 6.5   Fluid–fluid coexistence curves $k_B T/\varepsilon$ versus $\rho\sigma^3$ for a Kern–Frenkel fluid
with coverage $\chi = 0.8$ and range $\lambda = 1.5$. Data from the RHNC integral
equation (circles) are compared with Monte Carlo simulations (squares).
Adapted from reference 48.

$\chi = 0.8$ was selected to be half way between the fully occupied fluid and half
coverage that has peculiar behavior, as will be discussed shortly.

In Figure 6.5 we also report the coexistence lines and critical points using
Gibbs ensemble and grand-canonical Monte Carlo simulations, as outlined in
Sections 6.4.2 and 6.4.3. The former were used to evaluate coexistence in the
region where the gas–liquid free-energy barrier is sufficiently high to avoid
crossing between the two phases, whereas the latter was used to locate exactly
the critical point. The details of the analysis[48] (and for the corresponding two-
patch case[50]) can be found elsewhere.

The comparison between the RHNC integral equation and Monte Carlo
simulations for the Kern–Frenkel model with intermediate coverage $\chi = 0.8$
displayed in Figure 6.5 shows that the agreement is indeed comparable to the
square-well case, as anticipated.

The Kern–Frenkel model offers the possibility of continuously changing the
coverage interpolating from the isotropic square-well to the symmetric Janus-like
potential, when the coverage moves from $\chi = 1$ to $\chi = 0.5$ (see Figure 6.6a). To
investigate how the phase diagram of Janus particles arises, we calculate how the
gas–liquid coexistence is modified on progressively reducing χ. Figure 6.6b shows
Monte Carlo simulations (Gibbs ensemble for the coexistence lines and grand-
canonical μVT for the critical point) results for the gas–liquid phase coexistence
for several χ values, extending the original data of Kern and Frenkel.[20] A
progressive shrinking of the coexistence region to lower temperatures and
densities accompanies the decrease in the coverage. Consistently, as the coverage
decreases, both the critical temperature and critical density decrease (see
Figure 6.9). This is not surprising, in view of the fact that the coverage is a
measure of the attractive interactions intensity (by controlling the maximum

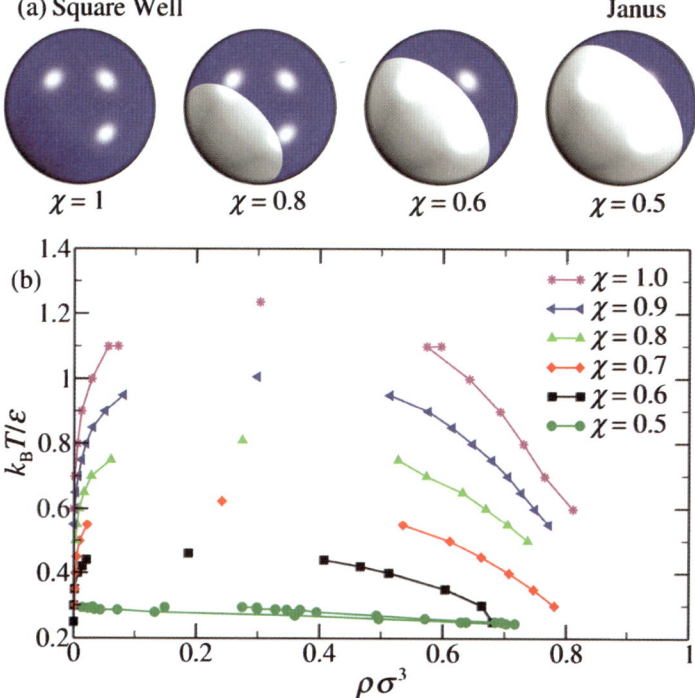

Figure 6.6 (a) Cartoon of a one-patch particle with the coverage (fraction of attractive surface, depicted in blue) changing from $\chi = 1.0$ (square well) to 0.5 (Janus). (b) Fluid–fluid coexistence curves $k_B T/\varepsilon$ versus $\rho\sigma^3$ for a Kern–Frenkel fluid with decreasing coverage from $\chi = 1.0$ (the square-well case) to the Janus limit $\chi = 0.5$. The range is always set to $\lambda = 1.5$. Data are from GEMC simulations for the coexistence lines and from grand-canonical Monte Carlo simulations for the critical points. This is the one-patch result; for the two-patches counterpart, see Figure 6.8. Adapted from reference 31.

number of nearest neighbor contacts[21,60]) that dictates the value of T_c and (perhaps more indirectly) ρ_c. It can be explicitly shown that this shrinking of the coexistence region is a non-trivial one and that it cannot be inferred by a simple scaling of either the temperature or the density. Up to $\chi = 0.6$ coverage, however, the morphology of the curve appears to be the standard one, with the gas and liquid coexistence lines widening on cooling.

The half-coverage $\chi = 0.5$ is known as the Janus limit and will be discussed in the next section, as it displays an interesting unconventional behavior.[28]

6.8.2 The Janus Limit

The value of $\chi = 0.5$ plays a special role in the Kern–Frenkel model with a single patch. This can be seen by following the change in the coexistence line locations as the coverage is reduced from the fully isotropic square-well fluid ($\chi = 1.0$) to the Janus case ($\chi = 0.5$), as discussed in the previous section.

In Figure 6.7a, the Janus case $\chi = 0.5$, already reported in Figure 6.6, is shown by magnifying the very narrow region where the transition occurs.

The difference from the standard phase diagram is clear. As the fluid is cooled to lower and lower temperatures, the coexistence region shrinks, in contrast to the standard case, and the two coexistence lines appear to approach one another at sufficiently low temperatures. The origin of this anomaly has far-reaching consequences that have already been discussed in details.[28,31] The crucial point is that at low temperature and density, monomers tend to aggregate into micelles and vesicles (see Figure 6.7b) bearing a well-defined number of particles (10, 41, ...) so that all favorable contacts are saturated inside the aggregate.[28,31] This means that, in all cases, the clusters always expose the hard-sphere parts to the external fluid and the condensation process is thus inhibited and eventually destroyed altogether. The specificity of these magic numbers can also be explained by means of a simple cluster theory,[61] and an even simpler approach[62] can be developed to mimic the onset of the re-entrant gas branch. Below the Janus limit, no evidence of fluid–fluid transition was found, in full agreement with the above interpretation.[28,31]

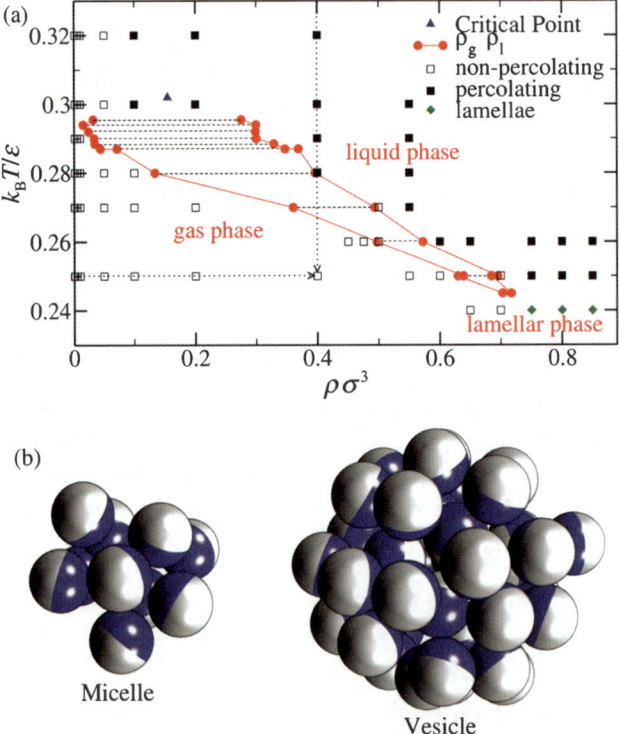

Figure 6.7   (a) The anomalous phase diagram of the Janus limit ($\chi = 0.5$). Adapted from reference 28. (b) Cartoon of the aggregates which form in the gas phase on cooling, micelles and vesicles. The blue surface is attractive, modeled *via* a square-well potential.

The unconventional shape of the phase diagram arises from the particular energetic and entropic balance associated with the transition from the micellar gas to the liquid phase. It has been found that, in contrast to the usual gas–liquid behavior, the potential energy per particle is higher in the liquid phase than in the micellar gas phase, a consequence of the greater energetic stability of the micelles and vesicles compared with the disordered liquid phase. Hence, despite the fact that the gas phase is stabilized by the translational entropy of the micelles, the coexisting liquid phase is more disordered than the gas phase. Such unconventional entropic stabilization of the liquid phase arises from the orientational entropy, since particles are orientationally disordered in the liquid phase whereas they are properly oriented in the micelles gas phase.

6.8.3 One *Versus* Two Patches

It is interesting to investigate the effects of having the same attractive square-well coverage split into two parts, at the opposite poles of the sphere so that they are symmetrically distributed. This is the two-patch case and again this case smoothly interpolates between the fully occupied isotropic square-well case $\chi = 1$ and the empty hard-sphere case $\chi = 0$.

The corresponding fluid-fluid (gas–liquid) phase diagram is depicted in Figure 6.8, which should be contrasted with the single-patch counterpart reported in Figure 6.6. Two main differences are noteworthy. First, the case $\chi = 0.5$ does not appear to play any particular role, at variance with the single-patch counterpart. Its coverage value corresponds to the so-called triblock Janus case and will be further discussed later.[25–27] This can be attributed to the fact that at higher valences it is not possible to saturate all favorable contacts and hence the condensation process always takes places in complete agreement with the previous interpretation of the Janus case. The second main difference is related to the fact that in the two-patch case, coverages below $\chi = 0.5$ exhibit a fluid–fluid transition, as displayed in Figure 6.8, where coverages as low as $\chi = 0.3$ are depicted. Below this value, the transition becomes metastable with respect to crystallization, as explained in detail elsewhere.[50]

The χ dependence of the critical point can be inferred by plotting the reduced critical densities and temperatures as a function of the different coverages, as shown in Figure 6.9.[50] Clearly, the two-patch densities and temperatures lie always above the one-patch counterparts, thus supporting the interpretation that it is easier to form a fluid with higher valence at a given coverage, a feature that is related to the existence of the fluid–fluid transition even for low coverages.

6.8.4 Evaluation of the Fluid–Fluid Coexistence Curves from Thermodynamic Perturbation Theory

As in the case of integral equation theory, the fluid–fluid phase diagram in the temperature–density plane can be computed even from Barker–Henderson (BH) thermodynamic perturbation theory,[54] as indicated in Section 6.7. This is

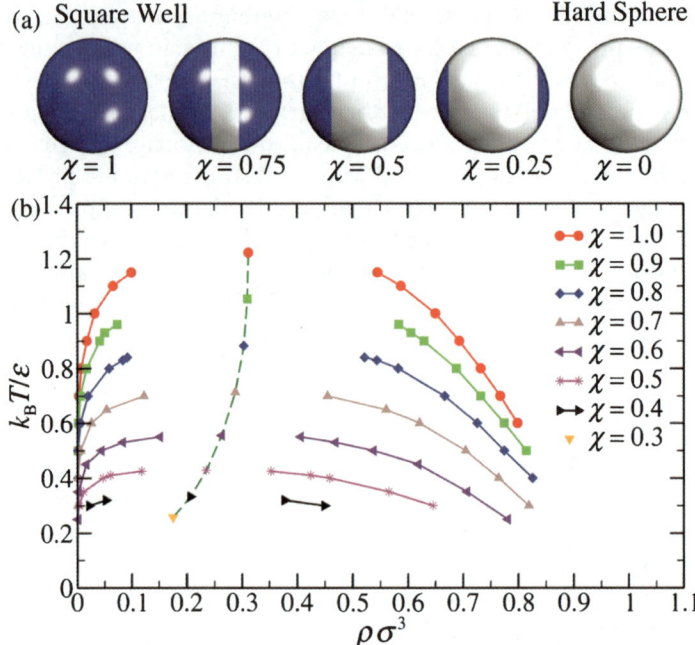

Figure 6.8   (a) Cartoon of two-patch particles, when the coverage (blue) varies from χ = 1 (square-well limit) to χ = 0 (hard-sphere limit). (b) Fluid–fluid coexistence curves $k_B T/\varepsilon$ versus $\rho\sigma^3$ for a Kern–Frenkel fluid with two patches and decreasing coverage from χ = 1.0 (the square-well case) to e χ = 0.3. The range is always set to λ = 1.5. Data are from GEMC simulations for the coexistence lines and from grand-canonical Monte Carlo simulations for the critical points. This is the two-patch counterpart of Figure 6.6. Adapted from reference 50.

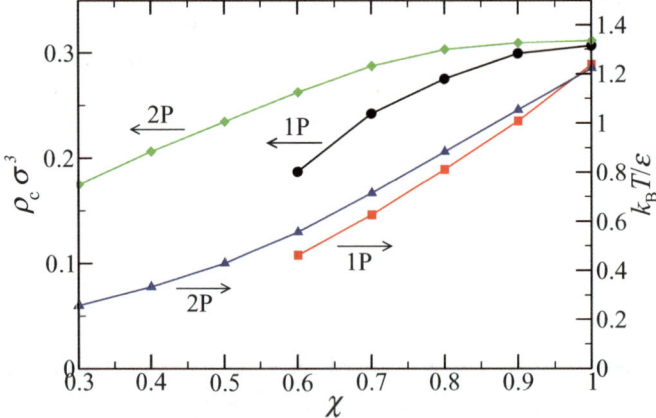

Figure 6.9   Comparison between the one- and two-patch coverage dependence of the reduced critical density (left axis) and of the reduced critical temperature (right axis). Adapted from reference 50.

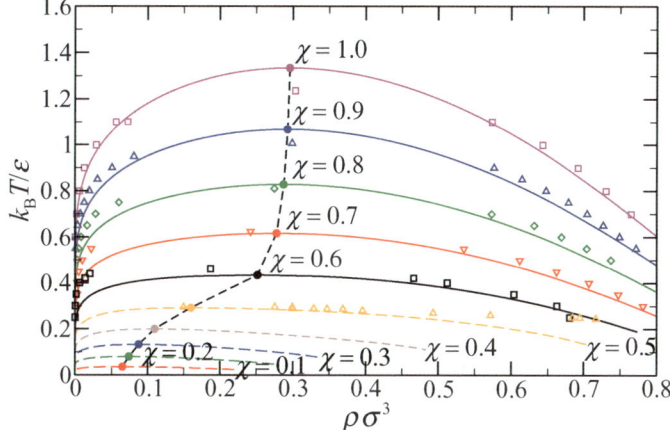

Figure 6.10 The fluid–fluid phase diagram from BH thermodynamic perturbation theory (continuous lines $0.6 \leqslant \chi \leqslant 1.0$, dashed lines $0.1 \leqslant \chi \leqslant 0.5$) contrasted with numerical simulations (open symbols). Closed symbols refer to the BH critical points. Adapted from reference 54.

shown in Figure 6.10 for the one-patch case and compared with the same Monte Carlo simulations used for comparison with integral equation theory.

Given the simplicity of the theory, the accuracy of the BH theoretical prediction is rather remarkable. On the one hand, it allows a prediction of the possible location of the true coexisting lines, even for coverages where Monte Carlo results are not yet available or in regions where crystallization prevents the evaluation of the location of the metastable gas–liquid critical point. Note that there are no restrictions on the applicability of the BH theory, either on thermodynamic parameters or on coverages, the only limitations being the convergence of the numerical non-linear solvers for eqns (6.44) and (6.45). On the other hand, the BH perturbation theory can be applied in its current form only to the one-patch case, as it lacks the possibility of discriminating the location of the patches. Notwithstanding these limitations, this approach is very promising as it even allows the computation of the fluid–solid branch, as discussed in the next section.

6.8.5 Fluid–Solid Coexistence

At higher densities, a fluid–solid transition is expected in a way similar to that occurring in the fully isotropic square-well case.[63,64] For the value of the square-well amplitude $\lambda = 1.5$, this transition occurs at densities that are well separated from the fluid–fluid counterpart. In the isotropic square-well scenario, this case has been studied several times with different methodologies.[63,64] Figure 6.11 reports one of those studies. by Young and Adler[63] (open squares), carried out using molecular dynamics techniques. In addition to the expected fluid–solid transition, with the solid being a face-centered cubic

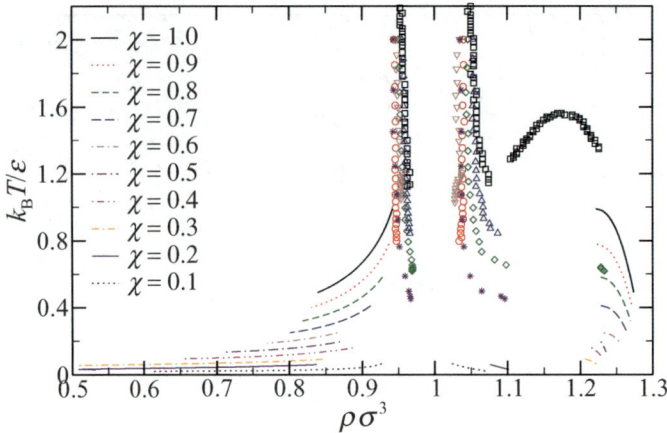

Figure 6.11   The fluid–solid phase diagram from BH thermodynamic perturbation
theory (continuous and dashed lines) contrasted with numerical simulations (open symbols). Closed symbols refer to the BH critical points. The
case $\chi = 1.0$ (open squares) corresponds to the Young and Adler
molecular dynamics results[63] and includes the FCC–FCC transition not
considered in the patchy cases. Adapted from reference 54.

(FCC) lattice, an additional FCC–FCC transition is visible as a large plateau
at higher densities (see Figure 6.11).

The other open symbols in Figure 6.11 represent results from the fluid–solid
transition of the Kern–Frenkel model with coverages from $\chi = 0.9$ down to the
Janus limit $\chi = 0.5$. In all cases, the final crystal is FCC with 12 nearest neighbors.
An additional FCC–FCC transition exists akin to that found in the square-well
case, corresponding to a transition from a more dilute to a denser FCC lattice,
which is possible for this range of the attractive well, but it will not be considered
here. The lines in Figure 6.11 represent results from thermodynamic perturbation
theory obtained by using the same approach as outlined above for the fluid–fluid
transition. In this case, a hint of the solid structural change is visible at higher
densities, but the exact coexistence lines could not be obtained owing to
limitations in the convergence of the numerical algorithm associated with the non-
linear solver in eqns (6.44) and (6.45). In spite of this drawback, thermodynamics
perturbation theory is able to capture the main features of the transition even for
lower coverages (as low as $\chi = 0.1$). It should be stressed, however, that
transitions to different crystal structures may be envisaged at low coverages and
this feature has not been accounted in the analysis reported in Figure 6.11.

6.8.6   Self-assembly in a Predefined Kagome Lattice

The subtle interplay between the fluid–fluid and fluid–solid transitions is one
of the most delicate issues in the framework of self-assembly patchy colloids,
especially in the presence of additional effects such as inhibiting clustering
transitions. This has already been hinted at earlier, but a very illuminating

example is provided by the phase diagram of the triblock Janus fluid, which has also been considered elsewhere in this volume from the experimental point of view.

Triblock Janus colloids are spherical colloidal particles decorated with two hydrophobic poles of tunable area, separated by an electrically charged middle band (triblock Janus).[25] The electric charge of the particles allows for a controlled switch of the interaction *via* addition of salt, which effectively screens the overall repulsion, offering the possibility of hydrophobic attraction between patches expressing itself. Once deposited on a flat surface, after the addition of the salt, particles organize themselves into a kagome lattice. The crystallization kinetics have been followed in real space in full detail.[25] The patch width in the experimental system, of the order of 65° corresponding to $\chi \approx 0.57$, allows for simultaneous bonding of two particles per patch, stabilizing the locally four-coordinated structure of the kagome lattice (see Figure 6.12).

A triblock Janus particle can be modeled *via* the two-patch Kern–Frenkel model in which the attractive region is split into two parts at the opposite poles of the sphere, whereas the repulsive part is concentrated in the middle strip at the equator. The square well mimics the short-range hydrophobic attraction, whereas the hard-sphere region represents the repulsive charge–charge interaction. Depending on the considered state point and on the patch amplitude, several possible phases are possible, as shown elsewhere[27] and summarized in Figure 6.13. In full agreement with the experiments, we observe at comparable interaction strengths the spontaneous nucleation of a kagome lattice. We also find at higher pressures spontaneous crystal formation in a dense hexagonal structure. Such ease of crystallization suggests that in this system crystallization barriers are comparable to the thermal energy at all densities. Interestingly, we find that for this model a (metastable) gas–liquid phase separation can be observed for large patch widths.

The ability to describe Janus triblock particles accurately with the Kern–Frenkel potential is particularly rewarding and provides strong support for the use of such a model for predicting the self-assembly properties of this class of patchy colloids. The possibility of numerically exploring the sensitivity of the

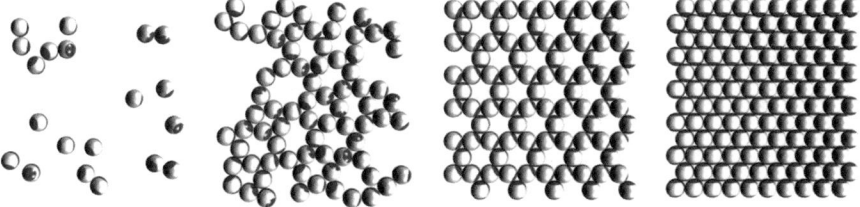

Figure 6.12   From left to right: snapshot of a gas, liquid, kagome lattice and hexagonal lattice. The kagome and the hexagonal crystals are formed at low and high pressures, respectively. Attractive patches are colored blue and the hard-core remaining particle surface gray. Particles are free to rotate in three dimensions but are constrained to move on a flat surface. Adapted from reference 27.

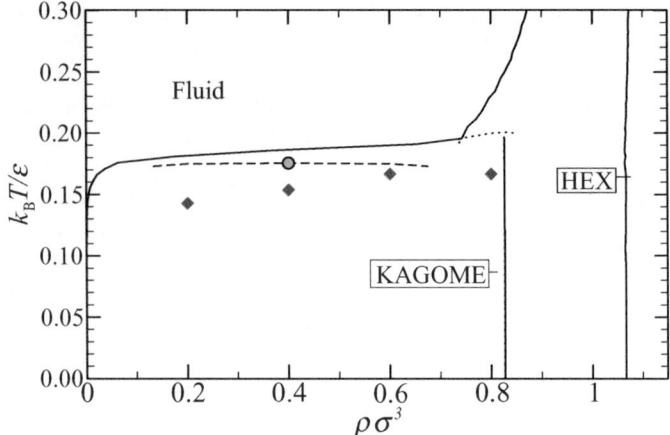

Figure 6.13   Phase diagrams in the $T-\rho$ plane for a wide ($\chi \approx 0.57$) patch model (see reference 27 for the short-range analog). Boundaries between stable phases are drawn as solid black lines and metastable phase boundaries are dotted. The orange points indicate the (metastable) gas–liquid critical point. The dashed line represents the metastable gas–liquid phase separation. Blue diamonds and red circles indicate the highest temperature at which spontaneous crystallization into the kagome and hexagonal lattice, respectively, was detected at the corresponding density. Adapted from reference 27.

phase diagram to the parameters (patch width and interaction range) entering in the interaction potential provides an important instrument and a guide to the design of these new particles to obtain specific structures by self-assembly.

6.9   Conclusions and Future Perspectives

In this chapter, we have discussed some of the main theoretical approaches that have been proposed in the last few years to address the computation of the phase diagram in the temperature–density plane of the Kern–Frenkel model, one of the paradigmatic models for patchy colloids. Three different approaches, namely Monte Carlo simulations, integral equations and perturbation theory, have been discussed and their performances contrasted against each other. We have attempted to present the main ideas behind each technique, along with some representative examples of applications. The main emphasis has been placed on the evaluation of the fluid–fluid (gas–liquid) and fluid–solid phase diagrams and their importance in the framework of the self-assembly processes. In fact, the exact location of the transition lines is a crucial ingredient necessary to implement bottom-up engineering of new-generation materials. It is remarkable that none of these techniques are really new, being part of the background given in any standard graduate course in statistical physics. Yet their implementation in the framework of patchy colloids has often (if not always) required significant improvements that have shed new light on the techniques themselves. This is true for the newly improved Monte Carlo schemes that have been devised in this field, but also for integral equation and perturbation theories that have a long and important history.

Several steps still need to be performed. We still need an accurate study of the role of the interaction range in selecting the most stable geometries. More specifically, we need to understand under which patch width and range conditions micelles and vesicles are the stable structures. Experiments[24] indicate that for Janus particles one-dimensional structures become preferred when the interaction range is only a few percent of the particle size. We also need to develop an accurate methodology to predict all possible crystal structures for different types of patterned particles (including Janus), their stability fields and the associated nucleation rates. Understanding self-assembly into ordered predefined structures can have relevant industrial applications.[65] The Janus paradigm (and its theoretical counterpart, the Kern–Frenkel model) can play an important role in this process.

The fact that notwithstanding its simplicity the Kern–Frenkel model is able to present such a complex and rich scenario in its phase diagram is related to the combined effect of two features. On the one hand, the short-range and reversible nature of the involved interactions allows a partial rearrangement, within a localized region in space, of the particles in a search for the optimal minimal energy configuration, a feature that would not be possible for stronger (irreversible) covalent longer range interactions. On the other hand, the specificity dictated by the patchy anisotropy is expedient in avoiding multiple degenerate configurations with similar energies, thus eliminating defects and polydispersity effects characteristic of isotropic colloids.

Qualitative and semiquantitative agreements with experiments can be obtained with the Kern–Frenkel model in some particular cases and it is hoped that the contribution of this book, including both theory and experiments in a well-chosen balance, will lead to the further strengthening of this very promising route.

Acknowledgements

The results presented in this chapter were obtained in collaboration with a number of colleagues, including Christoph Gögelein, Fred Lado, Julio Largo and Giorgio Pastore.

References

1. G. M. Whitesides and M. Boncheva, *Proc. Natl. Acad. Sci. U. S. A.*, 2002, **99**, 4769; G. M. Whitesides and B. Grzybowski, *Science*, 2002, **295**, 2418.
2. S. C. Glotzer, *Science*, 2004, **306**, 419.
3. S. C. Glotzer and M. J. Solomon, *Nat. Mater.*, 2007, **6**, 557.
4. A. Walther and A. H. E. Müller, *Soft Matter*, 2008, **4**, 663.
5. A. B. Pawar and I. Kretzchmar, *Macromol. Rapid Commun.*, 2010, **31**, 150.

6. A. Yethiraj and A. van Blaaderen, *Nature*, 2003, **421**, 513–517.
7. V. N. Manoharan, M. T. Elsesser and D. J. Pine, *Science*, 2003, **301**, 483–487.
8. Y.-S. Cho, G.-R. Yi, J.-M. Lim, S.-H. Kim, V. N. Manoharan, D. J. Pine and S.-M. Yang, *J. Am. Chem. Soc.*, 2005, **127**, 15968–15975.
9. D. J. Kraft, J. Groenewold and W. K. Kegel, *Soft Matter*, 2009, **5**, 3823–3826.
10. A. L. Hiddessen, S. D. Rotgers, D. A. Weitz and D. A. Hammer, *Langmuir*, 2000, **16**, 9744–9753.
11. G. Zhang, D. Wang and H. Möhwald, *Angew. Chem. Int. Ed.*, 2005, **44**, 1–5.
12. C. Mirkin, R. Letsinger, R. Mucic and J. Storhoff, *Nature*, 1996, **382**, 607–609.
13. V. T. Milam, A. Hiddessen, S. Rodgers and J. C. Crocker, *Langmuir*, 2003, **19**, 10317–10323.
14. E. Bianchi, R. Blaak and C. N. Likos, *Phys. Chem. Chem. Phys.*, 2011, **13**, 6397
15. J. P. Hansen and I. R. McDonald, *Theory of Simple Liquids*, Academic Press, New York, 1986.
16. C. G. Gray and K. E. Gubbins, *Theory of Molecular Fluids, Vol. 1: Fundamentals*, Clarendon Press, Oxford, 1984.
17. B. Smith and D. Frenkel, *Understanding Molecular Simulation: from Algorithms to Applications*, Academic Press, San Diego, CA, 2002.
18. M. P. Allen and D. J. Tildesley, *Computer Simulations of Liquids*, Clarendon Press, Oxford, 1987.
19. J. Lyklema, *Fundamentals of Interface and Colloid Science, Vol. I: Fundamentals*, Academic Press, London, 1991.
20. N. Kern and D. Frenkel, *J. Chem. Phys.*, 2003, **118**, 9882.
21. E. Bianchi, J. Largo, P. Tartaglia, E. Zaccarelli and F. Sciortino, *Phys. Rev. Lett.*, 2006, **97**, 168301.
22. H. Liu, S. K. Kumar and F. Sciortino, *J. Chem. Phys.*, 2007, **127**, 084902.
23. C. Gögelein, G. Nägele, R. Tuinier, T. Gibaud, A. Stradner and P. Schurtenberger, *J. Chem. Phys.*, 2008, **129**, 085102.
24. L. Hong, A. Cacciuto, E. Luijten and S. Granick, *Langmuir*, 2008, **24**, 621.
25. Q. Chen, S. C. Bae and S. Granick, *Nature*, 2011, **469**, 382.
26. F. Romano and F. Sciortino, *Nat. Mater.*, 2011, **10**, 171.
27. F. Romano and F. Sciortino, *Soft Matter*, 2011, **7**, 5799.
28. F. Sciortino, A. Giacometti and G. Pastore, *Phys. Rev. Lett.*, 2009, **103**, 237801.
29. J. Russo, J. M. Tavares, P. I. C. Teixeira, M. M. Telo da Gama and F. Sciortino, *Phys. Rev. Lett.*, 2011, **106**, 085703.
30. J. Russo, J. Tavares, P. Teixeira, M. da Gama and F. Sciortino, *J. Chem. Phys.*, 2011, **135**, 034501.
31. F. Sciortino, A. Giacometti and G. Pastore, *Phys. Chem. Chem. Phys.*, 2010, **12**, 11869.
32. W. Bol, *Mol. Phys.*, 1982, **45**, 605.
33. D. M. Tsangaris and J. J. de Pablo, *J. Chem. Phys.*, 1994, **101**, 1477.
34. J. Kolafa and I. Nezbeda, *Mol. Phys.*, 1987, **61**, 161.
35. A. Lomakin, N. Asherie and G. B. Benedek, *Proc. Natl. Acad. Sci. U. S. A.*, 1999, **96**, 9465.

36. R. Eppenga and D. Frenkel, *Mol. Phys.*, 1984, **52**, 1303.
37. V. I. Harismiadis, J. Vorholz and A. Z. Panagiotopoulos, *J. Chem. Phys.*, 1996, **105**, 8469.
38. A.Z. Panagiotopulos, *Mol. Phys.*, 1987, **61**, 813; A.Z. Panagiotopulos, N. Quirkem M,
39. N. B. Wilding, *Phys. Rev. E*, 1995, **52**, 602; for a review, see also N. B. Wilding, *J. Phys.: Condens. Matter*, 1997, **9**, 585.
40. C. Vega, E. Sanz, J. L. F. Abascal and E. G. Noya, *J. Phys.: Condens. Matter*, 2008, **20**, 153101.
41. D. A. Kofke, *J. Chem. Phys.*, 1993, **98**, 4149.
42. F. Lado, *Phys. Rev. A*, 1973, **8**, 2548.
43. F. Lado, *Mol. Phys.*, 1982, **47**, 283.
44. F. Lado, *Mol. Phys.*, 1982, **47**, 299.
45. L. Verlet and J. J. Weis, *Phys. Rev. A*, 1972, **5**, 939.
46. Y. Rosenfeld and N. W. Ashcroft, *Phys. Rev. A*, 1979, **20**, 1208.
47. F. Lado, *Phys. Lett.*, 1982, **89A**, 196.
48. A. Giacometti, G. Pastore and F. Lado, *Mol. Phys.*, 2009, **107**, 555.
49. A. Giacometti, F. Lado, J. Largo, G. Pastore and F. Sciortino, *J. Chem. Phys.*, 2009, **131**, 174114.
50. A. Giacometti, F. Lado, J. Largo, G. Pastore and F. Sciortino, *J. Chem. Phys.*, 2010, **132**, 174110.
51. J.A. Barker and D. Henderson, *Rev. Mod. Phys.*, 1976, **48**, 587.
52. D. Henderson and J. A. Barker, in *Physical Chemistry, an Advanced Treatise*, Vol. VIIIA, Academic Press, New York. 1971, p. 377.
53. R. Zwanzig, *J. Chem. Phys.*, 1954, **22**, 1420.
54. C. Gögelein, F. Romano, F. Sciortino and A. Giacometti, *J. Chem. Phys.*, 2012, **136**, 094512.
55. J. A. Barker and D. Henderson, *J. Chem. Phys.*, 1967, **47**, 2856.
56. E. Zaccarelli, G. Foffi, K. A. Dawson, S. V. Buldyrev, F. Sciortino and P. Tartaglia, *Phys. Rev. E*, 2002, **66**, 041402.
57. G. Foffi, E. Zaccarelli, S. V. Buldyrev, F. Sciortino and P. Tartaglia, *J. Chem. Phys.*, 2004, **120**, 8824–8830.
58. L. Vega, E. de Miguel, L. F. Rull, G. Jackson and I. A. McLure, *J. Chem. Phys.*, 1992, **96**, 2296.
59. F. del Río, E. Ávalos, R. Espíndola, L. F. Rull, G. Jackson and S. Lago, *Mol. Phys.*, 2002, **100**, 2531.
60. E. Bianchi, P. Tartaglia, E. La Nave and F. Sciortino, *J. Phys. Chem. B*, 2007, **111**, 11765–11769.
61. R. Fantoni, A. Giacometti, F. Sciortino and G. Pastore, *Soft Matter*, 2011, **7**, 2419.
62. A. Reinhardt, A. J. Williamson, J. P. K. Doyle, J. Carrete, L. M. Varele and A. A. Louis, *J. Chem. Phys.*, 2011, **134**, 104905.
63. D. A. Young and B. J. Adler, *J. Chem. Phys.*, 1980, **73**, 2430.
64. H. Liu, S. Garde and S. Kumar, *J. Chem. Phys.*, 2005, **123**, 174505.
65. F. Romano and F. Sciortino, *Nat. Commun.*, 2012, **3**, 975.

CHAPTER 7

Self-assembly of Dipolar and Amphiphilic Janus Particles

LIANG HONG*[a] AND ANGELO CACCIUTO*[b]

[a] The Dow Chemical Company, Midland, MI 48674, USA; [b] Department of
Chemistry, Columbia University, 3000 Broadway, New York, NY 10027, USA
*E-mail: lhong@dow.com or ac2822@columbia.edu

7.1 Introduction

Self-assembly of nanocomponents into micro- or mesoscopic aggregates is a
process of fundamental physical importance. From protein aggregation into
viral capsids[1–3] to self-organization of phospholipids into biological membranes[4] or nucleation of colloidal particles into perfectly ordered photonic
crystals,[5] self-assembly is the main mechanism leading to structure formation
in all those physical systems where thermal fluctuations play an important
role.[6–10] Uncovering the driving forces and the precise mechanisms leading to
self-assembly of biological or synthetic components at the nanoscale will
improve our understanding of important biological processes and can lead to
robust strategies for the development of materials designed to have specific
electronic, optical or mechanical properties.

Simple thermodynamic arguments have been put forward to explain the
equilibrium behavior of the aggregates formed via self-assembly.[11–14]
Unfortunately, the intrinsic dynamic nature of the problem and the crucial
role played by the physical properties of the single components, *i.e.* their bare
shape and pair interactions, make it very difficult for such theories to make
robust predictions on the structural properties of the aggregates.

RSC Smart Materials No. 1
Janus Particle Synthesis, Self-Assembly and Applications
Edited by Shan Jiang and Steve Granick
Published by the Royal Society of Chemistry, www.rsc.org

What is typically found in numerical and experimental studies is that self-assembly is indeed a very elusive process that demands a subtle balance between entropic and enthalpic contributions and precise engineering of the interactions between the nanoparticles. As a result, a large body of work has been carried out with the aim of gaining insight into how the geometry of the interparticle interactions and the bare shape of the particles themselves determine the kinetic pathways and structural morphologies of the final aggregates.

Most of the experimental work on self-assembly has historically focused on small molecules. Various techniques are used to characterize the morphological properties of self-assembled structures, such as X-ray analysis, atomic force microscopy and electron microscopy. However, these techniques are often limited by the spatial or time resolution to study interactions between small molecules especially at the single-molecule level. X-ray crystallography[15] or cryo-transmission electron microscopy[16] could provide excellent spatial resolution, but limited time resolution to study the dynamics of structure formation. Recent progress in single-molecule imaging techniques[17] could provide reasonable time resolution, but detailed structural information is still difficult to retrieve and fluorescent probes may interfere with the self-assembly behaviors.

The last decade has witnessed several breakthroughs in particle synthesis at the mesoscale. The reasons behind the recent focus on the synthesis of colloidal particles reside in the recognition that colloidal particles offer a flexible platform for the relatively cheap production of a potentially unlimited number of building blocks with tunable size, charge and chemical specificity. Furthermore, unlike their molecular counterparts, colloidal particles can be easily tracked via standard microscopes and it is relatively simple to alter their physical properties.

Historically, classical colloidal science dealt mostly with spherical particles interacting via isotropic potentials that depended only on particle separation. Using depletants such as polymers or smaller nanoparticles, it is possible to induce effective interactions between the colloids with range and strength controlled by the size and concentration of the depletant.[18] The competition between short- and long-range interactions, the latter due, for instance, to electrostatic interactions, has been shown to lead to micro-phase separation and consequent formation of cluster phases of tunable geometry and size.[19–21] Finally, theoretical isotropic pair potentials consisting of several minima and maxima as a function of particle separation[22] and the ultra-soft interaction between polymer nanoparticles such as star polymers and dendrimers are theoretically expected to lead to the formation of a variety of complex non-close-packed structures.[23]

Although the question of how macroscopic anisotropy can arise from local isotropic interactions is an interesting one, it is possible to encode explicit anisotropy at the single-particle level. Colloidal particles that are anisotropic in both shape and surface chemistry are now easily synthesized. Several routes can achieve this, for example, selective sintering, hybridization of DNA oligomers, electrodeposition,[24] emulsion encapsulation and shrinkage[25] and many more. The goal is eventually to be able to develop building blocks (colloidal molecules)[26] with tailored anisotropy in shape and chemistry whose interactions with other components in solution result in the spontaneous formation of desired macro- or microscopic aggregates. To this end, it is necessary to gather a sufficient understanding of the forward self-assembly process to then be able to sort out which interparticle interactions lead to what target structure. With this in mind, there is a clear necessity to explore a new dimension in the classical temperature–concentration phase diagrams: the geometry of the interactions; see reference 27 for a perspective on the subject and a review of the relevant literature.

In this respect, Janus particles represent what is possibly the simplest experimental and theoretical model where the effect of anisotropy in the pair potential can be studied in detail. By 'Janus particle' we usually refer to spherical colloidal particles (although non-spherical geometries have also been synthesized such as Janus rods, dumbbell structure, *etc.*), having the two hemispheres of the surface chemically treated independently to achieve different physical properties. The concept of Janus particles is quite generic and does not refer specifically to any particular pair of chemical properties; similarly, it can be extended to non-symmetric interaction geometries where the Janus balance (the locus separating opposite regions) does not lie on an equatorial line.

In this chapter, we summarize recent experimental and numerical results obtained with systems of dipolar and amphiphilic Janus spherical particles.[28,29] In the first case, the two hemispheres have opposite charge, representing in a way the simplest colloidal analog of a spherical dipole, whereas in the second case, one side is hydrophobic and the other is polar, representing a spherical surfactant. Apart from the simplicity of the system, what makes it particularly interesting is that the range of the interactions can be easily tuned by altering the salt concentration in solution, leading to complex phase behavior and dynamic heterogeneities. However, this simple symmetry breaking at the single colloidal level brings significant theoretical challenges. Separation between particles is no longer the only parameter determining the interaction for Janus particles. Instead, the interactions become separation and angular dependent and computer simulations become an attractive route to approach such a complex system.

It is also worth stressing that although Janus particles can be viewed as simple model systems of their molecular counterparts, the two systems are fundamentally different. For example, in small molecular systems such as surfactants, molecules often go through configurational changes when transitioning from a gas phase to a more ordered self-assembled structure. Furthermore, these configurational changes, typically involving C–C bond rotations, require small activation energies but lead to fast kinetics towards minimum energy states of the aggregates. In contrast, in colloidal systems, the individual particles can be considered as rigid, which implies that no morphological changes will affect the assembly process. Hence the kinetic pathway associated with the aggregation of these systems is expected to be significantly different and should rely on much slower dynamical processes such as hopping or sliding between the particles. This will be one of the main areas of discussions in this chapter.

The first two sections of this chapter detail synthesis strategies and numerical modeling employed to study the physical behavior of Janus particles. The subsequent two sections discuss experimental and numerical results on self-assembly of spherical dipolar and amphiphilic particles. The next section deals with extensions of the Janus model and is followed by a discussion of possible applications.

7.2 Numerical Methods: Modeling of Janus Particles

In this section, we discuss simple models of spherical Janus particles. We begin with the case of dipolar and then proceed to the case of amphiphilic Janus particles.

7.2.1 Dipolar Janus Particles

Dipolar Janus particles are characterized by two identical hemispheres having the same charge density but opposite sign. Given these geometric features, one would naively expect that a dipolar Janus particle could be described as a simple dipole oriented along the axis joining the two hemispheres, plus a generic isotropic excluded volume term to account for the spherical shape of the particle. This assumption, usually very reasonable when dealing with long-range interactions and typically interparticular separations much larger than the particles' diameter, becomes inadequate at very short distances. It turns out that the physically more interesting and relevant phase behavior occurs for moderately high salt concentrations for which the interaction range between the particles becomes just about 10% of the particle diameter and the dipolar assumption, which would naively predict head-to-tail chain formation, fails to characterize their phase behavior. The reason is that at such short ranges, we are basically dealing with a contact interaction that has a much more complex angular dependence than the dipolar one. For instance, when two oppositely charged hemispheres coming from two different particles are in contact, provided that the contact region is far from the location of the Janus boundary of either particle, one can completely ignore the presence of the other pair of hemispheres in each particle and treat the interaction as a rotationally invariant attraction that depends only on the number of charges near the contact region. However, as soon as one of the two particles exposes the Janus balance to the other, a complex angular dependence arises that unfortunately does not depend exclusively on the relative orientation of the particles, but also on the absolute orientation of one of them. The net result is a rather intricate multidimensional angular dependence of the pair potential.

To extract adequate pair potentials for the dipolar Janus character of these components, one can construct an explicit model for the Janus surface: a spherical particle decorated with 12 002 points charges distributed homogeneously over its surface (Figure 7.1e). The large number of surface particles ensures a smooth representation of the charge distribution on each hemisphere. At each point an effective charge is associated to match the estimated surface density in the experimental system. Two hemispheres are then identified and opposite charges are associated with them. Finally, two such particles are brought near each other and the direct/explicit pair potentials obtained by summing the contribution of every single charge in the system are computed as a function of the particle orientations and for different separations. This is a time-consuming procedure that gives great insight into the functional form of the pair potential but cannot be directly applied to full simulations of large number of Janus particles. Fortunately, the problem can be greatly simplified in the short-range/high salt concentration limit. In this limit, the overall curvature of the particles can be ignored and if one can simplify the charge–charge interactions with a square-well or a square-shoulder potential depending on the relative sign of the charges, the problem is reduced to a simpler geometric one.

In this representation, the absolute value of the interaction strength becomes trivially related to the number of charge involved in the interaction, *i.e.* the surfaces of the spherical cap $S = 2\pi Rh$, where R is the radius of the particle and h goes from zero, when the particles are further than the range of interaction R_{cut}, to R_{cut}, when they are in contact. This quantity can be actually tabulated and fitted to a simple functional form to account properly for all the cross-terms in the interactions.

To characterize the angular dependence of the potential and modulate the strength of the interaction by an angular factor $\alpha(n_1, n_2)$ dependent on the particle axis n_1 and n_2, one can now project the charge content of each of the two spherical caps involved in the interaction on two circular surfaces and measure the direct surface-on-surface charge overlap. To find these surfaces given two particles 1 and 2, one initially finds the axis joining their centers, $r_{12} = r_2 - r_1$, then one determines the plane normal to this axis and projects on to this plane the region of the charged surface of particle 1 that satisfies the constraint $n_1 \cdot r_{12} > 0$ and the region of the charged surface of particle 2 satisfying the constraint $-n_2 \cdot r_{12} > 0$. Figure 7.2 shows a schematic illustration of this procedure, which allows us to define $\alpha = [S(q_1 = q_2) - S(q_1 \neq q_2)]/S_{tot}$, where q_1 and q_2 are the charges on surfaces 1 and 2, $S(q_1 \neq q_2)$ is the overlapping area of charges having opposite sign and $S(q_1 = q_2)$ is the overlapping area of charges having the same sign and S_{tot} is the surface of the whole overlapping region. By construction $\alpha \in [-1,1]$. We have now created all the tools necessary to compute efficiently the pair interaction between two dipolar Janus particles modeled as simple spherical objects with an effective pair interaction. The excluded volume term is enforced via a hard core potential. Figure 7.1 shows the angular dependence of the effective potential for different orientations of the particles that involves a

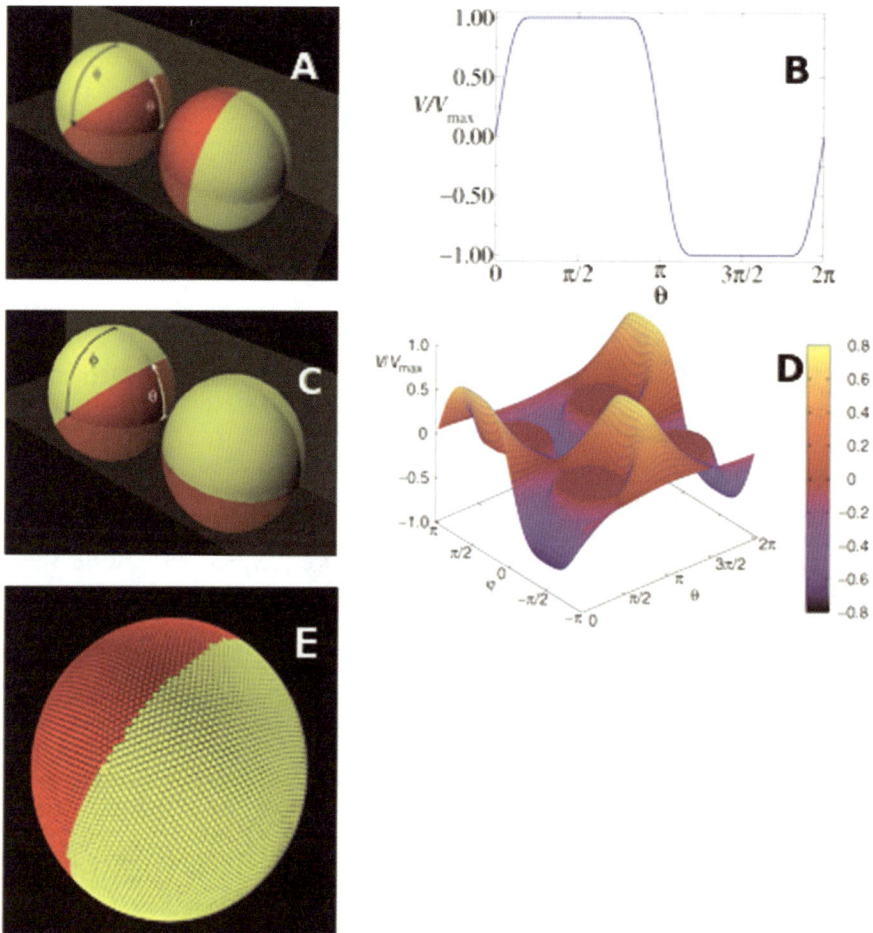

Figure 7.1 Two particles with bipolar charge, analyzed using simulations. The right-hand object is fixed; the angles (θ, φ) indicate orientation of the left-hand object. Red denotes positive charge, yellow denotes negative charge; these charges are equal in magnitude. (a) Charged hemispheres whose polar axes lie in the same plane. (b) Relative interaction energy for case (a), plotted against angular displacement for two particles at contact. The interaction potential switches from repulsive to attractive as the left-hand object rotates in the θ direction even when the particle–particle separation does not change. (c) General case where the polar axes fail to lie in the same plane. (d) For every orientation of the right-hand particle, a potential energy landscape must be determined. Thus, the parameter space become four-dimensional even at fixed surface separation, illustrating the rapid increase in complexity once one goes beyond the concept of isotropic potentials. The potentials shown were obtained numerically using a pair of composite particles having 6001 fixed and uniformly distributed 'nanoparticles' per hemisphere, representing the surface charges (e). Each pair of nanoparticles interacts via an attractive or repulsive square potential (representing their screened electrostatic interaction), according to the sign of both hemispheres. We find that the

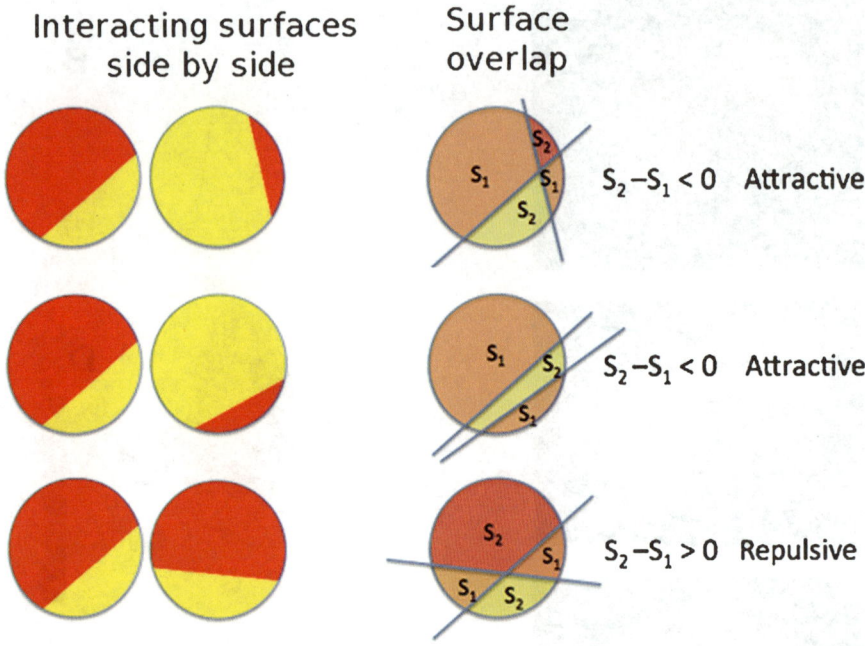

Figure 7.2 Sketch of how the sign and strength of interaction between two Janus dipolar particles facing each other are computed. The left column shows the two interacting spherical caps in three different scenarios. The right column shows the same surfaces on top of each other. S_1 corresponds to $S(q_1 = q_2)$, *i.e.* the total overlapping surface area of equally charged regions, and S_2 corresponds to $S(q_1 \neq q_2)$, *i.e.* the total overlapping surface area of differently charged regions. The relative size of the two contributions determines the total interaction.

complex multidimensional energy landscape. One can explicitly verify that variation of the number of surface particles does not change the shape of this potential, but merely rescales the contact energy of a pair of Janus particles. The magnitude of the colloidal potential quickly decays to zero with increasing surface-to-surface distance.

It should be noted that the flat region of the potential for large sections of the angular range is key to understanding the phenomenological behavior, in terms of both statics and dynamics, of the experimentally observed Janus clusters.

> shapes of the resulting clusters are insensitive to the precise range of the nanoparticle potential (and hence to the precise salt concentration), provided that this range is less than 30% of the colloid diameter. Since the equilibration time increases rapidly with decreasing interaction range, we have adopted a range of 10% of the particle size. The magnitude of the resulting *colloid potential* quickly decays to zero with increasing surface-to-surface distance. Reprinted with permission from reference 29. Copyright 2006 American Chemical Society.

Most of the numerical simulations described in this chapter were carried out using Monte Carlo simulations in the NVT ensemble with periodic boundary conditions. Potentials with contact interactions between 5 and $10k_BT$ were explored and systems with up to $N = 400$ particles were considered. The minimum-energy states were reached fairly quickly, but to ensure equilibration all runs were continued for a minimum of 1×10^6 Monte Carlo steps. Once the system had self-assembled into clusters of different sizes, the analysis of their structure was carried out and the results were compared with the experimental images and also with the known Lennard-Jones clusters of equal size.

To check for the stability of each cluster of size n, we performed numerical simulations where only n particles were placed within the simulation box and were allowed to self-assemble. By judiciously increasing the system temperature and allowing the clusters to sample all accessible configurations, it is possible to enumerate a great number of different structural rearrangements for each cluster size. Each of these clusters was then quenched to a minimum-energy state by decreasing the temperature using a standard simulated annealing technique. Finally, the energies of all resulting clusters were compared to find those with absolute minimum energy. While such an approach becomes inefficient for large n, as the number of accessible configurations grows exponentially with the cluster size, our analysis only included rather small clusters, giving us confidence that the structures presented below are indeed the most stable ones.

7.2.2 Amphiphilic Janus Particles

There are several ways of modeling the interactions between amphiphilic Janus particles. One of the main problems is that unlike the dipolar system, where one can use the superposition principle to compute the total potential as the sum of the interaction between every single charge in the system, the details of the hydrophobic interaction and in particular how it scales with the size of the hydrophobic surface are not fully understood. We expect it to be attractive and with a range that extends up to 100–200 nm from the particle surface.[30] Several functional forms for the pair potential have been put forward and usually involve a sharp exponential attraction at short distances and a somewhat weaker decay for larger separations; however, it is fair to say that we do not have a definite way of accurately treating the hydrophobic force between two surfaces in a pairwise manner. Given that extracting the pair potential by performing a full atomistic simulation with explicit water surrounding micron-sized particles would require a series a simulations with millions of water molecules, a typical choice is to describe hydrophobicity with a generic attractive potential and scan over a range of interaction strengths and ranges.

Within these limits, modeling the angular dependence of the amphiphilic Janus particles can be approached along the same lines as described for the dipolar case. In this case we used a hard-sphere potential for the excluded volume interactions and an isotropic Yukawa potential $V_E = A\exp[-\kappa(r - \sigma)]/r$ for $r < r_c$ and 0 otherwise, for the polar region, where r is the particle separation and σ is the diameter of the colloidal particle. The range r_c of this potential can be tuned by altering the Debye screening length κ. The hydrophobic part of the interaction is modeled using a generic attractive pair potential with exponential dependence on the particle separation and with angular modulation analogous to that described for the dipolar Janus case. The main difference is that now $\alpha \in$ [0,1] and is defined as $\alpha = S(H)/S_{tot}$, where $S(H)$ is the overlapping area of the hydrophobic regions $V_H(r, S_{eff}) = -B\alpha\exp[-\lambda(r - \sigma)]$, where λ is a constant determining the range of the interaction and B is the strength of the interaction.

All numerical data on Janus particles presented in this chapter were also obtained using Monte Carlo simulations in the *NVT* ensemble with the model described above. Because of this simplified model, we were able to study systems containing relatively large numbers of particles for a large number of steps, which is essential to ensure that stable configurations are reached. The minimum-energy states were reached fairly quickly, but again to ensure equilibration we continued all runs for a minimum of 1×10^6 Monte Carlo sweeps.

It should be stressed that the more interesting phase behavior turned out to occur for salt concentrations for which the Debye screening length is significantly smaller than the particle diameter. In this limit, the electrostatic potential can actually be ignored and one can treat the particle as a hard sphere with an angular-dependent short-range attractive interaction. The angular potential described above takes great care in describing the interaction potential near the Janus balance of the particles and we used it successfully to understand the experimental results obtained with amphiphilic Janus particles. However, its generalization to more complex interaction geometries can be cumbersome. For this reason, the preferred model of choice for amphiphilic particles is the patchy model. This model was first introduced by Kern and Frenkel[31] to understand the effect of surface anisotropy in simple models of protein crystallization and has been used extensively in models for particle self-assembly (see reference 32 for a recent review on the subject). It consists of a hard-sphere potential plus an attractive square-well potential multiplied by a step-function-like angular dependence and has the form

$$\psi(\hat{\mathbf{n}}_1, \hat{\mathbf{n}}_2, \hat{\mathbf{r}}_{12}) = \begin{cases} 1 & \text{if } \hat{\mathbf{n}}_1 \bullet \hat{\mathbf{r}}_{12} \geq \cos\theta_0 \text{ and } -\hat{\mathbf{n}}_2 \bullet \hat{\mathbf{r}}_{12} \geq \cos\theta_0 \\ 0 & \text{otherwise} \end{cases} \tag{7.1}$$

were $0 < \theta_0 < \pi$ is an angle describing the angular size of the hydrophobic region. Figure 7.3 shows a sketch of two Janus ($\theta_0 = \pi$) hydrophobic particles facing each other and forming the angles Ω_1 and Ω_2 with respect to the axis joining their centers. Although the patchy model describes incorrectly the complex angular potential landscape associated with the orientation of the Janus particles, it properly reproduces the equilibrium features of the experimental system, it is easy to code and is computationally extremely efficient. Several extensions of this potential for use in molecular dynamics

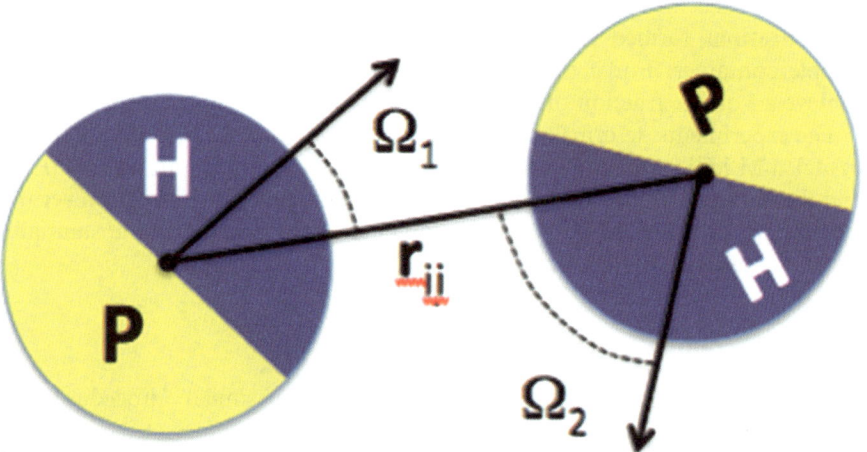

Figure 7.3   Sketch of two amphiphilic Janus particles at a distance r_{ij} from each other forming angles Ω_1 and Ω_2, respectively, with respect to the axis connecting their centers.

simulations have also been proposed. More details on patchy particles can be found in Chapter 6.

7.3   Experimental Methods

In this section, we discuss the experimental methods used to produce dipolar and amphiphilic Janus particles.

7.3.1   Dipolar Janus Particles

The zwitterionic colloids relevant for this chapter were produced using methods described by Hong *et al.*[29] A suspension of fluorescently labeled monodisperse colloidal particles is spread on a cleaned glass substrate such that a monolayer of colloids remains after the suspension liquid has evaporated. A thin (15 nm) gold film is then deposited using electron beam deposition. Monolayers of *N,N,N*-trimethyl(11-mercaptoundecyl)ammonium chloride are deposited from ethanol solution to produce positive charges and are washed multiple times with 1% solution of HCl in ethanol to remove the electrostatic adsorption. On the other hemisphere, negative charges results from carboxylic acid groups on the untreated side of carboxylate-modified polystyrene colloids. When this procedure is performed with care, the resulting particles present a nearly balanced opposite charge distribution on the two hemispheres, and this can easily be checked by measurements of the zeta potential. For example, the zeta potential of bipolar particles is 0.6 mV, whereas it is –41.6 mV for bare particles at the same salt concentration. At this point, the particles are ready to be suspended in aqueous solution and epifluorescence microscopy can be used to image regions of relatively high concentrations formed when the particles settle at the bottom of the glass sample container. For a thin volume of ~ 5 μm from the surface, one can achieve a volume fraction of $\sim 10^{-3}$.

The experiments described in this chapter were conducted in PBS buffer (pH 6) at 1 mM ionic strength, such that the Debye screening length was ~ 10 nm and the particles (F8819 from Invitrogen) had a diameter of 1 μm; however the particle size is not believed to be fundamental, provided that it substantially exceeds the screening length.

7.3.2 Amphiphilic Janus Particles

Amphiphilic Janus particles were created with a similar procedure. As described for zwitterionic particles, a suspension of (fluorescent) carboxy-late-modified polystyrene spheres was spread on a cleaned glass slide such that a monolayer of colloid remained after the suspension liquid had evaporated. A thin (15 nm) film of gold was then deposited using electron beam deposition on top of a titanium adhesion-promoting layer (2 nm). On to the gold hemisphere surfaces, monolayers of octadecanethiol (ODT) were deposited and washed multiple times first with 1% solution of HCl in ethanol and then with deionized water to remove non-specific adsorption. The other, untreated, hemisphere remained coated with negative charge from carboxylic acid groups on the parent spheres.

The resulting amphiphilic spheres were then removed from the surface by ultrasonication and used directly in control experiments to confirm the generality of the findings presented below. In most experiments the density of negative charge was increased by grafting DNA oligomers using a conjugation reaction.[33] In brief, the carboxylic acid groups on the particles surfaces were activated using 1-ethyl-3-(3-dimethylaminopropyl)carbodiimide hydrochloride (EDC) in 2-morpholinoethanesulfonic acid at pH 4.5 using 2-(*N*-morpholi-no)ethanesulfonic acid (MES) buffer solution, then these activated groups were allowed to react with the amino terminus of DNA. It was convenient to use the DNA heptamer 5′-/TAC GAG TTG AGA TTT TTT TTT T/iSP18/ 3AmM/-3′. After reaction, 2% of Tween 20 solution was used to wash away unreacted DNA molecules and the resulting particles were washed copiously. A control experiment showed that the other hemisphere remained hydro-phobic. Note that after DNA grafting, each of the 22 bases carried one negative charge, resulting in a very high surface charge density.

A new method was developed to prepare Janus particles based on Pickering emulsion approaches.[8] A typical process is sketched in Figure 7.4. Silica particles first dispersed in paraffin wax and then dispersed in water at elevated temperatures concentrate on the wax/water interfaces to minimize the surface energy. This dispersion is then cooled to room temperature so that the wax can provide sufficient protection for the part of surface area of the silica particles that it covers, thus allowing the exposed area to be selectively modified using

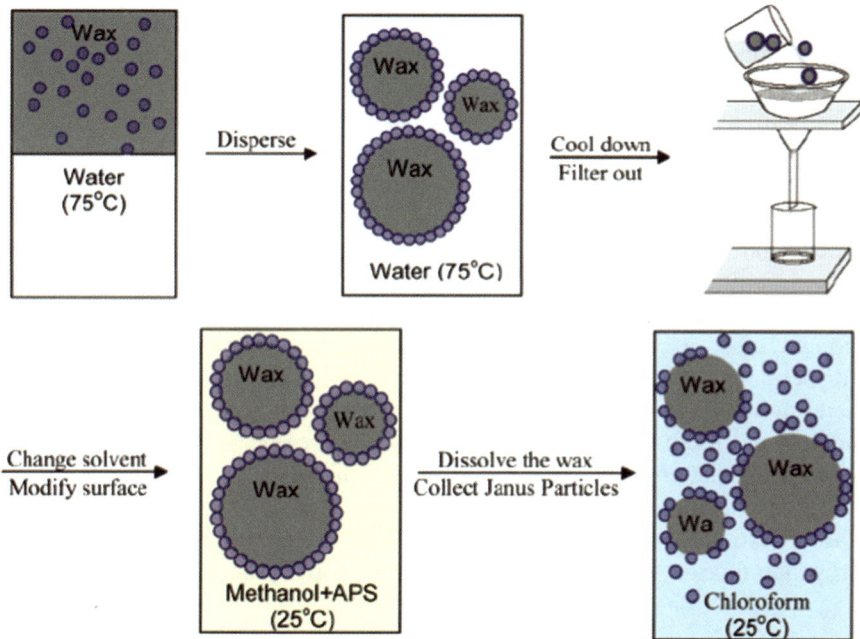

Figure 7.4 Schematic diagram of the procedure to create Janus particles by functionalizing particles adsorbed on an emulsion of water and oil, then cooling the sample so that the oil crystallizes to form a wax.[8] Reprinted with permission from reference 8. Copyright 2006 American Chemical Society.

silane chemistry. After the modification, the wax is dissolved to release the modified silica Janus particles. An advantage of this approach is that it could be used to control the Janus balance by controlling the three-phase contact angle on silica particles at wax/water interfaces. By selecting the appropriate water and wax phase with additives such as surfactants, three-phase contact angles can be controlled and produce particles with desired Janus balance.

7.4 Experiments on and Simulations of Janus Self-assembly

In the following sections we present the experimental results on the phase behavior and morphological properties of the different aggregates formed by dipolar and amphiphilic Janus colloidal particles in an aqueous solution. A comparison with numerical data is also given in some detail. Before we begin to describe the experimental details of these studies, it is useful to present some discussion considering the peculiar nature of the pair potential acting between these particles as extracted by numerical methods. Figure 7.1 illustrates the potential in all its complexity. To get a feeling for it, consider a polar axis pointing north–south through the hemispheres. When the axes of both particles lie in the same plane as illustrated in Figure 7.1a, we have a smooth interpolation of the interaction from repulsion to attraction depending on the relative orientation of the particles. Figure 7.1b shows the interaction potential relative to the configuration in Figure 7.1a. However, when the polar axes are not coplanar (Figure 7.1c), the interparticle potential will depend on the orientation of both particles as well as their separation, leading to a complex multidimensional parameter space. Figure 7.1d shows how the pair interaction depends on the two polar angles describing the orientation of one particle, for a *fixed* orientation of the second particle and at a fixed particle separation. The most important aspect of this analysis is that the interaction space can be thought of as a complex energy landscape in the true sense of the word. Furthermore, it is important to point out that one of the main features dictating the local dynamics observed in experiments, *i.e.* the high structural mobility of the clusters, is directly related to the presence of large gradient-free regions in the angular potential (typical for particles in the size range considered in this work at high salt concentrations). In fact, this particular feature allows particles to easily slide and wriggle over one another by Brownian motion, towards an efficient search for low-energy configurations.

7.4.1 Dipolar Janus Particles

As explained above, a crucial condition in the experimental setup is that the particle size of 1 μm greatly exceeds the electrostatic screening length of ~ 10 nm. Based on the manufacturer's specification and measurements of the zeta potential of unfunctionalized particles, one can estimate a surface charge on the order of one elementary charge per square nanometer. One of the main results, readily observable in epifluorescence images, is that at low densities, particles self-assemble into well-defined aggregates whose morphology is strongly dependent on the number of particles forming that cluster. This result is indeed very different from the linear aggregation that one would naively expect by mapping Janus dipolar particles to a system of dipoles. The different geometries as a function of particle number are shown in the Experiment column in Figure 7.5. Clusters larger than 12 particles have not been observed in experiments because of the low particle concentration considered so far; however, a low-energy configuration for a larger number of particles can be obtained numerically. A 13-particle cluster, for instance, is also shown in Figure 7.5.

Figure 7.6 shows a few examples of these dynamic rearrangements within small clusters. Specifically, it shows local cluster rearrangements in (a) a tetramer, (b) a hexamer and (c) a heptamer. A sketch of the particle moves is also shown on the right.

Numerical simulations of zwitterionic Janus particles, modeled according to the interaction potential previously described, generated exactly the same sequence of clusters as a function of particle number that was observed experimentally. The results are also shown in Figure 7.5, Simulation column. Because it is not easy to access directly the orientation of the particles in the

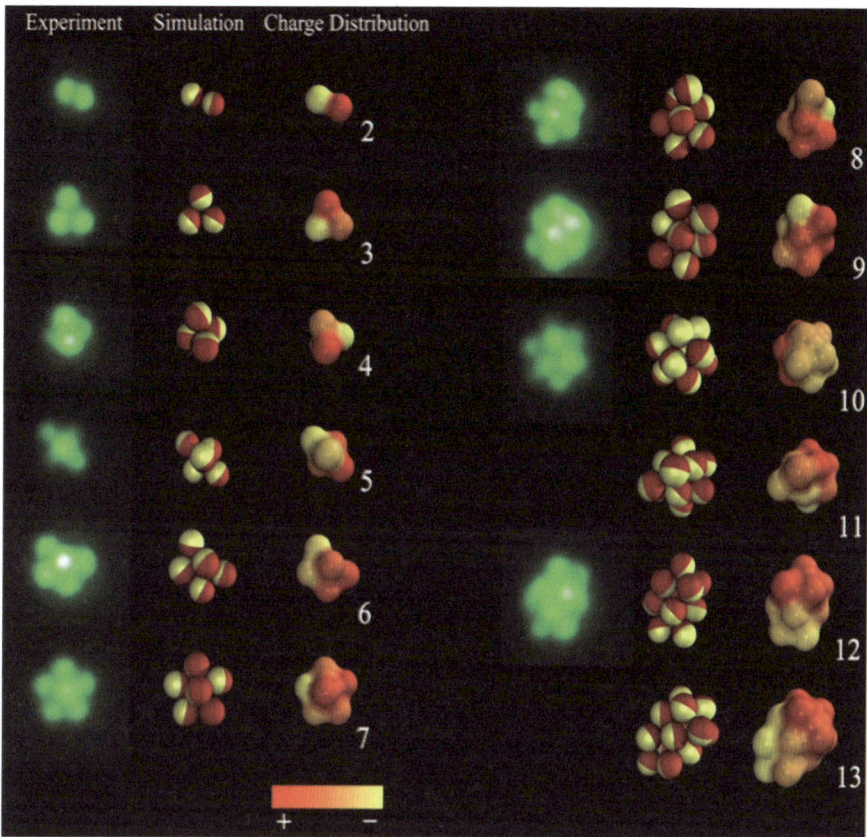

Figure 7.5 Comparison of experimental epifluorescence images and Monte Carlo computer simulations of the self-assembled structures of particles with near-equal positive and negative charges on the two hemispheres (denoted by red and yellow colors). Clusters form with monotonically decreasing energy as the number of colloidal particles increases. Their computed structures (Simulation column) agree quantitatively with the observed structures (Experiment column). The charge distribution in these clusters is also computed; the color goes smoothly from red to yellow depending on the inner product of the vector from the center of mass of the cluster to each colloid and the axial vector (pointing to the positive side) of this colloid (Charge Distribution column). In the charge distributions, a wrapping has been added to emphasize the outer surface. As described in the text, experimental measurements of zeta potential showed it to be <1 mV, confirming that charge on two hemispheres was nearly balanced. Reprinted with permission from reference 29. Copyright 2006 American Chemical Society.

experiments, the typical charge distribution of the clusters was extracted from the numerical data.

For dimers, one finds an ensemble of degenerate low-energy configurations, where the hemispheres of opposite sign face each other, but the large gradient-

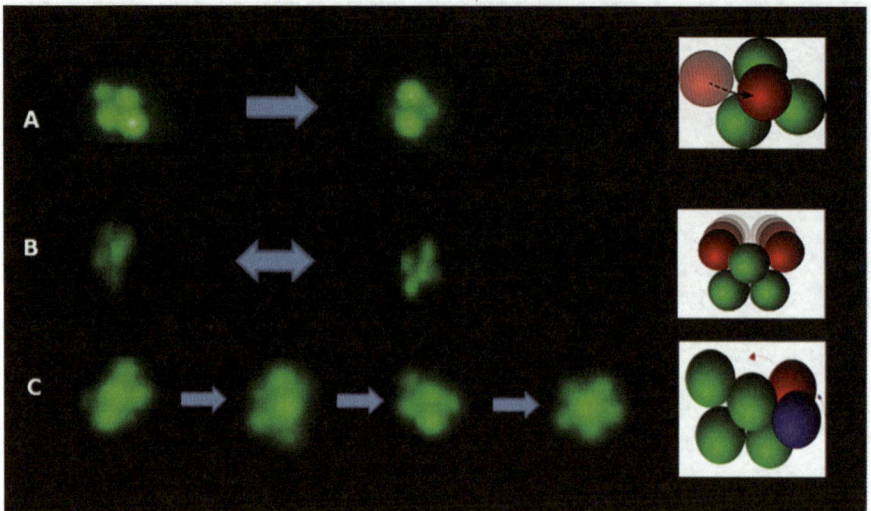

Figure 7.6   Examples of dynamic rearrangements within small clusters, showing local
cluster rearrangements in (a) a tetramer, (b) a hexamer and (c) a
heptamer. The particle moves are shown on the right.

free angular regions in the pair potential allow for significant wiggling of the
two particle axes. Clusters larger than $n = 2$ are subject to stronger angular
constraints; typically one finds that the polar axes of the hemispheres twist in
space as neighboring particles circle the structure. As a result, the charge
distribution is not equally distributed over the cluster surface. By projecting
the charge of each particle over the surface of a sphere encapsulating the entire
cluster, it is possible to see that indeed the clusters retain the overall dipolar
character of the single components. Figure 7.5, Charge Distribution column,
shows the result of these calculations for all the cluster sizes considered.

Another intriguing result that can be readily computed in numerical
simulations is that the total energy of the clusters is a decreasing function of
their size. The overall charge asymmetry coupled with the trend in cluster
energy indicates that if experiments were performed at higher concentrations,
smaller clusters would fuse together to produce larger ones. If we think of the
clusters as particles in their own right whose self-assembly would lead to larger
mesoparticles, this scenario suggests a hierarchical (or perhaps fractal) multi-
stage self-assembly process for larger clusters. Unfortunately, the low particle
concentrations prevented us from probing this scenario in experiments.

It should be stressed that the morphological properties of the clusters formed
by Janus particles presented in Figure 7.5 are very different from those observed
in several studies of either magnetic[34] or electric dipoles[35] where linear or
circular aggregates are routinely observed. The main reason for this resides in
the very short range of the electrostatic interactions for the relatively large
nanoparticles considered in our study. Linear aggregation is expected whenever
the range of the interaction becomes larger than the particle diameter. It is

however instructive to compare the zwitterionic Janus clusters to those formed by isotropic microspheres aggregating within an evaporating liquid emulsion droplet. Pine and coworkers have shown that in this case the resulting clusters pack in a way that minimizes the second moment of the mass distribution[36].

Unlike zwitterionic Janus particles that spontaneously assemble into clusters of different sizes starting from a dilute solution due to the anisotropic nature of their mutual interaction, the stabilization of isotropic microspheres is due to the van der Walls interactions felt by the colloids when mechanically brought into contact with each other by the shrinking/evaporating liquid droplets[36]. Apart from the orientational degrees of freedom that are obviously not a factor in isotropic particles, the positional ordering of the particles in these clusters was indistinguishable from that reported in Figure 7.5 for $n=2$–5 and $n=7$. Systematic differences are observed for clusters containing more than 7 particles, but for $n=7$ to $n=10$ dipolar Janus particles reproduce the clusters formed by unconstrained Lennard-Jones particles. Finally, zwitterionic clusters formed by 6, 11 and 13 particles are not matched by any other cluster formed by isotropic particles as the former tend to aggregate into less symmetric structures.

Apart from the intriguing geometric properties of some of these clusters and their overall dipolar character, what is probably their most fascinating feature concerns the dynamic intricacies leading to their formation. Real-time imaging shows that aggregation of zwitterionic particles proceeds according to a complex hierarchy of different sub-processes. The particles initially assemble into small clusters of three to four components and these clusters coalesce at a later time into larger clusters. The typical time scales involved in the process are of the order of seconds to minutes, suggesting that the interactions between these particles are relatively weak. Estimates from numerical simulations suggest binding energies of the order of 5–$10k_\mathrm{B}T$. This experimental approach will allow comparison with prior computer-based calculations of how patchy particles associate[37].

The snapshots corresponding to the tetramer and hexamer and a heptamer are shown in Figure 7.6 and illustrate how two or more particles rearrange to minimize the energy of the clusters. In this case, equilibrium is reached on a time scale that ranges from 10 s to a few minutes and could be even longer and involve more particles for clusters of larger size. For a suspension at much larger particle volume fraction, one can envision greatly extended relaxation time scales, which may lead to unusual structural and dynamic behavior compared with the formation of large clusters of isotropic particles. This scenario is supported by numerical simulations.

As we shall see in the next section, this breathing dynamics of the clusters between different conformations is very similar to that observed in amphiphilic particles, but the reaction kinetics are still not well understood. Unlike the amphiphilic case where charged surfaces repel each other and limit the size the clusters, zwitterionic clusters, in principle, do not have such a limitation. Furthermore, in the amphiphilic system the electrostatic repulsion is mostly decoupled from the hydrophobic interaction, so one can tune the repulsive forces by controlling the ionic strength while maintaining the attractive hydrophobic interactions in a range of a few $k_B T$. For zwitterionic particles, the electrostatic interactions are more convoluted, making controlled experiments on the cluster dynamics much more difficult.

7.4.2 Amphiphilic Janus Particles

Amphiphilic Janus particles present a structural and dynamical behavior that is as rich and diverse as that observed in systems of zwitterionic particles. However, unlike zwitterionic particles for which aggregation is expected at all salt concentrations as one forms linear aggregates in the low-salt *dipolar* limit and finite size zwitterionic clusters in the Janus high-salt limit, amphiphilic particles present the opposite trend with ionic strength. This is illustrated explicitly in Figure 7.7, where the structural properties of the aggregates formed by amphiphilic particles are shown for different salt concentrations. In deionized water, the electrostatic interparticle repulsion is so strong that no aggregate is seen in experiments or in simulations (Figure 7.7a). When the ionic strength is raised so that the electrostatic screening length is reduced to about 10 nm (Figure 7.7b), experiments and simulations concur in finding small clusters. They rearrange dynamically (see below) and contain various numbers of particles, $N = 2$–12, but they always retain a compact shape. When the electrostatic screening length is further reduced and becomes significantly smaller than the range of the hydrophobic interaction, colloidal aggregates with a significantly different morphology are observed (Figure 7.7c). In this regime, the number of particles in each aggregate is larger by at least an order of magnitude and their morphology resembles that of extended, branched rods. Because real-time microscopy also reveals shape fluctuations as the rods change contour by Brownian motion, we refer to them as 'worm-like' clusters.

In this latter case, it is fair to assume that the electrostatic interactions have been screened to such an extent that the hydrophilic region of the particles plays no role in the observed phase behavior. Indeed, Monte Carlo simulations performed at similar volume fractions and interaction strengths show that, when a hard wall replaces the amphiphilic hemisphere of the particles, analogous linear structures do indeed form. In this regime, the pair potential of amphiphilic Janus particles, and in particular its angular dependence, presents features that are similar to those computed for dipolar Janus particles; it is therefore reasonable to expect intricate dynamics underlying their aggregation. Indeed, real-time imaging reveals such multi-step processes as particles self-assemble.

To gain more insight into how the morphology of the self-assembled structures depends on the solution's ionic strength, it is useful to estimate numerically the energy cost associated with the electrostatic interaction as a function of the relative rotation between two particles. Clearly, in the low-salt limit, the electrostatic repulsion between the particles is minimized when they are placed head-on with the particles' axes parallel to the axis connecting their centers. In the opposite limit, at very high salt concentration, we expect a very

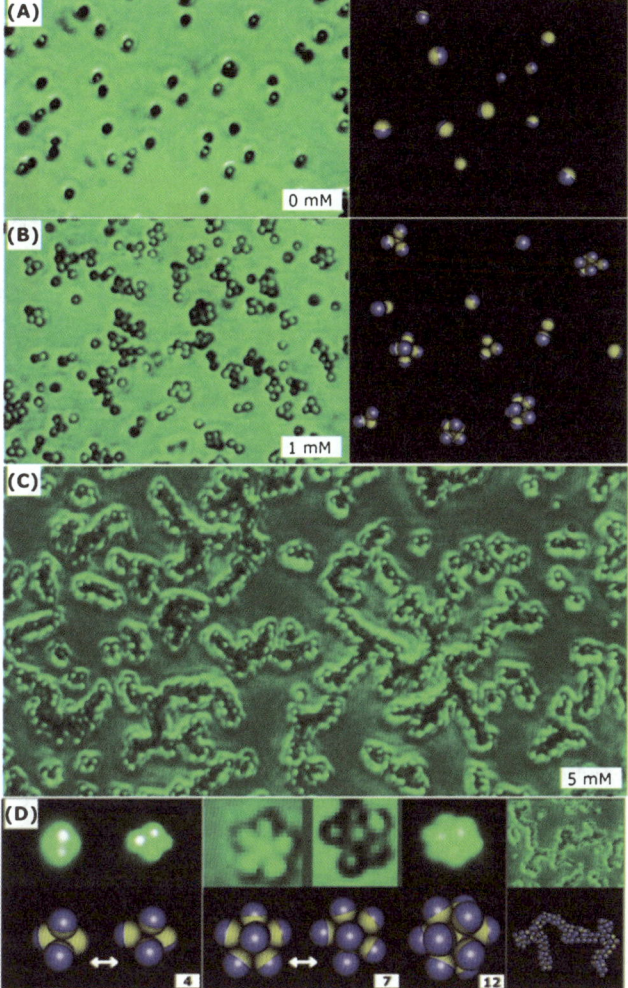

Figure 7.7 By tuning the salt concentration, particles hydrophobic on one hemisphere and charged on the other (denoted for epifluorescence experiments by images with a green background and for simulations by red and yellow colors) assemble into clusters of different size and shape. (a) Discrete particles (deionized water, no salt added); (b) small clusters in equilibrium with one another, icosahedra and smaller (1 mM KNO₃); (c) long, branched strings (5 mM KNO₃) (d) comparison of experimental epifluorescence images and Monte Carlo simulations with excellent agreement. Experimentally, clusters with the same number of particles (N) interconvert dynamically between different shapes, as confirmed from simulations for tetramers, N = 4 (left), and heptamers, N = 7 (middle). The icosahedral structure, N = 12, is also confirmed (middle) and the assembly of small clusters into worm-like, branched strings is confirmed to occur as the range of electrostatic repulsion, relative to hydrophobic attraction, decreases with increasing salt (right). In experiments, the diameter of the silica particles is 1 μm. Reprinted with permission from reference 28. Copyright 2008 American Chemical Society.

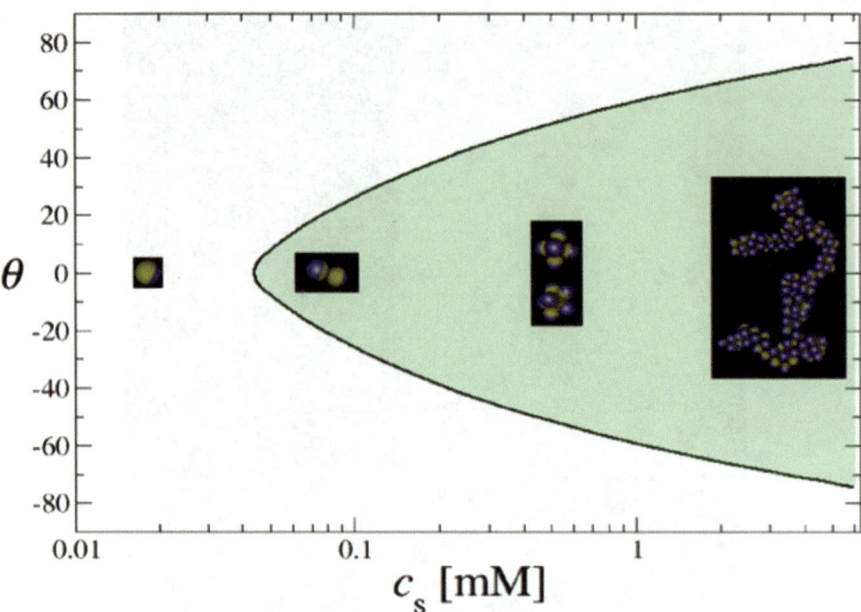

Figure 7.8   Region of 'permitted' tilt angles between two charged hemispheres held at
a center-to-center distance of 1.020 μm as a function of salt concentration.
The rotational boundary is determined by requiring the electrostatic
energy cost per simultaneous rotation of the particles of an angle θ to be
smaller then $1k_BT$, As a reference, $\theta = 0$ refers to the case where the
hydrophobic hemispheres face each other and the particles' polar axes are
parallel to the vector joining the centers of the particles. Reprinted with
permission from reference 28. Copyright 2008 American Chemical
Society.

limited (if any at all) angular dependence in the pair potential due to the
hydrophilic region. By placing two particles next to each other and changing
the relative orientation of their axes, it is possible to calculate numerically the
angular contribution of the electrostatic repulsion to the total energy for
different screening lengths. Figure 7.8 shows the $1k_BT$ line of permitted angles,
θ, and its dependence on the morphology of the self-assembled structures (see
Figures 7.9 and 7.10 and their captions for detailed information on the
calculation of this diagram.)

As expected, at low salt concentration the electrostatic repulsion is so strong
that the particles do not form any aggregate. As the salt concentration increases
and the electrostatic screening becomes more significant, particles can approach
each other to a sufficiently close distance to feel their hydrophobic hemispheres.
This also coincides with a weakening of the angular dependence of the potential
and leads to the formation of dumbbells. The onset screening length for this
particle size at the reported surface charge density is around 0.1 mM, which in
qualitative agreement with the experiments. At even higher concentration,
thermal fluctuations allow the particles to explore a wider range of angles when

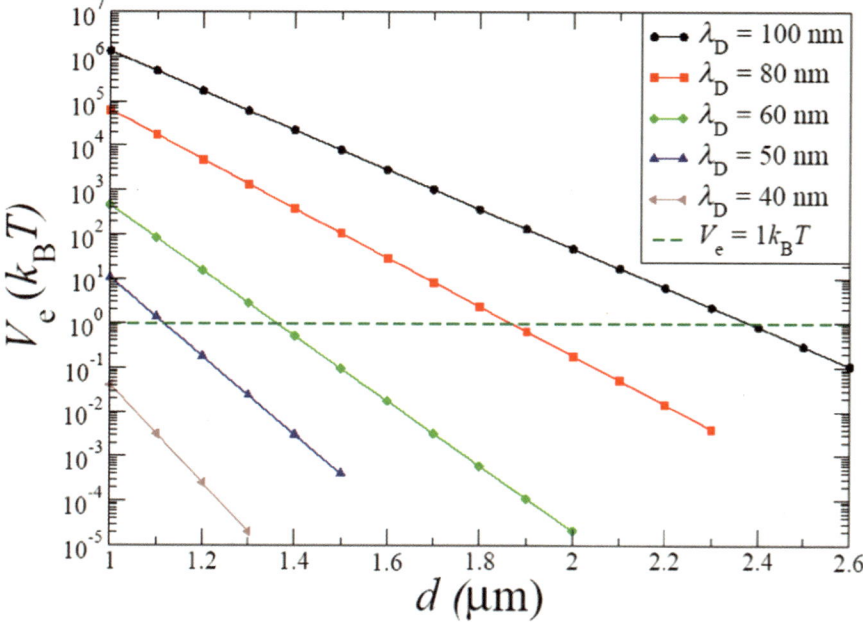

Figure 7.9   To calculate the phase diagram, the atomistic electrostatic energy between two hemispherical charged particles whose hydrophobic sides are facing each other, as a function of interparticle separation for different values of the Debye length, was calculated; the charge density was chosen according to the experimental conditions. d is the separation between two centers of the particles; λ is the Debye length, which was chosen based on experimental conditions; the dashed line represents $1k_{\mathbf{B}}T$, where the repulsive energy is at the same level of thermal energy. This figure shows that when the Debye length is longer than 30 nm ($C_s < 0.1$ mM), the repulsive energy is greater than the thermal energy, which means that Janus particles are suspended, but when the Debye length is shorter than 30 nm, the particles prefer to aggregate at this orientation.

in contact. Small clusters containing up to 12 particles dominate this region. Finally, at even higher concentrations, when the angular gradients are too weak to impose a preferred orientation between any two particles facing their respective hydrophobic hemispheres, linear aggregates form as a result of a complex multi-stage fusion of smaller clusters.

The images in Figure 7.7d show how small clusters interconvert dynamically between different shapes for $N = 4$, 7 and 12. The striking agreement between experiments and simulations suggests that the structures observed experimentally are equilibrated and indicates that typical interaction strengths between the hydrophobic regions are again of the order of only 5–$10k_{\mathrm{B}}T$. Interestingly, some clusters, for example the dodecamer ($N = 12$), has a single stable structure whereas two distinct configurations were observed both experimentally and numerically for the tetramer ($N = 4$). For the heptamer ($N = 7$), while the compact variant (left in Figure 7.7d) is the most stable; the more 'open'

$$d = 1.02 \ \mu m$$

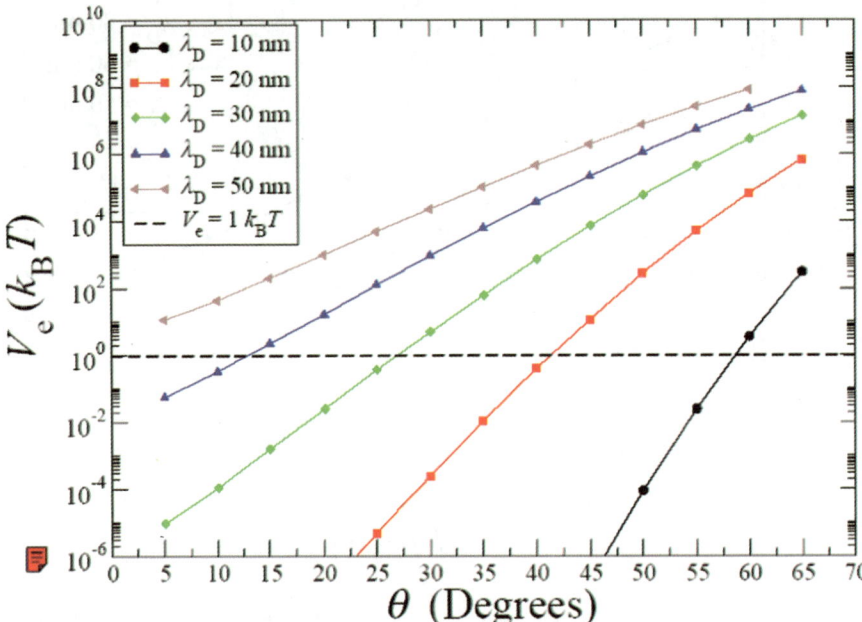

Figure 7.10   Based on Figure 7.4, at shorter Debye length, particles prefer to aggregate, but the relative orientations of these particles can vary due to the Brownian rotational dynamics. However, the rotation is also hindered by the repulsion from the charged hemispheres. Here we calculate the atomistic electrostatic energy between two hemispherical charged particles kept a distance of 1.02 μm as a function of the tilt angle θ. We simultaneously tilt both particles by an angle θ, as in Figure 7.4, for different values of the Debye length. We can deduce a phase diagram (Figure 7.8) based on the combination of Figures 7.4 and 7.5 to understand the self-assembly evolution of amphiphilic Janus particles.

arrangement on the right, also observed both experimentally and numerically, has an important role in the dynamic sub-processes leading to the formation of the larger worm-like structures. As in the case of zwitterionic particles, at high salt concentrations small clusters of amphiphilic Janus particles can dynamically breathe between several competing structures, thus facilitating the integration of extra particles in the cluster, and favoring its growth.

The dynamics leading to the formation of the self-assembled structures are of particular interest.[38] Figure 7.11 shows the intricate kinetic pathways of amphiphilic Janus particles forming structures containing up to seven components at an NaCl concentration of 3.8 mM. This diagram, confirmed by experimental observations, highlights how monomer addition, cluster fusion and isomerization are the main modes of cluster formation. Typically, after the system has equilibrated, the structures still go through a process of isomerization. For instance, the capped trigonal bipyramid (CTBP) structures

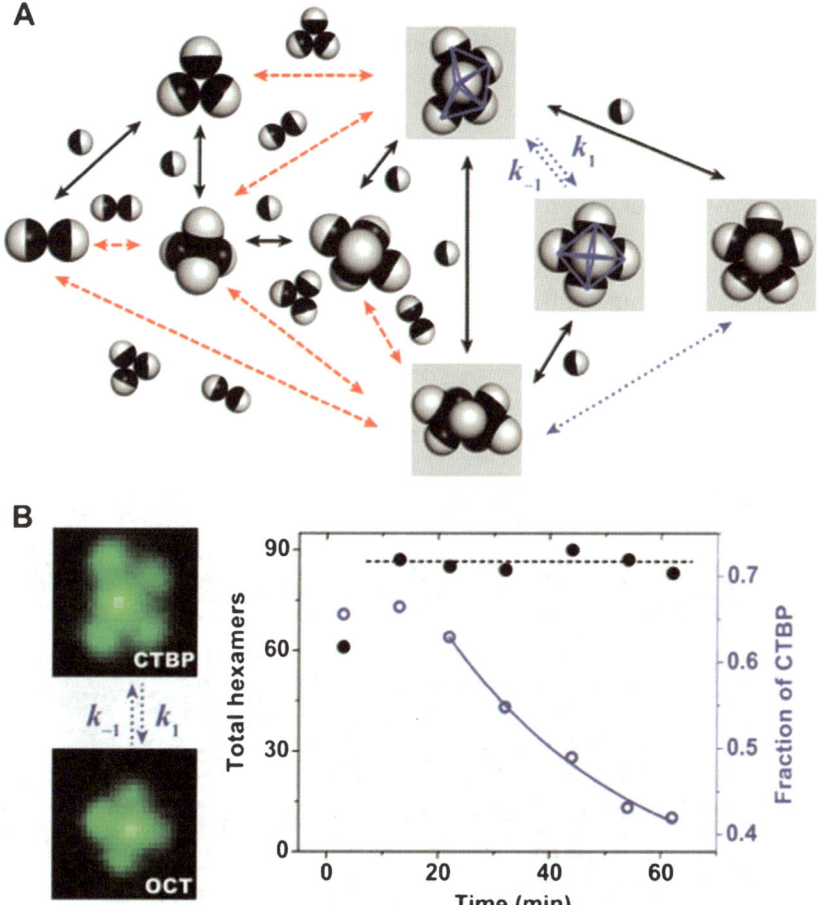

Figure 7.11   Clusters formed from Janus spheres with one hydrophobic hemisphere. (a) Network of reaction pathways, all of which we have observed in experiments at 3.8 mM NaCl. Reaction mechanisms of monomer addition, cluster fusion and isomerization are denoted by black, red and blue arrows, respectively. Isomers with $N = 6$ and 7 elemental spheres are highlighted in boxes. (b) A study of isomerization between two types of $N = 6$ clusters, the capped trigonal bipyramid (CTBP) and the octahedron (OCT). Here and in all other images, the Janus spheres have a diameter of 1 mm. After initiating the cluster process by setting the NaCl concentration at 3.8 mM, the partition of Janus spheres between clusters of different size equilibrates after 20 min but isomerization continues. Once the total number of hexamers (black filled circles) has stabilized, isomerization (fraction of CTBP, blue open circles) is consistent with first-order reaction kinetics in time t, $d[OCT]/dt = k_1[CTBP] - k_{-1}[OCT]$, time constant 34 min, $k_1/k_{-1} = 2.2$ and $k_1 = 0.02$ min^{-1}. Here the calculation is based on the ensemble behavior of many clusters, among which individual ones can follow different reaction pathways.[42] Reprinted with permission from *Science*, 2011, **331**, 199.

form first as they are much easier to reach by monomer addition or a cluster fusion process; however, over time they gradually isomerize to more stable octahedral (OCT) structures. Kinetic data support a reversible first-order reaction.

In contrast to the assembly of molecular surfactants, which is characterized by sequential additional of individual particles, the fusion kinetics through smaller clusters locking on to one another are rather unique to the assembly processes of these particles. Figure 7.12 illustrates in detail the typical multi-stage pathway corresponding to the formation of large linear clusters obtained from numerical simulations. Figure 7.12a shows how heptamers form from the fusion of a trimer and a tetramer. Figure 7.12b shows how during the fusion of a heptamer with another cluster, such as a pentamer ($N = 5$), the former imprints its topology (a pentagon-shaped arrangement of five colloids flanked by two additional particles) on the latter and how subsequent fusion of this

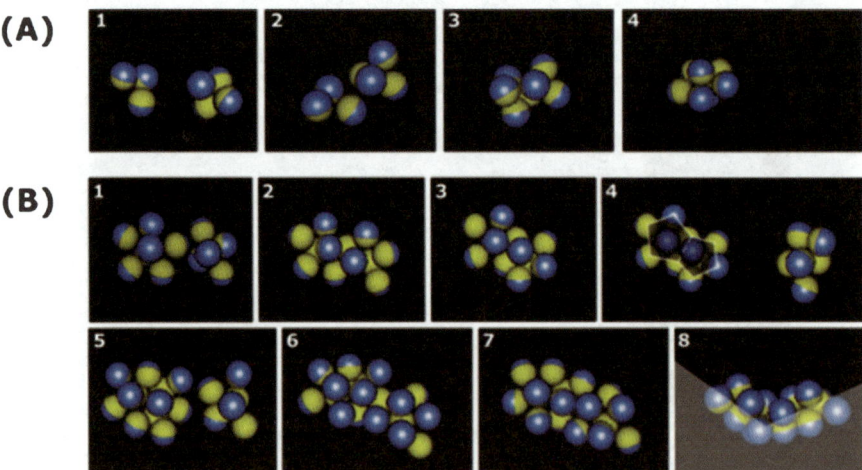

Figure 7.12   Monte Carlo simulations of the polymerization of clusters of amphiphilic spheres into worm-like strings. Assembly proceeds in sequence from left to right. Part (a) illustrates that, as a trimer ($N = 3$) and tetramer ($N = 4$) approach, particles within these clusters reorient until they fuse together in a lock-and-key structure to form a heptamer ($N = 7$). Part (b) shows similar fusion of a heptamer and a pentamer (images 1–4), followed by fusion with a hexamer (images 4–7), leading to a kink in the worm-like string (image 8). As the small clusters fuse, individual colloids rearrange to maintain fivefold rings and reorient such that their hydrophobic sides are directed towards the interior of the worm. Specifically, when the $N = 5$ cluster merges with an $N = 7$ cluster, two pentagons are formed that are not coplanar [part (b), image 4]. When the resulting aggregate incorporates another $N = 6$ cluster, the individual particles rearrange once more to form a third pentagonal structure [part (b), images 6 and 7]. Note that large structures grow through cluster aggregation and not *via* the stepwise addition of discrete particles. Reprinted with permission from reference 28. Copyright 2008 American Chemical Society.

aggregate with smaller cluster leads eventually to the worm-like strings observed in Figure 7.12c. A common feature of the linear aggregates is the presence of kinks. They typically form when a hexamer ($N = 6$) meets a string end formed from a heptamer and a pentamer. These configurations tend to be fairly stable and the fusion of such clusters requires the latter to distort as the particles in the hexamer attempt to arrange in a five-fold ring (image 8). Figure 7.12b (images 5–7) shows snapshots from computer simulations illustrating the scenario described above.

7.5 Off-balance Amphiphilic Janus Particles

While the dividing surface between hydrophobic and hydrophilic regions in Janus particles is set at the equator, it is also interesting to consider particles in which this boundary is located at an arbitrary latitude on the particle's surface to understand how its exact location may affect the structure and dynamics of the aggregates. The behavior in a few special cases is well understood. For instance, we know that when the hydrophobic region covers a small area, particles can only assemble into dumbbells. We know that when the dividing surface is placed to slightly larger values, so that more particles can share the same hydrophobic surface, particles condense into small, stable clusters (this would be equivalent to the Janus case at intermediate salt concentration). We have just described the structures resulting from self-assembly of Janus particles (half coverage) and, of course, in the limit of full coverage we recover an isotropic potential, which is known to lead to the formation of an FCC crystal. Changing the location of the Janus balance has a great influence on the structure of the aggregates. We go from small isotropic clusters to long linear (one-dimensional) aggregates, to full-scale three-dimensional crystals as one moves the location of the Janus balance from one pole to the other.

Recent numerical simulations[39] have shown that a wide variety of novel structures become accessible on altering the balance between hydrophobic and hydrophilic regions on the particles surface in the high-salt limit. Figure 7.13 shows at the top a diagram indicating the structural diversity obtained for different values of the size of the hydrophobic region described in terms of the angle $0° < \theta_{max} < 180°$ and the strength of the hydrophobic interaction ε. What follows is a description of the phase diagram reproduced from reference 39. At low binding energy the system is in a gaseous state. For small values of θ_{max} and moderate values of ε (Figure 13a) particles aggregate into small clusters of 4–13 particles (mesoparticles) including icosahedral structures like those seen in the experiments described above. For $\theta_{max} \approx 90°$, self-assembly yields the well-known worm-like extended structures (Figure 1 b)). As the angular location of the hydrophobic region increases from $90°$ to $180°$ one finds, in order of appearance, self-connected worm-like structures (Figure 13c), flat-crystalline bilayers (Figure 13d), faceted hollow cages (Figure 13e, also observed in equilibrium studies of the same system[40]), dense fluid clusters (Figure 13f) and finally FCC crystals (Figure 143g).

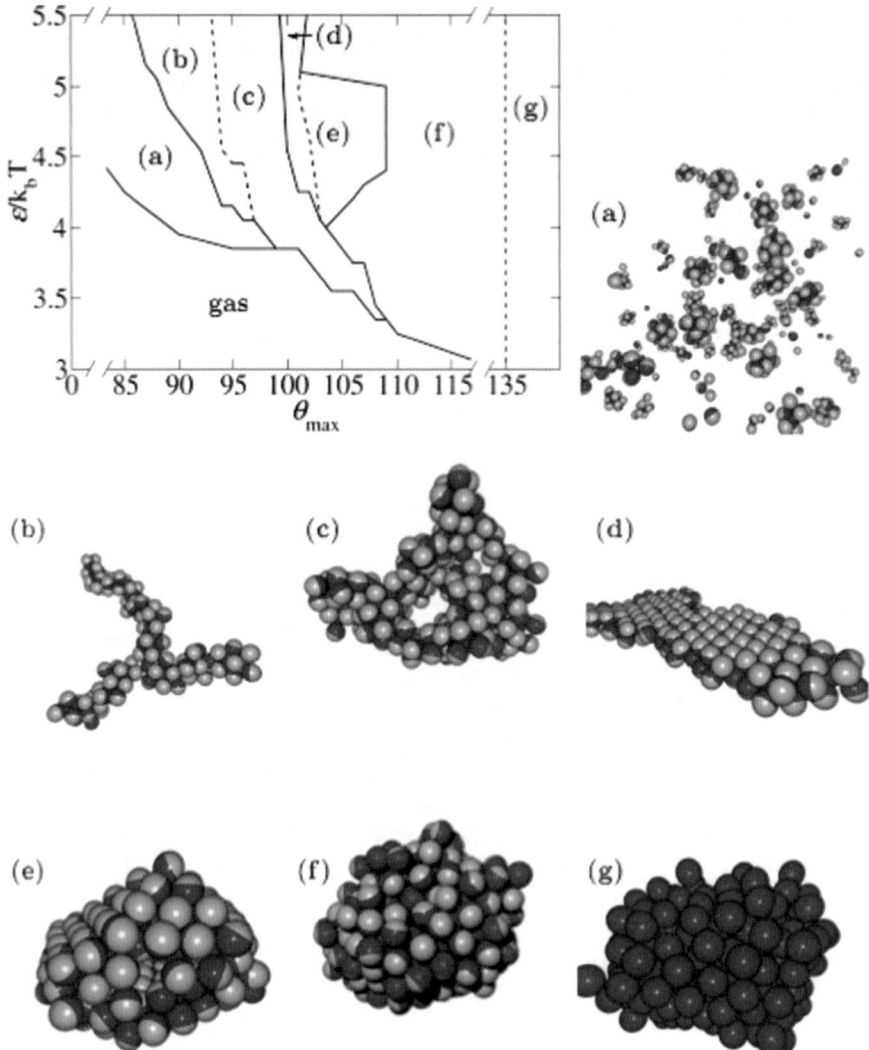

Figure 7.13 Top: self-assembly diagram of amphipathic colloidal particles as a function of binding energy ε and size of hydrophobic region θ_{max}. Region (a) is populated by small micellar-like clusters containing 4–13 particles. Region (b) contains branched worm-like aggregates. In region (c), one finds connected worm-like aggregates. Region (d) delimits flat hexagonal bilayers. Region (e) is populated by faceted hollow cages. In region (f), one finds dense fluid amorphous aggregates. Finally, in region (g), particles arrange in large clusters with FCC order.[39] Reprinted with permission from *Phys. Rev. E*, 2009, 80, 021404.

The morphology of the structures relative to phase (d) and (e) is interesting as each bears a clear resemblance to the structures that are typically found in self-assembly of lipid molecules or block copolymers. In a way they can be thought of as the colloidal analogs of bilayers and vesicles. However, the most intriguing aspect of the problem again concerns the dynamics of their formation. As also observed in the experiments on symmetric Janus particles described above, none of these structures arise via the classical particle-by-particle growth mechanism predicted by the classical self-assembly pathway. We find instead that a rather elaborate two- to three-step hierarchy of processes is usually required before the final structure is equilibrated. The first step involves the formation of a meso-structure (typically a cluster with large inner-mobility) whose morphology depends on the specific value of θ_{max}, and it is the fusion of these clusters that results in the subsequent formation of either extended worm-like aggregates or larger fluid clusters that, once beyond some threshold size, spontaneously reorganize into structured faceted vesicles or FCC crystals.

Let us consider the complex three-step mechanics leading to the formation of phases (c), (d) and (e). Following the discussion in reference 39, the worm-like clusters stable at $\theta_{max} = 90°$ are the common precursor to the formation of most of the structures obtained for $\theta_{max} > 90°$. Self-connected worm-like aggregates form as a result of the improved flexibility of the worm-like structures. In fact, larger values of θ_{max} imply that the particles can more easily rotate about their axes, leading to an expanded space of available structural configurations. The first accessible and stable conformations to the worm-like clusters due to a slight increase in θ_{max} are those leading to the interconnection of the branches. The formation of the bilayers phases (d) is also preceded by the branched-interconnected aggregates that in this case initially acquire an overall flat aspect ratio to maximize the number of contacts with the neighboring particles and eventually minimize the total energy by closing any exposed gap in the structure to form a crystalline bilayer *via* a complex coalescence of multiple coplanar loops (see the sequence in Figure 7.14). Finally, the faceted cages form, not as a result of self-assembly of misoriented disjoined bilayers, but *via* a mechanism involving multiple steps. For sufficiently large values of the hydrophobic region, the flexibility of the intermediate structures is further increased, leading to their prompt folding into small isotropic fluid clusters. Although small clusters tend to remain fluid, large ones morph into faceted cages via a mechanism similar to that described for the formation of bilayers. Fusion of small fluid clusters, as

Figure 7.14 Sequence of three snapshots showing the mechanism of bilayer formation *via* coalescence of three coplanar loops.[43] Reprinted with permission from *Phys. Rev. E*, 2009, **80**, 021404.

(1)   (2)   (3)

(4)   (5)   (6)

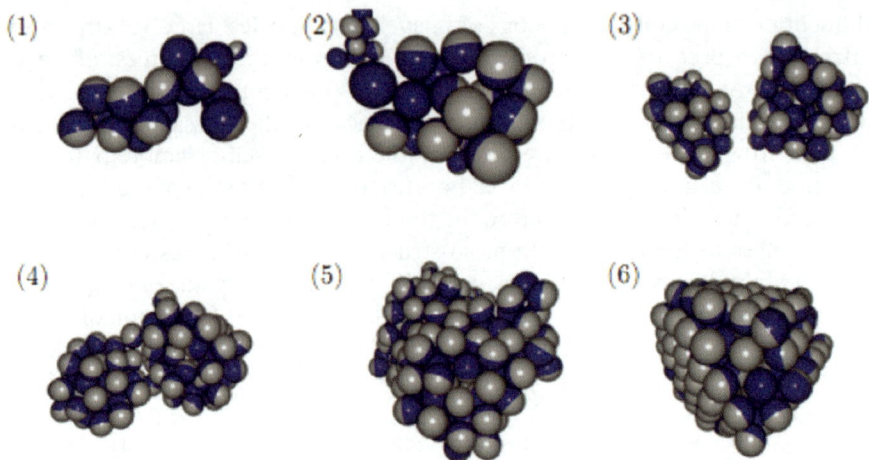

Figure 7.15   Sequence of six snapshots (1 → 6) showing faceted cage formation *via* initial fast folding of short worm-like clusters (1 → 2) and subsequent fusion of fluid blobs (3 → 6).[43] Reprinted with permission from *Phys. Rev. E*, 2009, **80**, 021404.

illustrated in Figure 7.15, is, however, the main mechanism through which large blobs, which eventually turn into faceted cages, are generated.

Systematic experiments on Janus particles with precisely controlled Janus balance have not been carried out, however the results of these molecular dynamics simulations suggest that not only is the structure of the aggregates very sensitive to the specific location of the Janus interface, but also the self-assembly kinetics seem to be greatly dependent on the details of the interparticle interaction.

7.6   Conclusion

We have reviewed a few studies on the self-assembly of Janus particles. Specifically, we considered the simple cases of zwitterionic and amphiphilic microscopic spherical components. We discussed how going beyond isotropic interactions by adding what is possibly the simplest angular dependence to the pair potential can lead to a very rich physical behavior, not only at the level of the morphology of the aggregates formed by these particles, but also at the level of the dynamics of their self-assembly and the complex hierarchies of the different sub-processes associated with it. We also reported on the effectiveness of different numerical models used to mimic the behavior of Janus particles and illustrated the main synthetic routes for their fabrication.

One of the main advantages of working with micron-size particles is that, unlike their molecular counterparts, it is possible to visualize in detail most of the self-assembly pathway using modern optical microscopy technology and estimate the long relaxation times associated with the kinetic processes. With the help of numerical simulations, it is then possible to draw an accurate picture of the phenomenological behavior presented by these particles and relate it to the complexity of their pair potential at different salt concentrations. Although the exploratory research presented in this chapter represents a first step towards a systematic understanding of the physical properties associated with this new class of nanocomponents, we expect that both structure and dynamics will also be very much dependent on the geometry (the bare shape) of the particles.

One obvious direction for further experimental work concerns the role of the Janus balance. As the ratio between the size of the two differently coated surfaces changes, the interaction potential will also change, leading to different cluster structures and ultimately phase behavior in the system. As shown before, this has already been predicted numerically. It would also be interesting to study more complex particles that include multiple patches having more than two chemical properties, going towards a level of complexity similar to that of biological macromolecules such as proteins. Another obvious direction for this research would be to consider non-spherical colloidal particles with definite symmetries such as rods or oblate spheroids or to create a Janus balance by altering the geometry of one side of the particle. Finally one can envision combining alterations in both shape and surface chemistry to provide an even broader number of nano- or micro-components to design materials with novel physical properties.

Note that so far most of the experimental work has been focused on the dilute regime; however, computer simulations predicted an equally intriguing phase behavior at higher concentrations. As of today, no experimental confirmation of the growing numerical work on the subject has been provided. Furthermore, it is fair to say that the details of the kinetic pathways and rearrangement mechanisms are not well understood even for spherical colloids. For instance, we expect the rotational dynamics of the particles to play an important role in the rearrangement and breathing dynamics, but the exact mechanisms of such rotation are still unclear and so is its link to the nature of the free energy difference between the different states.

Finally, it should be stressed that Janus colloidal particles, apart from been an ideal system to understand fundamental physical questions concerning the role of directionality and anisotropy in the process of self-assembly, also has some interesting potential applications. For example, these amphiphilic particles could be considered as large surfactants and, because the energy cost associated to their removal from a water/oil interface is orders of magnitude larger than that required by small molecular surfactants, it would result in significantly more stable emulsions.[41] These particles could also be used for surface modifications; for instance, Janus nanoparticles were applied to textiles to prepare water-repellent fibers.[42] Consumer electronics has also been a field where Janus particles have found some use; for instance, they have been exploited for some display applications where controlling the orientation of the particles allows for switchable black–white displays.[43] Further, Janus particles have some potential uses in nanomedicine and could be used in drug delivery and for combined imaging and magnetolytic therapy.[44]

Acknowledgements

The authors would like to thank S. Granick and E. Luijten for insightful discussions and mentorship.

References

1. B. Alberts, A. Johnson, J. Lewis, M. Raff, K. Roberts and P. Walter, *Molecular Biology of the Cell*, 5th ednGarland Science, New York, 2008.
2. F. H. C. Crick and J. D. Watson, *Nature*, 1956, **177**, 473.
3. A. Klug and D. L. D. Caspar, *Adv. Virus Res.*, 1960, **7**, 225.
4. E. Sackmann, *Can. J. Phys.*, 1990, **68**, 999.
5. G. Subramanian, V. N. Manoharan, J. D. Thorne and D. J. Pine, *Adv. Mater.*, 1999, **11**, 1261.
6. G. A. DeVries, M. Brunnbauer, Y. Hu, A. M. Jackson, B. Long, B. T. Neltner, O. Uzun, B. H. Wunsch and F. Stellacci, *Science*, 2007, **315**, 358.
7. M. Li, H. Schnablegger and S. Mann, *Nature*, 1999, **402**, 393.
8. L. Hong, S. Jiang and S. Granick, *Langmuir*, 2006, **22**, 9495.
9. H. Weller, *Philos. Trans. R. Soc. London A*, 2003, **361**, 229.
10. E. K. Hobbie, B. J. Bauer, J. Stephens, M. L. Becker, P. M. McGuiggan, S. D. Hudson and H. Wang, *Langmuir*, 2005, **21**, 10284.
11. J. N. Israelachvili, D. J. Mitchell and B. W. Ninham, *J. Chem. Soc., Faraday Trans. 2*, 1976, **72**, 1525.
12. T. Hu, R. Zhang and B. I. Shkovskii, *Physica A*, 2008, **387**, 3059.
13. D. Leckband and J. Israelachvili, *Q. Rev. Biophys.*, 2001, **34**, 105.
14. R. Nagarajan and E. Ruckenstein, *Langmuir*, 1991, **7**, 2934.
15. A. McPherson, *Introduction to Macromolecular Crystallography*, Wiley, Hoboken, NJ, 2003.
16. J. Frank, *Three-Dimensional Electron Microscopy of Macromolecular Assemblies*, Oxford University Press, New York, 2006.
17. B. Huang, W. Wang, M. Bates and X. Zhuang, *Science*, 2008, **319**, 810; E. Betzig, G. H. Patterson, R. Sougrat, O. W. Lindwasser, S. Olenych, J. S. Bonifacino, M. W. Davidson, J. Lippincott-Schwartz and H. F. Hess, *Science*, 2006, **313**, 1642.
18. S. Asakura and F. Oosawa, *J. Chem. Phys.*, 1954, **22**, 1255.
19. P. N. Segrè, V. Prasad, A. B. Schofield and D. A. Weitz, *Phys. Rev. Lett.*, 2001, **86**, 6042.
20. A. D. Dinsmore and D. A. Weitz, *J. Phys. Condens. Matter*, 2002, **14**, 7581.
21. F. Sciortino, S. Mossa, E. Zaccarelli and P. Tartaglia, *Phys. Rev. Lett.*, 2004, **93**, 055701.
22. S. Torquato, *Soft Matter*, 2009, **5**, 1157.
23. C. N. Likos, *Soft Matter*, 2006, **2**, 478.

24. S. C. Glotzer and M. J. Solomon, *Nat. Mater.*, 2007, **6**, 557.
25. V. N. Manoharan, M. T. Elsesser and D. J. Pine, *Science*, 2003, **301**, 483.
26. A. Van Blaaderen, *Science*, 2003, **301**, 470.
27. S. C. Glotzer and M. J. Solomon, *Nat. Mater.*, 2007, **6**, 557.
28. L. Hong, A. Cacciuto, E. Luijten and S. Granick, *Langmuir*, 2008, **24**, 621.
29. L. Hong, A. Cacciuto, E. Luijten and S. Granick, *Nano Lett.*, 2006, **6**, 2510.
30. J. N. Israelachvili, *Intermolecular and Surface Forces*, Elsevier, Amsterdam, 2011.
31. N. Kern and D. Frenkel, *J. Chem. Phys.*, 2003, **18**, 9882.
32. E. Bianchi, R. Blaak and C. Likos, *Phys. Chem. Chem. Phys.*, 2011, **13**, 6397.
33. C. De Michele, S. Gabrielli, P. Tartaglia and F. Sciortino, *J. Phys. Chem. B*, 2006, **110**, 8064.
34. M. Klokkenburg, R. P. A. Dullens, W. K. Kegel, B. H. Erné and A. P. Philipse, *Phys. Rev. Lett.*, 2006, **96**, 037203.
35. K. Van Workum and J. F. Douglas, *Phys. Rev. E*, 2006, **73**, 031502.
36. V. N. Manoharan, M. T. Elsesser and D. J. Pine, *Science*, 2003, **301**, 483.
37. K. Van Workum and J. F. Douglas, *Phys. Rev. E*, 2006, **73**, 031502; J. P. K. Doye, M. A. Miller, P. N. Mortenson and T. R. Walsh, *Adv. Chem. Phys.*, 2000, **115**, 1; D. J. Wales, *Int. J. Mod. Phys. B*, 2005, **19**, 2877; Z. Zhang and S. C. Glotzer, *Nano Lett.*, 2004, **4**, 1407; C. De Michele, S. Gabrielli, P. Tartaglia and F. Sciortino, *J. Phys. Chem. B*, 2006, **110**, 8064; D. Zerrouki, B. Rotenberg, S. Abramson, J. Baudry, C. Goubault, F. Leal-Calderon, D. J. Pine and M. Bibette, *Langmuir*, 2006, **22**, 57.
38. Q. Chen, J. Whitmer, S. Jiang, S. C. Bae, E. Luijten and S. Granick, *Science*, 2011, **331**, 199.
39. W. Miller and A. Cacciuto, *Phys. Rev. E*, 2009, **80**, 021404.
40. F. Sciortino, A. Giacometti and G. Pastore, *Phys. Rev. Lett.*, 2009, **103**, 237801.
41. L. Hong and S. Granick, *US Patent Application*, 20100305219A1, 2010.
42. A. Synytska, R. Khanum, L. Ionov, C. Cherif and C. Bellmann, *ACS Appl. Mater. Interfaces*, 2005, **3**, 1216.
43. N. Takasi, T. Torii, T. Takahashi and Y. Takizawa, *Adv. Mater.*, 2006, **18**, 1152.
44. M. Yoshida, K.-H. Roh, S. Mandal, S. Bhaskar, D. W. Lim, H. Nandivada, X. Deng and J. Lahann, *Adv. Mater.*, 2009, **21**, 4920; S.-H. Hu and X. Gao, *J. Am. Chem. Soc.*, 2010, **132**, 7234.

CHAPTER 8

Self-assembly of Janus Particles Under External Fields

ILONA KRETZSCHMAR*[a], SUMIT GANGWAL[b],
AMAR B. PAWAR[c] AND ORLIN D. VELEV[d]

[a] Department of Chemical Engineering, The City College of New York, 140
Street and Convent Avenue, New York, NY 10031, USA; [b] Xanofi Inc., 7516
Precision Drive, Raleigh, NC 27617, USA; [c] Momentive Performance
Materials, 769 Old Saw Mill River Road, Tarrytown, NY 10591, USA;
[d] Department of Chemical and Biomolecular Engineering, North Carolina
University, 911 Partners Way, Raleigh, NC 27695, USA
*E-mail: kretzschmar@ccny.cuny.edu

8.1 Introduction

Janus particles are the simplest example of new types of anisotropic particles
that have been synthesized and studied thanks to the recent advances in
colloidal engineering. The self-assembly of anisotropic particles has attracted
much attention in the last decade.[1,2] Entire journal issues have been dedicated
to reports and reviews on the preparation and assembly of particles anisotropic
both in shape and surface material.[3,4] Multiple reviews have been dedicated to
Janus particle synthesis and assembly.[5–9]

More recently, the manipulation of Janus particles with electric and
magnetic fields has garnered interest among the colloidal research community.
The use of external fields for the assembly of particles into well-defined
structures has a long history.[10,11] Field assembly has been found to allow the
formation of thermodynamically stable or metastable structures, has permitted

RSC Smart Materials No. 1
Janus Particle Synthesis, Self-Assembly and Applications
Edited by Shan Jiang and Steve Granick
© The Royal Society of Chemistry 2012
Published by the Royal Society of Chemistry, www.rsc.org

the determination of particle zeta potentials and has been applied in electrorheological and magnetorheological fluids. For particles to interact with a field, either the field needs to have a gradient, as is the case in convective flow fields, or the field has to cause a non-symmetric dipolar interaction between particles. The advantage of using external fields is that one can precisely tune forces exerted on particles by the field and also the resulting field-induced particle–particle interactions.[12,13] Further, the parameters affecting the colloidal assembly can be controlled by external electronics, thereby allowing the simple adjustment of the driving forces for the assembly. Finally, external field assemblies discussed in this chapter are characterized by their straightforward and simple experimental study and fabrication without the need for expensive or difficult techniques.

Figure 8.1 summarizes schematically the particle–field interactions and presents examples of observed particle assembly behavior for isotropic particles in convective flow, electric DC, electric AC and constant magnetic fields, which are briefly reviewed in the first subsection of this chapter. Next, the behavior of Janus particles in convective flow fields, DC and AC electric fields and magnetic fields is covered for the case where uniaxial fields are used, culminating in the discussion of first reports on the assembly of Janus particles in crossed AC electric and magnetic mixed biaxial fields. Owing to the focus on external fields, we will not discuss here many other interesting examples of Janus particle interactions and dynamics, such as (i) the interaction of gold-coated silica Janus particles modified with homeotropic DMOAP surfactant in the field of a nematic crystal 5CB in which a novel boojum-ring configuration has been observed,[14,15] (ii) gold-capped Janus particles in a defocused laser beam, where the interaction of the gold cap with the laser trap causes self-thermophoresis,[16,17] and (iii) Janus particles with reactive caps that have been shown to create their own concentration gradients, leading to a phenomenon called self-diffusiophoresis.[18–21] The chapter is concluded with a brief Future Outlook section that discusses potential and realized applications of field-addressable Janus particles.

8.1.1 Convective Flow and Uniaxial Electric/Magnetic Fields

Convective flow fields are often established during the drying of colloidal solutions. The simplest case of a convective flow field is established when a small droplet containing particulate matter is allowed to dry. The three-phase contact line formed by the substrate, the surrounding air and the drying liquid is pinned due to the presence of defects in the substrate and particles in the solution. The schematic in Figure 8.1a(i) shows a cross-section of the edge of a drying droplet at various drying times. As the evaporation proceeds from the droplet surface, the volume of the droplet decreases. If the contact line moves freely over the substrate without pinning, then the volume decrease would lead to receding of the contact line in order to maintain the most favorable three-phase contact angle. However, since the contact line is pinned in the systems of

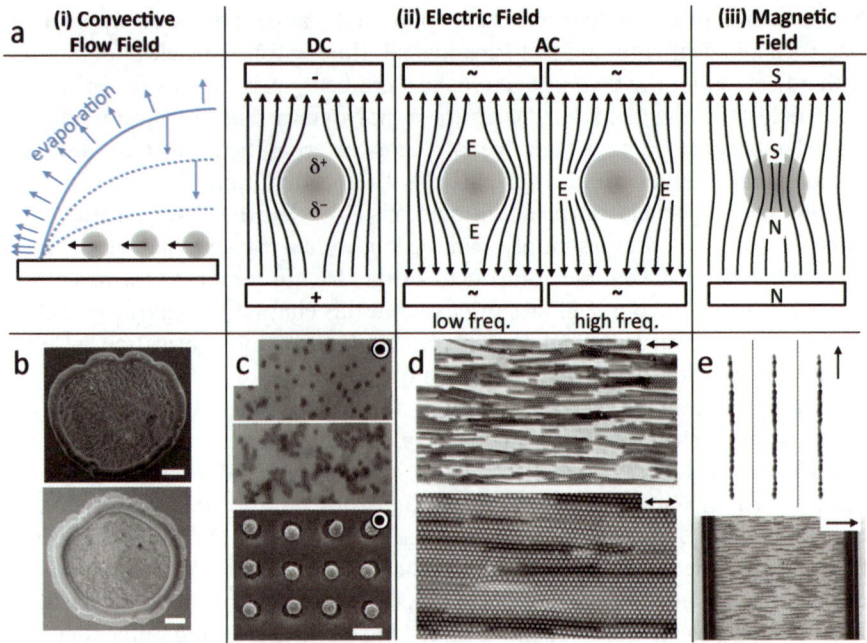

Figure 8.1 Particle–field interaction and assembly behavior for isotropic particles in convective flow, electric DC and AC fields and magnetic fields. (a) Schematic representation of particle–field interaction (from left to right): (i) convective flow: particles are carried to the rim of a drying droplet due to convective flow towards the pinned contact line; (ii) electric field: particles align their induced charge dipole opposite to the external DC field direction, experience maximum field intensity at the poles pointing towards the electrodes in low-frequency AC field [positive dielectrophoresis (DEP)] and at the left and right side in high-frequency AC field (negative DEP); and (iii) magnetic field: particles align their magnetic dipole parallel to the external field direction. (b) Scanning electron microscopy images of convectively assembled particulate residues left after drying of two aqueous droplets with 2.4 μm sulfated polystyrene particles (scale bar: 0.5 μm). (c) Top and middle: electrophoretic deposition (EPD) in a 2 V DC field of 4 μm polystyrene particles in 10^{-5} M KCl at time $t = 90$ and 900 s, respectively. Bottom: 0.5 μm streptavidin-coated, fluorescent polystyrene microbeads assembled into 0.6 μm wells at a pitch of 1.2 μm (scale bar: 1 μm). (d) 1.4 μm latex spheres assembled at $V_{AC} \geq 75$ V and $f_{AC} = 0.6$ kHz at $t = 5$ s (top) and 15 s (bottom). (e) Top: 0.902 μm PS particles with 40% Fe_3O_4 content in water at $\phi < 0.08\%$ in a magnetic field of $H_0 = 0.0054$ T. Bottom: 0.32 μm diameter particles at $\phi = 0.005$ after 3 min of application of a pulsed magnetic field ($H_0 = 1480$ A m^{-1}, $f = 2.0$ Hz). Field directions: radially outwards (b), out-of-plane (c), horizontal (d), vertical (e, top) and horizontal (e, bottom). Reprinted with permission from (c, top/middle) reference 29, (d) reference 37 and (e, top) reference 39. Copyright 1996, 2004 and 1995 American Chemical Society. (c, bottom) Reproduced from reference 32. (e, bottom) Reprinted with permission from reference 38. Copyright 1997 American Physical Society.

interest here, a convective flow field (black arrows) from the center of the drop towards the rim is established in order to replenish the fluid at the rim of the drop. Particles are trapped by the flow and advected to the rim of the droplet leading to particulate residues with a distinct ring structure, known as the 'coffee ring' effect.[22,23] Figure 8.1b shows scanning electron microscopy images of two such ring structures obtained after drying of two aqueous droplets of a 2.4 μm sulfated polystyrene particle suspension. A review on forces guiding the self-assembly in wet coating technology with specific focus on dip coating, spin coating and convective coating can be found elsewhere.[24,25]

When a DC electric field is applied to conducting or charged particles in an electrolytic medium, their electron density or ionic cloud, respectively, is polarized as illustrated in the first schematic in Figure 8.1a(ii).[12,26] The polarization causes the particle to move towards the electrode with the polarization opposite to that of the particle. This phenomenon, termed electrophoresis, was discovered in 1807 by Ruess,[27] and is characterized by the electrophoretic mobility, μ. The electrophoretic mobility is the proportionality factor between the electric field intensity and the particle velocity. The particle mobility increases with the zeta potential, ζ, and decreases with the viscosity of the medium, η. It can be described either with the Hückel equation, eqn (8.1a), for particles with radii smaller than the Debye length ($1/\kappa$) or the Helmholtz–Smoluchowski equation, eqn (8.1b), when the particle radii are much larger than the Debye length.[28]

$$\mu = \frac{2\varepsilon\varepsilon_0\zeta}{3\eta} \quad (r \ll 1/\kappa) \tag{8.1a}$$

$$\mu = \frac{\varepsilon\varepsilon_0\zeta}{\eta} \quad (r > 1/\kappa) \tag{8.1b}$$

Using eqn (8.1), where ε and ε_0 are the dielectric permittivity of the medium and vacuum (8.854 $\times$ 10^{-12} C^2 N^{-1} m^{-2}), respectively, one can determine an object's zeta potential by observation of the object's mobility as a function of field intensity and solution viscosity. In the extreme case, particles move until they reach the electrode, a process called electrophoretic deposition (EPD). The top and middle panels of Figure 8.1c depict an electrode surface submerged horizontally in 10^{-5} M aqueous KCl solution containing 4 μm polystyrene particles at time t = 90 and 900 s, respectively.[29] Particles are allowed to settle by gravity and form close-packed 2D colloidal layers upon application of a 2 V DC potential. The effect is rationalized by the existence of electroosmotic flows arising from the interaction of the external electric DC field with the equilibrium charge on the particles.[30,31] EPD can also be used to assemble particles into well-defined patterned arrays. The bottom panel of Figure 8.1c shows 0.5 μm streptavidin-coated, fluorescent polystyrene microbeads that have been assembled into 0.6 μm wells at a pitch of 1.2

μm.[32] A major limitation in using DC fields besides the occurrence of electroosmotic flows is the electrolysis of water at higher potentials, which limits the range of applied voltage that can be used for electrophoretic experiments. Both of these side effects can be avoided when AC electric fields are used.

In AC electric field assembly, particle interactions are determined by the presence of the dielectrophoretic (DEP) force.[12,33–35] In contrast to the polarization observed in DC electric fields, the dipoles induced in particles under AC electric field exposure interact with the gradient of the inhomogeneous field. The DEP force is dependent on particle volume (r^3) and the gradient of the field squared, ∇E^2:

$$F_{DEP} = 2\pi\varepsilon_1 \text{Re}|K(\omega)|r^3\nabla E^2 \qquad (8.2)$$

where ε_1 is the dielectric permittivity of the surrounding medium and $\text{Re}|K(\omega)|$ is the real part of the Clausius–Mossotti function, K:

$$\text{Re}|K(\omega)| = \frac{\varepsilon_2 - \varepsilon_1}{\varepsilon_2 + 2\varepsilon_1} + \frac{3(\varepsilon_1\sigma_2 - \varepsilon_2\sigma_1)}{\tau_{MW}(\sigma_2 + 2\sigma_1)^2(1 + \omega^2\tau_{MW}^2)} \qquad (8.3a)$$

$$\tau_{MW} = \frac{\varepsilon_2 + 2\varepsilon_1}{\sigma_2 + 2\sigma_1} \qquad (8.3b)$$

where ε_1 and σ_1 are the dielectric permittivity and conductivity of the surrounding medium, respectively, ε_2 and σ_2 are the dielectric permittivity and conductivity of the particle, respectively, and τ_{MW} is the Maxwell–Wagner charge relaxation time. Two distinct frequency regimes exist: (i) the low-frequency regime, in which $Re(K) > 0$ and particles are attracted to the high field intensities, and (ii) the high-frequency regime where $Re(K) < 0$ and particles are repelled from high field intensity regions. The two regimes are delineated by a crossover frequency, $\omega_c = \tau_{MW}^{-1}$, at which K changes its sign. The second and third schematics in Figure 8.1a(ii) depict locations of high field intensities on a dielectric particle experiencing positive DEP (attraction to high fields) in which the field intensity is high at the particle's poles pointing towards the electrodes. At high frequencies, the particle experiences negative DEP, *i.e.* it is repelled from high fields and highest field intensities are detected at the equator of the particle. The $Re(K)$ and thus the crossover frequency can be determined experimentally using a two-step method recently proposed by Honegger *et al.*[36] in which the particles have first to be positioned precisely on activated electrodes according to the applied frequency followed by particle velocity measurements in the pure DEP regime. At higher particle concentrations, isotropic microspheres are likely to interact and form chains.[37] The chaining force can be determined using:

$$F_{chain} = -C\pi\varepsilon_1 r^2 K^2 E^2 \tag{8.4}$$

where C is a coefficient ranging from 3 to $>10^3$ depending on the distance between particles and chain length. Note that the chaining force is always positive and attractive regardless of the relative polarizability of the particles and the medium at the given AC field frequency. Figure 8.1d depicts the combination of chaining and DEP behavior observed for 1.4 μm latex spheres assembled at $VAC \geq 75$ V and $f_{AC} = 0.6$ kHz at $t = 5$ s (top panel) and 15 s (bottom panel).[12,37] Initially, the particles form multiple parallel chains, which are then attracted to the electrode by DEP and merge into a 2D colloidal crystal due to lateral interactions between the chains.

Structures formed in the presence of electric DC and AC fields have the advantage that they can be reversibly assembled and disassembled, *i.e.* annealed, by simple removal and reapplication of the field. However, the major limitation of electric field assembly is the presence of electrolytes in commonly used aqueous media. The electrolyte limits the range within which a voltage can be applied, and can lead to current-induced heating and bubbling; when the electrolyte is drained the electric field-induced forces will dissipate quickly and the structures formed will break apart, causing difficulties in deposition of particle structures for further use. Magnetic fields present an interesting alternative to electric fields when magnetizable or magnetic particles are available.

When a magnetizable particle is exposed to a magnetic field, a magnetic dipole moment, μ, aligned with the field is induced, as shown schematically in Figure 8.1a(iii). The strength of the induced dipole can be described by:

$$\mu = \frac{4}{3}\pi r^3 \mu_0 \chi H_0 \tag{8.5}$$

where r is the radius of the particle, μ_0 is the magnetic permeability of vacuum ($4\pi \times 10^{-7}$ N A^{-2}), χ is the magnetic susceptibility of the particle and H_0 is the applied magnetic field.[38] When two particles with such an aligned induced dipole approach each other, they interact *via* dipole–dipole attraction as described by the dipole–dipole interaction potential $U(r,\theta)$ given by the equation:

$$U(r, \theta) = \frac{(\mu^2/4\pi\mu_0)}{r^3}(1 - 3\cos^2\theta) \tag{8.6}$$

where μ is the induced magnetic dipole moment and θ is the angle between the line connecting the particles' centers and the magnetic field direction. The dipole–dipole attraction is maximized when the two dipole moments are aligned with the field and placed in a head-to-tail configuration, *i.e.* $\theta = 0°$, leading to the chaining of particles.[39] This chaining process has been employed in magnetorheological fluids[40] and in the fabrication of well-controlled porous structures.[41] A useful dimensionless dipole strength parameter, λ, that relates the dipole–dipole energy to the thermal energy can be defined by:

$$-\lambda = \frac{U_{max}}{\kappa_B T} = \frac{\pi \mu_0 r^3 \chi^2 H_0^2}{9 \kappa_B T} \qquad (8.7)$$

where k_B is the Boltzmann constant and T is the absolute temperature. Particles will start to aggregate at a critical λ_c. For two particles, a critical λ_c of 1 indicates the equivalence of magnetic and thermal forces, whereas one expects $\lambda_c < 1$ for systems of more than two particles as each of the particles contributes to the magnetic interaction.[39] The top panel in Figure 8.1e shows chains formed by 0.902 µm PS particles with 40% Fe_3O_4 loading in water at a volume fraction of $\phi < 0.08\%$ in a magnetic field of $H_0 = 0.0054$ T.[39] The λ value for the system is reported to be 76, indicating that the magnetic forces are dominating the assembly. The bottom panel of Figure 8.1e depicts the formation of aggregates from 0.32 µm diameter droplets comprised of 10 nm Fe_2O_3 particles suspended in octane at a volume fraction of $\phi = 0.005$ after 3 min exposure to a pulsed magnetic field ($H_0 = 1480$ A m^{-1}, $f = 2.0$ Hz).[38] The droplets are found to form ellipsoidal aggregates aligned with the direction of the external magnetic field.

The situation is slightly different for the case of intrinsically magnetic particles. Their assembly has been reviewed both for colloidal solutions[42–44] and for confined assembly.[45] They assemble even in the absence of a magnetic field owing to their intrinsic magnetic dipole and the resulting dipole–dipole interactions described by the equation:

$$U_{ij} = -\frac{3\left(m_i \cdot \hat{r}_{ij}\right)\left(m_j \cdot \hat{r}_{ij}\right) - m_i m_j}{4\pi \mu_0 r_{ij}^3} \qquad (8.8)$$

where m_i and m_j are the magnetic dipole moments of the two interacting particles and $\hat{r}_{ij}$ is the distance vector between the centers of the two particles.[46] An interesting recent review discusses the spontaneous assembly of magnetic colloids at interfaces and the driving of these assemblies to non-equilibrium configurations by means of an applied alternating magnetic field.[47] Another recent assembly theme uses a magnetized ferrofluid to assemble diamagnetic and paramagnetic particles into highly reproducible, rotationally symmetric arrangements of, for example, Saturn rings (axial quadrupoles) and flowers (axial octupoles).[48]

8.1.2 Biaxial Combinations of Electric and Magnetic Fields

As the assembly behavior of particles in uniaxial electric and magnetic fields is becoming increasingly understood, researchers have started to explore the use of biaxial fields for improved control of colloidal assembly. Figure 8.2 reports on two experiments: the assembly of PMMA particles in a biaxial AC electric

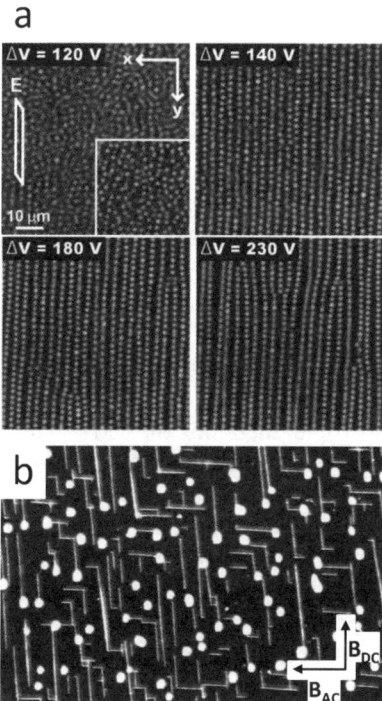

Figure 8.2 (a) Sheet structures formed by 2.0 μm PMMA particles at $\phi = 0.20$ imaged with a confocal microscope at the center of the sample space 11 min after switching on of the varying external voltages indicated in the images with a frequency of 4 Hz and the biaxial electric field aligned with the yz-plane. Reprinted with permission from reference 49. Copyright 2009 Wiley-VCH Verlag GmbH. (b) Structures formed by paramagnetic particles under an AC–DC combined in-plane biaxial magnetic field at time $t = 5$ min and $f = 10$ Hz. Reprinted with permission from reference 50. Copyright 2011 American Chemical Society.

field (Figure 8.2a)[49] and the assembly of paramagnetic particles under a biaxial DC–AC field (Figure 8.2b).[50] In the case of biaxial AC electric fields, 2.0 μm PMMA particles at a volume fraction of $\phi = 0.20$ are found to form sheet structures that are aligned with the yz-plane in which the biaxial electric field resides. Monte Carlo simulations show good agreement with the experimental observations and also yield phase diagrams for both uncharged and charged colloidal particles.[51] Unsurprisingly, a hexagonally close packed (*hcp*) structure is found to be the lowest energy structure in the uncharged case. For the charged case, the coexisting liquid becomes inhomogeneous for high field strengths with a dimensionless dipole strength parameter $\gamma > 16$ [electric field analog to λ, eqn (8.7)] and a structure with fluid-like layers is formed. When a biaxial AC–DC combination of fields is applied to a suspension of paramagnetic particles, chains that oscillate with the magnetic field are formed at low frequencies (<3 Hz). At higher frequencies, the chains coagulate into disk-like clusters as shown in Figure 8.2b.[50] The combination of mixed fields has also been explored for the case of a convective flow field and AC electric field.[52–54] For example, Schöpe[53] reported the use of an AC electric field parallel to the substrate and crystal growth direction during the convective assembly of a particle monolayer. The oscillating field causes a periodic shearing perpendicular to the growth direction leading to fewer crystal defects (content of bad spots is reduced by a factor of 3 and dislocations by a factor of 1.5), larger single-crystal domains (average grain size increased by a factor of 6) and faster colloidal assembly. Ristenpart *et al.*[52] applied AC electric fields perpendicular to the substrate and found a strong dependence of colloidal quality on the field frequency. Kleinert *et al.*[54] used an AC electric field as a means of speeding up the convective assembly process and depositing larger, better organized crystal domains.

The behavior of isotropic metallic, dielectric and magnetic particles in applied electric and magnetic fields can be summarized in that particles exposed to a field are polarized to various extents and in varying patterns. Dipole–dipole interactions lead to chain formation followed by formation of 2D and 3D crystal structures at longer times. When dielectric properties are combined with metallic and/or magnetic properties, the competition between the polarizability of the different components can lead to distinct and interesting assembly behaviors. Janus particles provide such a situation and the following section briefly introduces the fabrication of such particles and the experimental techniques used to study their behavior in external fields.

8.2 Janus Particle Preparation and Cell Set-up for Field Assembly

In this section, the materials, preparation methods and assembly cells used for the field exposure studies of Janus particles as discussed in this chapter are summarized. In subsection 8.2.2, we discuss specifically the preparation of dielectric particles with metal caps. The reader is directed to other reviews[5,7,9,55–58] for more information on capillary fluid flow, nanosphere lithography, templating, colloidal cluster, particle lithography and synthetic methods that have been used for Janus particle preparation.

8.2.1 Materials

The specific experiments discussed in this chapter use Janus particles made of surfactant-free sulfate-terminated polystyrene (PS) particles purchased from Interfacial Dynamics Corporation (IDC) with diameters of 2.4, 4, 5.7 and 8 μm. The sulfate termination provides the particles with a negative surface charge. Particles are generally washed three times with deionized (DI) water before use, using a centrifugation–decanting–sonication cycle to remove any surfactants or electrolyte contaminations. The non-ionic surfactant Tween 20 was purchased from Acros Organics, tungsten baskets, gold wire and titanium sponge from Ted Pella, Inc., tungsten boats, chromium pellets and iron pellets from Kurt J. Lesker Co., and glass slides and solvents from Fisher Scientific. A 3:1 Ar–O_2 mixture used for the preparation of iron oxide caps was purchased from Airgas, Inc.

8.2.2 Janus Particle Preparation

Janus particles are prepared in a three-step procedure. First, a 16 wt% suspension of washed isotropic particles is used to prepare a monolayer of particles using a convective assembly process on an acid-cleaned glass slide.[59] Next, the monolayers are placed inside a metal evaporator for metal deposition. Two types of thermal evaporators are used: (i) a Cooke Vacuum Products (Model FPS-41) and (ii) a Ted Pella (Cressington 308 R) system, the evaporation geometries of which are depicted in Figure 8.3a and b, respectively. Figure 8.3c depicts the evaporation geometry used for the preparation of patchy particles, *i.e.* particles with metallic patches covering less that 50% of the particle surface. In the case of gold-capped Janus particles,

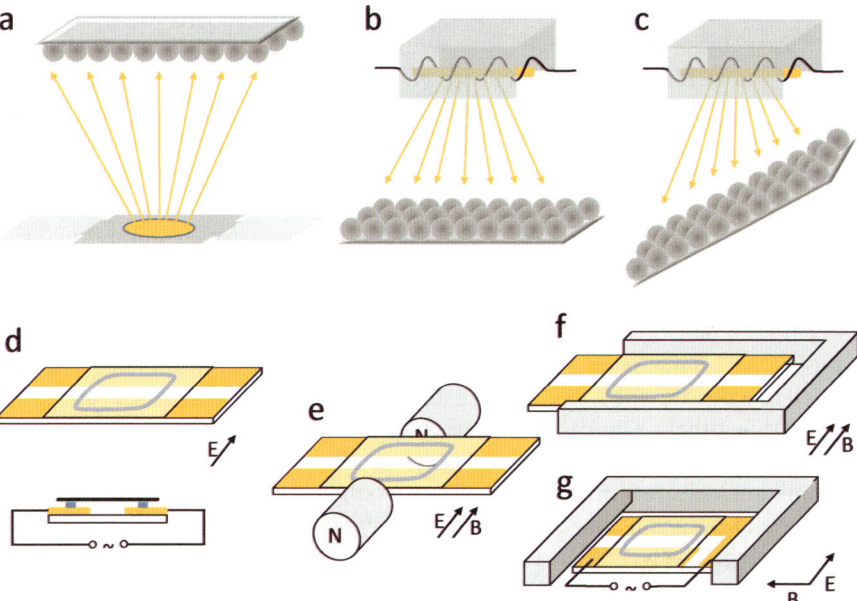

Figure 8.3 Evaporation and cell geometries used in Janus and patchy particle fabrication and assembly. Thermal evaporation set-up in (a) a Cooke Vacuum Product system and a Ted Pella Cressington 308 R system in (b) normal and (c) glancing angle geometry. (d) Cell set-up for AC electric field assembly (top) and cross-sectional view (bottom). (e) Cell set-up for magnetic field assembly. (f) Cell set-up for field assembly in parallel AC electric and magnetic fields. (g) Cell set-up for field assembly in perpendicular AC electric and magnetic mixed biaxial field. E and B indicate electric and magnetic field directions, respectively.

a thin adhesion layer of either 10 nm Cr or 5 nm Ti is deposited first followed by deposition of 20–30 nm of Au. In the case of Fe and iron oxide (Fe_xO_y) caps, no adhesion layer is required and caps of 8, 17, 34 and 50 nm are deposited without (Fe) and in the presence of a 3:1 Ar–O_2 mixture ($p = 5 \times 10^{-3}$ mbar). Finally, the glass slides are placed upside down inside a Petri dish with DI water and are sonicated briefly to remove the Janus particles. In some cases, a small amount of Tween 20 is added as a non-ionic surfactant to prevent particle aggregation. Particle suspensions are stored in a refrigerator and sonicated for 5 min prior to use.

8.2.3   Assembly Cells for Field Assembly

Various modifications of experimental cells are depicted in Figure 8.3d–f. Typically, the cell consists of a glass slide coated with 100 nm thick parallel gold electrodes (with 5 or 10 nm Ti or Cr adhesion layer, respectively) separated by an electrode gap of 3–5 mm, a thin spacer (hydrophobic marker, hydrophobic tape or silicon well) and a microscope cover glass slide. Janus particle suspensions of a few microliters are added to the cell. The cell shown in Figure 8.3d has a gap of 3 mm and a hydrophobic marker spacer 2–3 particle diameters thick and is used for studies of gold-capped Janus particle suspensions in the presence of an AC electric field.[60,61] The AC electric field ranging from 1 to 90 V with frequencies of 1–200 kHz is produced by an Agilent 33120A 15 MHz function generator connected to an RG-91 ramp generator/amplifier and applied to the two gold electrodes. A 1 µF capacitor is added to the circuit to filter any DC component of the applied field. The cell geometry in Figure 8.3e is used for magnetic field-assisted assembly of iron-capped Janus particles and has an electrode gap of 3 mm and a hydrophobic marker spacer of ~2–3 particle diameters.[62] A constant magnetic field is applied *via* a pair of aligned cylindrical permanent magnets (Magkraft, $d = 6.4$ mm, $l = 19.2$ mm, $B_R = 1.29$ T and $H_C = 11.9$ kOe) placed on either side of the cell at a distance of 20 mm from each other and their north and south poles facing each other. The cell geometries depicted in Figure 8.3f are used for magnetic field assembly of various iron oxide-capped Janus particles (top without gold electrodes) and also for the study of these Janus particles in parallel and perpendicular AC electric–magnetic mixed biaxial fields (top and bottom, respectively).[63,64] The electrode gap is 5 mm and a hydrophobic tape from 3M is used as a cell spacer. The magnetic field is provided by a permanent U-shaped magnet with a magnetic field of 0.08 T. The set-up for AC electric field generation is identical with that used for the cell shown in Figure 8.3d.

8.3   Field Assembly of Janus Particles

In the following four subsections, the behavior of Janus particles in four types of fields is discussed: (i) a convective flow field caused by evaporation of water from a drying droplet with a pinned contact line, (ii) a uniaxial AC electric field at low and high frequencies, (iii) a constant uniaxial magnetic field and (iv) perpendicular AC electric and magnetic mixed biaxial fields. Each subsectionr is subdivided into three sections discussing theoretical considerations for the specific field-induced particle–particle interactions, presenting results from experiments and detailing molecular dynamics or two-dimensional (2D) COMSOL simulations of the process dynamics and driving forces.

8.3.1 Janus Particles in Convective Flow Fields

Evaporation of a solvent from a particle-laden, sessile droplet results in deposition of the particulate matter at the rim of the droplet, a phenomenon mentioned above and known under the terms 'coffee stain' or 'coffee ring' problem.[22] Figure 8.4a shows an optical microscopy image of a section of such a rim formed when a 0.5 µL droplet of aqueous 2.4 µm PS particle suspension is dried under ambient conditions.[65] Negatively charged spherical particles form a close-packed layer at the rim when dried on a negatively charged substrate because of the convective flow field inside the droplet. The convective

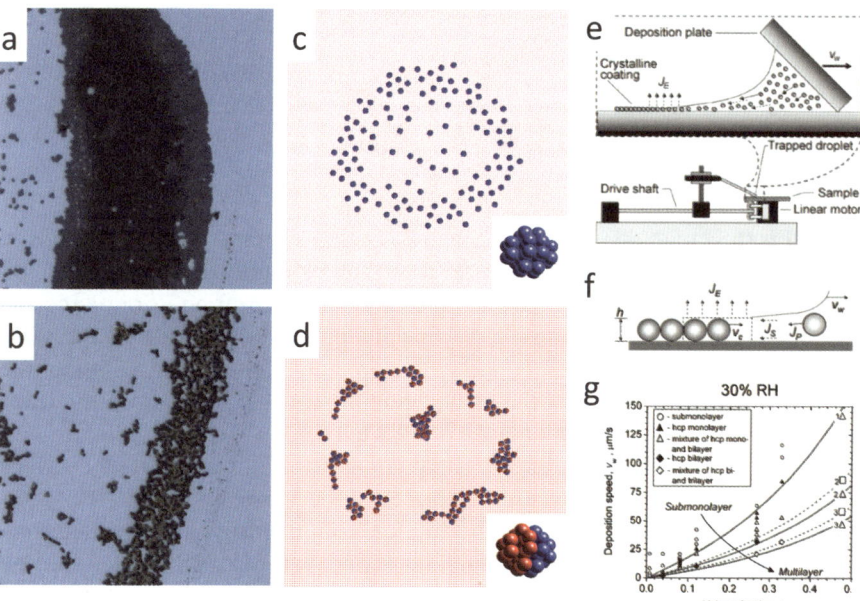

Figure 8.4   Optical images of particle residues at $10 \times$ original magnification left after drying of 0.5 µL of aqueous 2.4 µm PS (a) and gold-capped Janus PS particle (b) suspensions.[65] Top view of particle residue obtained in molecular dynamics simulation using Lennard-Jones type interactions of 2 nm isotropic (c) and Janus particles (d) in a 3.4×10^{-11} nL droplet.[72] Convective flow field assembly: (e) set-up, (f) schematic of the fluid balances and forces during the assembly and (g) operational phase diagram at 30% relative humidity. Parts (d)–(f) reprinted with permission from reference 59. Copyright 2004 American Chemical Society.

flow field is believed to originate from the combination of contact line pinning
and enhanced evaporation at the edges of the droplet.[22] More recently, the
effects of particulate shape[66,67] and surface anisotropy[65] have been studied,
indicating that both parameters affect the deposition pattern and structure
formed. In the following section, we briefly review the theory developed to
describe the convective flow field,[23,68–71] present the results of molecular
dynamics simulations and experiments describing the behavior of Janus
particles during droplet drying[65,72] and introduce an assembly method based
on convective flow fields used for the rapid preparation of colloidal sub-,
mono- and multilayers.[59]

The image in Figure 8.1a(i) schematically depicts a cross-sectional view of a
droplet periphery as the liquid evaporates for the case of a pinned contact line
and non-uniform evaporation of the solvent, *i.e.* evaporative flux is highest at
the rim. As the droplet volume shrinks (solid–dashed–dotted), an outward flow
is established to replenish the solvent at the edge of the droplet. The driving
force for the outward flow is the tendency of the air/liquid interface to
minimize its energy. Assuming an axisymmetric drop and conservation of the
fluid, Deegan *et al.*[23] showed that the vertical averaged radial flow, v, the
position of the air/liquid interface, h, and the evaporation flux, J_S, can be
related as follows:

$$v[(r, t)] = \frac{1}{\rho r h} \int_0^r dr \, r \left[J_S(r, t) \sqrt{1 + \left(\frac{\partial h}{\partial r}\right)^2} + \rho \frac{\partial h}{\partial t} \right] \qquad (8.9)$$

where r is the radial distance from the drop center, t is time and ρ is the density
of the liquid. Equation (8.9) indicates that there is a non-zero radial velocity,
$v(r,t)$, if there is a mismatch between the rates of evaporation $[J_S(r,t)]$ and the
rate of change of the interface $(\partial h/\partial t)$. Specifying h using the pressure term $(\nabla p
= 0)$ of the Navier–Stokes equation (hydrostatic limit) and J_S using the
Laplace eqn $(\nabla^2 u = 0)$ under the assumption of steady-state evaporation leads
to a boundary problem analogous to that used for a charged conductor with
the following boundary conditions: (i) air is saturated along the droplet surface
$(u = u_s = $ constant), (ii) the current normal to the substrate is zero (the vapor
does not penetrate the substrate) and (iii) far from the drop, u converges to u_∞.
Using these assumptions and the boundary conditions, Deegan *et al.*[23] showed
that the evaporation current diverges near the contact line of the droplet and
provide the following analytical solution for the evaporative flux:

$$j(r, \Theta) = j_0 \left(1 - \frac{r^2}{R^2}\right)^{-\lambda(\Theta)} \qquad (8.10a)$$

$$\lambda(\Theta) = \frac{1}{2}\left(1 - \frac{\Theta}{\pi - \Theta}\right) \qquad (8.10b)$$

where R is the radius of the droplet's base and Θ is the contact angle. Since Deegan *et al.*'s seminal work,[22,23] others have furthered the understanding of this problem and its effect on the deposition pattern both theoretically[68–71] and experimentally.[73–76]

8.3.1.2 Evaporative Drying of Particle and Janus Particle-laden Sessile Droplets

In analogy with Deegan *et al.*'s experiments, Guzman *et al.*[65] studied the assembly of isotropic and gold-capped Janus PS particles during evaporative drying. Figure 4a and b show optical microscopy images of a rim section obtained after 0.5 μL of an aqueous suspension of 2.4 μm isotropic and gold-capped Janus PS particles, respectively, had been dried under ambient conditions on a silicon wafer with a native oxide layer. The isotropic PS particles are found to show the typical *hcp* step structure at the rim. In contrast, the gold-capped Janus PS particles form chain-like structures close to the rim, leading to a partially close-packed structure with voids at the rim of the droplet. Comparison of optical images with scanning electron microscopy images confirms that the golden color belongs to Janus particles with their caps facing towards the substrate, while the black color indicates gold caps facing up. The chain formation and more loosely packed rim residue are rationalized by considering the attractive gold–gold, gold–PS and gold–SiO_2 and repulsive PS–SiO_2 and PS–PS interactions. It is likely that attractive interactions lead to preferential 'sticking' of gold caps to other gold caps and polystyrene cores. Near the rim, the height of the liquid film reaches the dimensions of the particle diameter and the particles are more likely also to experience 'sticky' interactions with the substrate surface, leading to arrest of the chains. In the case of isotropic polystyrene particles, all close-range interactions in the system are repulsive, leading to non-sticky particle collisions and easy sliding of particles over the SiO_2 surface. Eventually, convective forces and drying overcome the repulsive interactions, leading to close packing. Colloidal probe atomic force microscopy data confirm the proposed attractive and repulsive interactions.[65]

8.3.1.3 Simulation of Particle and Janus Particle-laden Evaporating Sessile Droplets

In order to understand the difference in evaporation-driven deposition behavior, Chen *et al.*[72] performed molecular dynamics (MD) simulations of isotropic and Janus particle-laden nanosized droplets on a planar substrate. The simulations were performed using standard MD techniques[77–79] assuming Lennard-Jones type interactions:

$$V\left(r_{ij}\right) = 4\epsilon \left[\left(\frac{\sigma}{r_{ij}}\right)^{12} - \left(\frac{\sigma}{r_{ij}}\right)^{6} \right] \qquad (8.11)$$

where ϵ, σ and m are non-dimensional energy, length and mass scales, respectively. In a simulation, a drop consisting of 72 236 fluid atoms in a hemispherical cap of radius 20 nm with 119 randomly placed particles (each comprised of 32 atoms in a face-centered cubic lattice) is placed on a solid monolayer of substrate atoms. The droplet is then allowed to equilibrate and evaporate. The droplet shrinks monotonically due to evaporation, while the particles first settle on the substrate and then are advected to the rim of the droplet. Figure 4c and d display the residues obtained from 3.4×10^{-11} nL droplets with 119 isotropic and Janus particles, respectively. The insets show enlarged images of the 32 atom-containing isotropic and Janus particles. The Janus nature of the particles is implemented by making the atoms in only one of the two hemispheres of a particle attractive to other particle atoms, *i.e.* red and blue atoms attract each other; while all particle atoms attract the fluid and wall in the same way. In good agreement with the experimental data (Fig 8.4b), the Janus particles form chain-like structures, whereas the isotropic particles, which carry a uniform negative charge of 8, remain isolated. Note that only chain structures with head-to-tail particle orientation are observed, whereas no such preference is observed in the experiments. The rationale for this behavior is the fact that attractive gold–gold interactions (red–red) are neglected in the calculations. The discrepancy in deposition pattern between Figure 8.4a and c can be attributed to the difference in droplet size (500 *versus* 3.4×10^{-11} nL) and the resulting short evaporation times in the simulation. If a droplet with a larger number of fluid atoms were to be used in the simulation, the structure in Figure 8.4c is likely to converge to the close-packed structure observed in Figure 8.4c; however, a larger number of fluid atoms would also linearly increase the simulation run time.

8.3.1.4 Rapid Convective Assembly for Monolayer Fabrication

Prevo and Velev[59] developed a method for the assembly of monolayers utilizing convective flow fields. Figure 8.4e depicts the convective assembly set-up consisting of a syringe pump and two glass slides kept at a well-defined angle. Positioning of a microliter droplet of a particle suspension with known volume fraction in the wedge formed by the two glass slides followed by motion of the angled glass slide by the syringe pump leads to the rapid deposition of particle sub-, mono- and multi-layers owing to the controlled dragging of the meniscus, as shown in Figure 8.4f. Dimitrov and Nagayama[80] proposed eqn (8.12) for the description of the colloidal assembly rate, v_c:

$$v_c = \frac{K\phi}{h(1-\varepsilon)(1-\phi)} \tag{8.12a}$$

$$K = j_e \beta l \tag{8.12b}$$

where j_e is the solvent evaporation flux, β is the hydrodynamic parameter, l is the drying length, ϕ is the particle volume fraction and h is the colloidal crystal thickness. Equation (8.12) balances the volumetric fluxes of the solvent and the accumulation of particles in the drying region under the assumption of steady-state assembly. Figure 8.4g depicts an operational 'phase' diagram at 30% relative humidity (RH), where deposition speed is plotted against volume fraction.[59] Higher deposition speeds generally lead to less packed layers, whereas higher volume fractions lead to thicker colloidal crystals being deposited. It would be interesting to test this deposition method with Janus particle suspensions in the presence of AC electric fields similarly to the recently reported combination of convective assembly and AC fields by Kleinert *et al.*[54] However, the current low production rates of Janus particles have so far prevented such experiments.

8.3.2 Janus Particles in Electric Fields

The focus of this section is on AC electric field behavior of metallodielectric Janus particles with a dielectric polystyrene core and a conducting gold cap on one hemisphere. Experiments were performed by Gangwal and co-workers[60,61,81] in the low[60] and high AC frequency range[61] and with particles that have caps smaller than 50% and discontinuous caps.[81] Several reviews on electrokinetic phenomena have included sections on Janus particles. Velev *et al.*[26] reviewed particle-localized AC and DC electric-field manipulation and electrokinetics, whereas Bazant and Squires[82] and Daghighi and Li[83] reviewed induced-charge electrokinetic phenomena. One aspect that is not specifically discussed in this section is the motion of particles in channels, which the reader can find in papers by Honegger *et al.*[84] and Daghighi *et al.*[85]

8.3.2.1 *Theory of Induced-charge Electroosmosis, Induced-charge Electrophoresis and Dielectrophoresis for Janus Particles*

The zeta potential of a Janus particle has been a subject of investigation,[86] but has not yet been measured accurately, because Janus particles will align their dipole with the direction of the field in the presence of an electric field and in some cases even show transverse motion with respect to the field.[60] Either of these two processes will lead to erroneous zeta potential measurements. The fact that asymmetry causes rich effects in electrokinetics was recognized and explored by Long and Ajdari.[87] In a series of studies, Squires and Bazant[31,82,88] researched this problem specifically for metallodielectric Janus particles. They determined that the induced charge electroosmotic (ICEO) flow around an asymmetric conducting or polarizable particle causes particle motion either along or transverse to the applied field and named the phenomenon induced charge electrophoresis (ICEP). Generally, ICEO flow is only observed in a band of driving AC frequencies that can be evaluated by the characteristic timescales of the system charging and polarization:

$$\tau_e^{-1} \geq \omega \geq \tau_p^{-1} \qquad (8.13a)$$

$$\tau_e = \frac{\lambda L}{D} \qquad (8.13b)$$

$$\tau_p = \frac{\lambda R}{D} \qquad (8.13c)$$

where τ_p is the characteristic resistor–capacitor (RC) time for the induced screening cloud with λ as the Debye length and D as the ionic diffusion coefficient, and τ_e represents the charging time of electrodes with $2L$ being the electrode separation. Compared with standard linear electrophoresis, ICEO flow is longer ranged and can be especially important for dense suspensions.

Figure 8.5a shows a schematic diagram of a gold-capped Janus polystyrene particle in one half cycle of an AC electric field in the stable configuration, *i.e.* the plane between the metallic gold (black half) and dielectric polystyrene core (gray half) of the particle is aligned with the electric field direction. A stronger polarization of the electric double layer on the gold-capped hemisphere occurs when the field is applied and induces a stronger ICEO slip flow compared with the ICEO slip flow on the polystyrene hemisphere from the poles of the particle towards the equator. This flow imbalance results in an overall motion of the particle normal to the electric field direction with the dielectric polystyrene hemisphere facing forward in the direction of motion. The effect has been reported experimentally and characterized by Gangwal *et al.*[60] The ICEP velocity with which a Janus particle will move is approximated by:

$$U_{ICEP} = \frac{9}{64} \frac{\varepsilon r E_0^2}{\eta} \left(\frac{1}{1+\delta} \right) \qquad (8.14a)$$

$$\delta = \frac{C_D}{C_S} \qquad \delta \propto \sqrt{c_0} \qquad (8.14b)$$

where ε and η are the permittivity and viscosity of the bulk solvent, respectively, r is the radius of the particles, δ is the differential capacitance ratio, C_D and C_S are the linear differential and linear surface capacitance, respectively, and c_0 is the electrolyte concentration. Equation (8.14) was derived by Squires and Bazant[88] using the following general approach. First, the Laplace equation is solved for a steady-state electric field subjected to the no-flux boundary condition. Next, the induced zeta potential is determined and the total charge condition is enforced. This step is followed by determination of the slip velocity, u_s, and solving of the Stokes equations resulting in the linear ICEP velocity in eqn (8.14). More recently, Kilic and

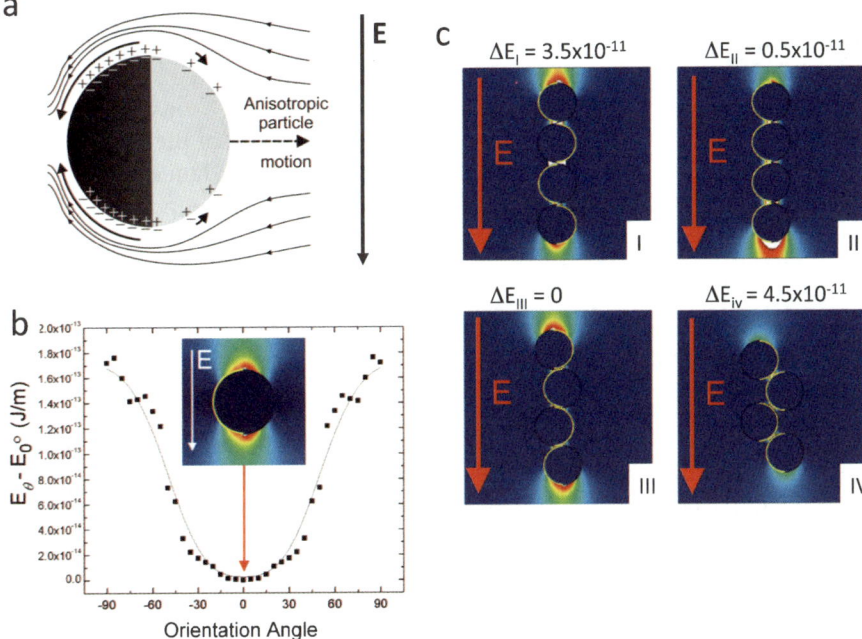

Figure 8.5 (a) Schematic diagram of a gold-capped Janus polystyrene particle in one half cycle of an AC electric field in the stable configuration. The black side indicates gold and gray indicates polystyrene. The electric double layer is more strongly polarized on the gold side than on the polystyrene side, resulting in a stronger ICEO slip on the gold side and an overall ICEP motion in the direction of the dielectric side. Reprinted with permission from reference 60. Copyright 2008 American Physical Society. (b) 2D COMSOL simulation of the effect of particle orientation on the calculated potential difference for various Janus particle orientations in an AC electric field. (c) 2D COMSOL simulation of the electric energy density contours around different particle configurations of four gold-capped Janus polystyrene particles in one half cycle of an AC electric field. The potential energy difference (ΔE) between the four structures is given relative to configuration III. Parts (b) and (c) reprinted with permission from reference 61. Copyright 2008 American Chemical Society.

Bazant[89] investigated the behavior of Janus particles near a wall and found that a Janus particle is always attracted to the wall of a channel or cell. The attraction is a result of the torque originating from the presence of ICEP, which rotates the insulating end of the particle towards the wall.[89] As discussed in Section 8.1.1, the Maxwell–Wagner theory can be employed to predict the DEP force acting on a polarized particle. The frequency-dependent Clausius–Mossotti factor [eqn (8.3a)] determines the orientation of the induced dipole with respect to the applied external AC electric field. At the critical field frequency, ω_c, eqn (8.3a) becomes equal to zero and indicates the transition between positive (attracted to high fields) and negative DEP (attracted towards lower fields). A gold-capped Janus particle will experience positive DEP in a broad AC frequency range due to the infinite conductivity of the gold cap. However, surface modification of the cap with molecular or insulating layers of various thicknesses will strongly affect the conductance of the gold half and result in negative DEP at low frequencies.[90] In the case of a Janus particle in an electrolyte, both ICEP and DEP forces are at work. They oppose each other and are summarized under the term dipolophoresis.[91]

8.3.2.2 Simulation of Metallodielectric Janus Particles in AC Electric Fields

In order to understand the assembly behavior of metallodielectric Janus particles in AC electric fields, the DEP force acting on a single particle and the field intensity distribution of the system need to be calculated.[61] If the assembling particles are more polarizable than the surrounding medium, they experience a positive DEP force in the presence of an AC electric field, which drives them towards high field intensity areas in an electric gradient, while at the same time minimizing the total electric energy of the system, W_e:

$$W_e = \int_V w_{es} dV \qquad (8.15)$$

where w_{es} is the local electric density, which can be calculated for particles in an aqueous medium using the equation:

$$w_{es} = \frac{1}{2} \varepsilon \varepsilon_0 E^2 \qquad (8.16)$$

Simulations on a 2D rendition of a Janus particle with selected orientations by placing a metal-dielectric circle inside a parallel plate capacitor were performed by Gangwal *et al.*[61] The field is kept constant, whereas the orientation of the circle is changed by rotation. The total stored electric energy of the system is calculated by integration over the area of the circle and given in units of J m^{-1}. The 2D Janus particle system is comprised of four subdomains: (i) the aqueous surrounding medium ($\sigma_w = 10^{-4}$ S m^{-1} and $\varepsilon_r = 78$), (ii) a thin conductive, counterionic shell on the polystyrene side of the particle ($\sigma_{ionic} = 0.2$ S m^{-1} and $\varepsilon_r = 78$), (iii) the dielectric polystyrene core of the particle ($\varepsilon_{PS} = 2.55$) and (iv) the metallic section ($\sigma_m = 4 \times 10^7$ S m^{-1} and $\varepsilon_r = 10^9$). Note that owing to its low electrical conductivity, a constant bulk permittivity value is used for the polystyrene core instead of the complex permittivity. The dielectric permittivity for each subdomain is calculated using the complex permittivity, $\tilde{\varepsilon}$, given by Morgan *et al.*[92] as follows:

$$\tilde{\varepsilon} = \varepsilon_r \varepsilon_0 - \frac{i(\sigma)}{\omega} \qquad (8.17)$$

where ε_r is the relative permittivity, ε_0 is the permittivity of vacuum, σ is the electrical conductivity and ω is the AC field frequency. Figure 8.5b depicts the results of the 2D COMSOL simulation for a Janus circle with varying orientations in an AC electric field of $f_{AC} = 10$ kHz.[61] The effect of metal orientation angle on the calculated potential difference $(E_\theta - E_0°)$ is plotted and reveals that an orientation in which the plane between the metallic and dielectric parts aligns with the field direction is the most energetically favorable configuration (Figure 8.5b, inset). An orientation where the metal section faces either of the two electrodes is $(E_{\theta = 90°} - E_0°) = 1.7 \times 10^{-13}$ J m^{-1} higher in energy at a field frequency of 10 kHz. Multiplication of the $(E_{\theta = 90°} - E_0°)$ value by the particle radius (2.5 μm) yields a value of $\sim 100 k_B T$, indicating that a Janus particle in an AC electric field will be arrested at an energy that is likely to prevail the thermal fluctuations and hydrodynamic disturbances in the system.

With the Janus particle orientation established, Gangwal *et al.*[61] increased the number of Janus particles in their simulations from one to four particles to study two straight and two staggered chain configurations. The four configurations tested are depicted in Figure 8.5c. In configurations I, III and IV the caps are facing in alternating directions, whereas in configuration II all caps face to the left. Configurations I and II represent straight particle chains and configurations III and IV are staggered chains. Further, the amount of cap contact increases from configurations I and II to III and is highest in configuration IV. Energetically, configuration III ($\Delta E_{III} = 0$) is the most favorable, followed by II ($\Delta E_{II} = 0.5 \times 10^{-11}$), I ($\Delta E_I = 3.5 \times 10^{-11}$) and IV ($\Delta E_{IV} = 4.5 \times 10^{-11}$). The low energy of configuration III implies that the gold surface dominates the particle assembly at medium and high field intensities, *i.e.* the particles are first oriented by the field as shown in Figure 8.5b, then they are attracted by their poles. Next, the gold caps connect slightly off-center and the longest dipole aligned with the electric fields lines and a narrow conductive gold lane is formed along the center of the staggered particle chain.[61] Interestingly, configuration II is only $\Delta E_{II} = 0.5 \times 10^{-11}$ above configuration III, indicating that particles within chains might be able to rotate around the chain axis, enabling easy chain rotation when two preformed chains connect. However, these chains were not observed experimentally. On the other hand, only slight positional displacement along the field direction, resulting in less (configuration I) and more cap overlap (configuration IV), leads to a dramatic increase in chain energy. Overall, the metal polarization appears as the leading effect in the assembly process and orientation.

8.3.2.3 Experimental Observations of Metallodielectric Janus Particles in AC Electric Fields

The behavior of metallodielectric particles, *i.e.* gold-capped polystyrene particles, in AC electric fields was thoroughly studied and characterized by Gangwal and co-workers[60,61] for both low[60] and high AC electric frequen-

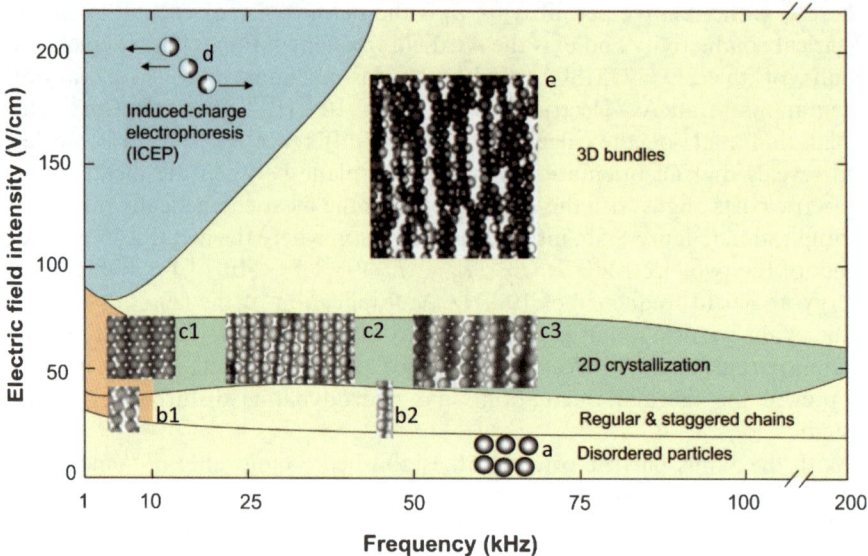

Figure 8.6 Diagram displaying the dynamic and structural response of gold-capped Janus polystyrene particles to AC electric field intensity *versus* field frequency. Reprinted with permission from reference 61. Copyright 2008 American Chemical Society.

cies.[61] Their findings are summarized in the phase diagram shown in Figure 8.6, in which the structural response is given as a function of AC electric field frequency and field intensity.[61] Five assembly regimes are observed (bottom to top): (i) the disordered particles region, (ii) the formation of regular and staggered chains region, (iii) the 2D crystallization region, (iv) the ICEP motion region and (v) the 3D bundles region.

At electric field strengths below 25 V cm^{-1}, the field-induced dipole in the particles is not strong enough to overcome the thermal energy of the system and the particles remain disordered, performing Brownian motion. Gangwal *et al.*[61] reported that most of the particles appear to orient with the gold-capped side facing up and away from the bottom glass slide at these frequencies (Figure 8.6, a). An increase in the field strength to the 25–40 V cm^{-1} range results in a DEP force that is strong enough to result in particle assembly into linear, regular chains at low frequencies (f_{AC} <10 kHz, Figure 8.6, b1) and staggered chains at higher frequencies (f_{AC} > 10 kHz, Figure 8.6, b2). Particles in the regular chains formed below 10 kHz are found to orient such that the gold cap faces upwards. The staggered chains assemble in a two-step process. Initially, when the AC field is turned on, particles rotate to align the plane between the cap and polystyrene with the field line as shown in the inset of Figure 8.5b. Next, the polarized particles are attracted towards each other and form staggered chains, where the caps point alternately left and right and the length of the chain is oriented parallel to the field lines as shown in configuration III depicted in Figure 8.5c. In contrast to what is observed for gold nanoparticle chains,[93] the staggered Janus particle chains fall apart upon removal of the AC field. The van der Waals interaction of the gold point contacts in the Janus particle chains is not large enough to overcome the repulsive interaction of the nearby polystyrene core.

Dipolar attractions between chains at electric field strengths ranging from 40 to 75 V cm^{-1} lead to the next regime in which 2D crystallization is observed. Formation of three types of 2D crystals is detected. At low frequency (1–10 kHz), regular 2D crystals (Figure 8.6, c1) form as a result of association of regular chains (Figure 8.6, b1) in which the gold cap is observed to face upwards. In the 10–200 kHz range, two additional types of crystals are observed. One of the 2D crystal structures (Figure 8.6, c2) is comprised of Janus particle chains formed at lower fields with configuration III shown in Figure 8.5c. The Janus particle chains assemble into a close-packed lattice with narrow parallel metallic lanes. By characterizing the staggered chains and 2D crystallization region using the order and polarization parameters for dipolar liquid crystals, Gangwal *et al.*[61] determined an order parameter of $S = 0.98$ and a polarization parameter of $P = -0.003$, which indicates that the 2D Janus crystal in Figure 8.6, c2 represents a well-aligned antiferroelectric lattice. The third 2D crystal is observed less frequently and characterized by broader lanes formed when two sets of chains associate such that each gold cap is in contact with its four neighboring gold caps near the equator and the poles (Figure 8.6, c3).

An increase in the field to above 75 V cm^{-1} leads to induced charge electrophoretic motion of the Janus particles normal to the field direction at frequencies between 1 and 40 kHz (Figure 8.6, d). The Janus particles move with the polystyrene side pointing forward normal to the field direction (Figure 8.5a) and are found to be dynamically attracted to the bottom of the glass cell, in good agreement with the hydrodynamic attraction observed for swimming microorganisms.[94] Careful investigation[60] of the particle motion in this frequency range shows that the particle velocity increases linearly with the electric field intensity squared ($E_0^2 = 10\,000$–$100\,000$ V^2 cm^{-2}) and the particle diameter ($d = 4$–8 μm) and decreases with the electrolyte concentration ($c_0 = 0$–0.3 mM), in good agreement with eqn (8.14a). As predicted by eqn (8.13), ICEP mobility is observed to decrease near the upper and lower characteristic frequencies of $\tau_e^{-1} = 20$ Hz and $\tau_p^{-1} = 12$ kHz when experiments are performed in experimental cells with 60–80 μm heights. In cells with smaller heights (10–15 μm), ICEP motion persists up to 40 kHz at high field strength. Two discrepancies with the ICEP theory discussed in Section 8.3.2.1 are found, however: (i) an increase in salt concentration leads to an increase in the critical field intensity required for inductance of particle motion and (ii) extrapolation of the electrolyte concentration velocity correlation predicts a zero velocity at an electrolyte concentration of 10 mM. A possible explanation for both observations could be an increased particle–wall interaction at higher electrolyte concentrations, as the Debye lengths of the cell glass surface and the particles are reduced. Finally, at medium to high frequencies (5–200 kHz) and high fields (>75 V cm^{-1}), the formation of 3D bundles is driven by high DEP and particle chaining forces (Figure 8.6, e). These 3D bundles are mostly isolated from each other and stretching towards the two electrodes. Particles within the bundles are able to rotate around their axis and also have their gold caps facing up.

Zhang and Zhu[90] reported that modification of the gold cap with dielectric coatings results in a strong alteration of the dielectrophoretic behavior of metallodielectric Janus particles. Polystyrene particles of 0.915 and 3.8 μm coated with 30 nm gold were further modified with 1-octanethiol (C$_8$), 1-dodecanethiol (C$_{12}$) and 1-hexadecanethiol (C$_{16}$) through formation of a self-assembled monolayer.[95] Coating of the gold cap with thiolated self-assembled monolayers (SAMs) increases the zeta potential from -52.37 mV for the unmodified polystyrene particle to -39.41 mV for the thiol-modified gold-capped Janus particles, indicating that the gold surface is coated with an insulating SAM layer. Unmodified gold-capped Janus particles exhibit positive DEP in the entire frequency range owing to the high conductivity of the gold cap. The Janus particles modified with a thiol coating show a critical DEP frequency, ω_c, that increases with increasing solution conductivity, σ_m, and length of thiol, l_{thiol}, and decreases with increasing particle radius, r:

$$\omega_c \sim \frac{1}{RC} \sim \frac{l_{\text{thiol}}\sigma_m}{\varepsilon_{\text{thiol}}r} \tag{8.18}$$

where $\varepsilon_{\text{thiol}}$ is the dielectric constant of the thiol layer. In the case of $\omega_c < 1/RC$, the electric field cannot penetrate the SAM coating, leading to domination of interfacial polarization of the double layer in the DEP behavior, whereas for $\omega_c > 1/RC$ the field penetrates the thiol layer, leading to a positive DEP behavior that is dominated by the gold cap.

8.3.2.4 *Experimental Observations of Patchy Particles in AC Electric Fields*

Since the gold cap is found to dominate the assembly behavior strongly, it is interesting to investigate the effect of a reduced gold cap size or a non-continuous gold cap on the particle assembly behavior in AC electric fields as it allows the pre-programming of the DEP force. Particles with caps smaller than 50%, so-called patchy particles, can be prepared by using the evaporation geometry shown in Figure 8.3c.[96] Particles with discontinuous caps can be prepared by addition of an inversion step using a partially cured poly-dimethylsiloxane thin film.[97] Gangwal *et al.*[81] investigated the AC field assembly of 2.4 and 5 μm diameter single-patch particles coated with an 11% gold patch and two-patch particles with a 25% patch on each pole. Upon exposure to an AC electric field, the following distinctive behavior is observed. Particles with a single 11% patch form chains at $f_{\text{AC}} < 200$ kHz. With increase in the frequency above 200 kHz, the particle suspension behavior changes dramatically, leading to the formation of parallel and staggered chains aligned with the field direction and unprecedented linear chains normal to the field direction. In the linear chains, particles orient such that their patches are pointing in the same direction. The explanation of this chaining in two directions is based on the unique quadrupolar polarization of the patchy particles. The smaller size of the gold patch leads to induced dipoles of similar magnitude but opposite direction in the gold and polystyrene parts of the particles at high frequencies. The resulting quadrupolar interaction permits chain formation perpendicular to the field direction.

The two-pole patchy particles are found to assemble into regular chains below 50 kHz and into staggered and diagonal chains oriented at an angle of 45° with respect to the field direction at frequencies above 50 kHz. Careful inspection of the patch orientation indicates that at high frequencies the patches are oriented such that they point left and right with respect to the field direction. This patch orientation and the large enough size of the patches allow staggering configurations to occur on both sides of a particle, resulting in diagonal chains. At higher concentrations, the assembly motive of the two-pole patchy particles can be used to form interesting 2D crystal assemblies with rhombic structures.

8.3.3 Janus Particles in Magnetic Fields

In this section, we focus on Janus particles with magnetizable caps made of iron or iron oxide and their assembly behavior upon exposure to a static magnetic field.[62,63] A rich variety of magnetic Janus particles have been prepared by microfluidic synthesis,[98,99] *via* microemulsions[100,101] and through the deposition of a magnetic or other magnetizable materials. For example, nickel caps of various thicknesses have been deposited on polystyrene particles with diameters ranging from 0.5 to 10 μm.[102–105] In one case, where 4.7 μm silica particles were used, the nickel film was capped with an additional 20 nm gold protection layer.[105] Further, 50 nm cobalt caps have been evaporated on 10 μm fluorescent polystyrene particles,[106] and 6/90/6 nm Ag/Fe/Ag stacks have been deposited on 1.86 μm silica microspheres.[107] Silica particles of 4.75 μm with [Co/Pt]$_8$/Pt layers represent an interesting system as this material has been characterized to have a permanent dipole normal to the cap.[108,109] The assembly of particles with off-center magnetic moments normal to the cap has been studied using analytical calculations and Monte Carlo simulations.[110] Fully magnetic, fluorescent particles with diameters ranging from 200 nm to 4.4 μm have been coated with 100 nm of either Al or Au for the study of Brownian-modulated optical nanoprobes.[111]

8.3.3.1 *Dipolar Interactions Between Magnetic Janus Particles*

Currently, a unified theory for the interaction of magnetic Janus particles does not exist. Generally, when a magnetizable object is exposed to an external magnetic field, it develops a magnetic dipole that aligns with the magnetic field direction (see Section 8.1.1). In the case of an asymmetric object, the magnetic dipole develops along the longest axis of the magnetizable material in order to maximize the field interaction. In a particle with a magnetizable cap, the longest axis of magnetization is parallel to the plane between the cap and the particle, leading to the alignment of the Janus particle with the field direction such that the magnetic cap is oriented sideways in a vertical field. Once the particle caps have been magnetized and aligned with the magnetic field, particles are attracted to each other because of dipolar interactions. The dipole–dipole interaction [eqn (8.6)] leading to the assembly of magnetic doublets is strongest when the dipole moments are aligned in a head-to-tail arrangement ($\theta = 0$). The off-center nature of the dipole moment in magnetic Janus particles yields such a head-to-tail alignment when the Janus particles are in a staggered configuration as shown in Figure 8.7a.[62] The caps are facing alternatingly left and right and are touching at the center of the chain, forming a chain structure with a central magnetic lane and maximized magnetic moment (broadly similar to the staggered Janus particle assembly observed in

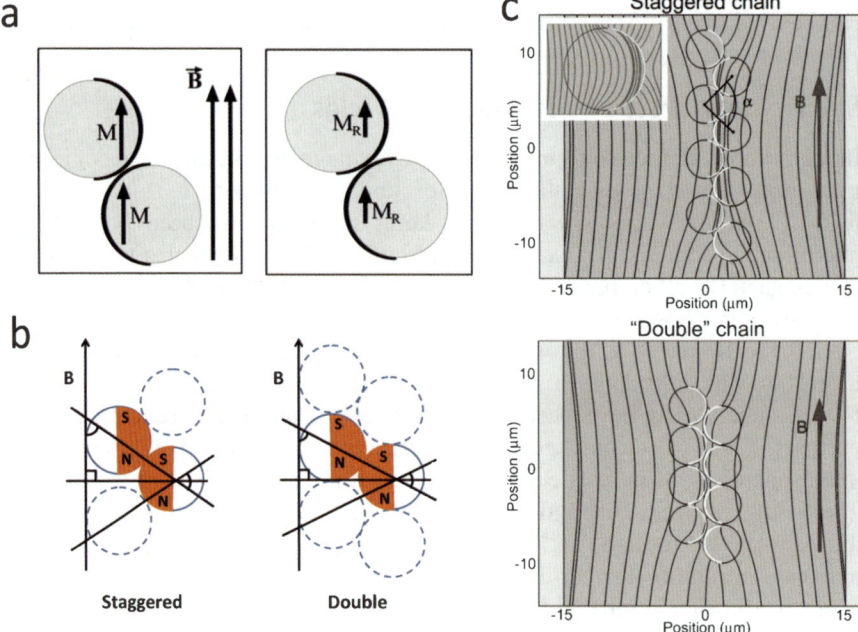

Figure 8.7 (a) Interaction of magnetizable Janus particles in a magnetic field. (b) Schematics of pre-oriented doublets formed upon application of the external magnetic field leading to staggered (left) and double chains (right). (c) 2D COMSOL simulation of the magnetic energy density contours in an applied magnetic field around different magnetic Janus particle configurations. Parts (a) and (c) reproduced from reference 62. Part (b) reprinted with permission from reference 63. Copyright 2012 American Chemical Society.

AC electric fields). In the specific case of magnetization of $Fe_{1-x}O$- and Fe_3O_4-capped Janus particles, two types of doublets are formed with the orientation shown schematically in Figure 8.7b. Their pre-orientation at 43° and 52° with respect to the magnetic field leads to staggered and double chains, respectively,[63] with three-particle angles of 94° and 76°, respectively. The dotted blue circles indicate where additional particles can connect to the pre-formed doublet. Analysis of the magnetic moments in these pre-oriented doublets using eqn (8.6) leads to an attractive interaction for the doublet that is precursor for staggered chains ($\theta = 43°$), while the doublet leading to the double chain structure seems to be at an angle ($\theta = 52°$) where the dipole–dipole force between the two particles is close to zero [*i.e.* $(1 - 3\cos^2\theta) = 0$ for $\theta = 54°$]. As both particle types have the same cap thickness, one would assume that their magnetic moments are located in similar off-center positions and it is not clear at this point what drives this particular doublet pre-orientation. More theoretical analysis is needed to understand the pre-orientation and the resulting stability of the double chain structure.

8.3.3.2 Simulation of Magnetic Field Assembly of Janus Particles

COMSOL simulations using a 2D projection of a Janus particle, *i.e.* a circle with a layer of magnetic material on one half, are used to evaluate the magnetic field distribution and magnetic energy density around a magnetic Janus particle and assemblies of magnetic Janus particles by solving Maxwell's equations. In analogy with the electric field equations [eqns (8.15) and (8.16)], the total magnetic energy, W_B, and the local magnetic energy density, w_B, can be defined as follows:

$$W_B = \int_V w_B \mathrm{d}V \tag{8.19}$$

$$w_B = \frac{1}{2}\frac{B^2}{\mu_0} \tag{8.20}$$

where μ_0 is the permeability of free space ($4\pi \times 10^{-7}$ N A^{-2}). The system is comprised of three subdomains with the following electric conductivities, σ_i, and relative permeabilities, μ_i: (i) water medium ($\sigma_w = 1 \times 10^{-4}$ S m^{-1}, $\mu_w = 1 - \chi_m = 1 - 9.04 \times 10^{-6}$), (ii) dielectric polystyrene ($\sigma_{PS} = 1 \times 10^{-16}$ S m^{-1}, $\mu_{PS} = 1 - \chi_m = 1 - 8.21 \times 10^{-6}$) and (iii) a 34 nm iron shell ($\sigma_{Fe} = 1 \times 10^7$ S m^{-1}, $\mu_{Fe} = 7.00$), where χ_m is the magnetic susceptibility of the materials. The particle diameter used in the simulation is 4 μm. Smoukov *et al.*[62] found that the exact shape of the cap is important because it is thinner at the equator than at the poles of the particle owing to the directional nature of the evaporation process. A reasonable cap shape is achieved by the use of two circles with a diameter of 4 μm off-set by the cap thickness. The inset in the top panel of Figure 8.7c shows that the energetically most favorable orientation of a magnetic Janus particle in a magnetic field is identical with that of a Janus particle in an AC electric field, *i.e.* the plane between the magnetic section and dielectric section is aligned with the field direction. A 90° rotation of an individual Janus particle with a 34 nm cap in a field of 0.15 T results in a total system energy increase of $\sim 10^6 k_B T$. This energetic penalty is found to decrease with decrease in cap thickness. In comparison, the same rotation of a Janus particle in a 100 V cm^{-1} field at 10 kHz requires only $100 k_B T$. Figure 8.7c shows the magnetic energy density contours in an applied magnetic field around magnetic staggered (top) and double (bottom) Janus particle configurations comprised of eight 4 μm polystyrene particles with 34 nm iron caps in a homogeneous magnetic field of 0.15 T (123 400 A m^{-1}).[62] The staggered configuration with a three particle angle of $\alpha = 90°$ is very similar to that predicted for gold-capped Janus particles (Figure 8.5c). Shifting of particle centers into a double chain configuration in which the three-particle angle, α, is 60° is found to be $\sim 28\,000 k_B T$ higher in energy per particle. Overall, the three-particle angle and the energetic difference between structures are strong functions of the cap material's susceptibility and magnetic permeability.

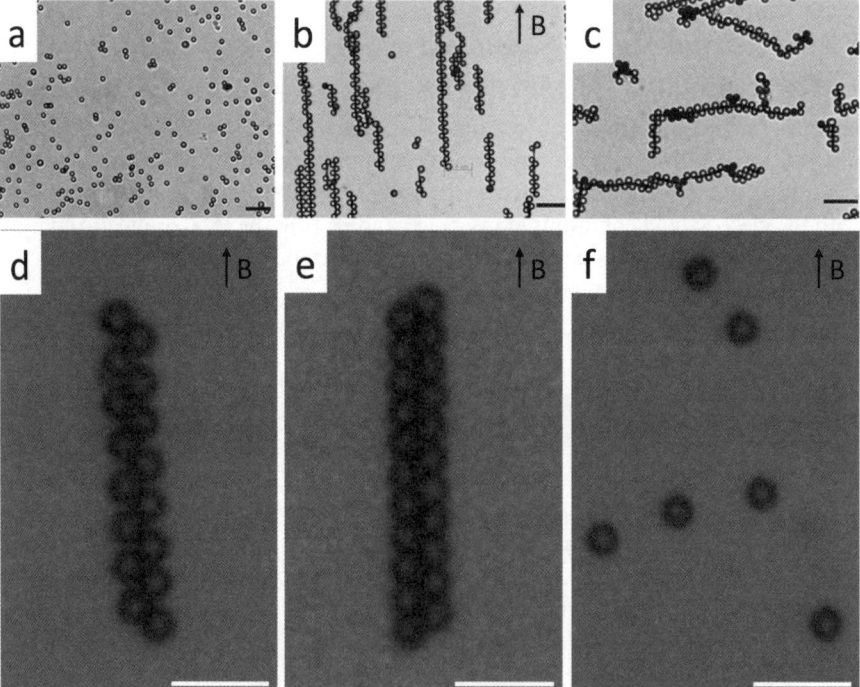

Figure 8.8   Optical micrographs of iron and iron oxide-capped Janus particle assemblies. Iron Janus particles: (a) prior to application of the field, (b) in a 0.15 T field and (c) after removal of the magnetic field (scale bar: 20 μm). Iron oxide Janus particles with (d) $Fe_{1-x}O$ cap, (e) Fe_3O_4 cap and (f) Fe_2O_3 cap in a field of 0.08 T (scale bar: 10 μm). Parts (a)–(c) reproduced from reference 62. Parts (d)–(f) reprinted with permission from reference 63. Copyright 2012 American Chemical Society.

8.3.3.3 Magnetic Field Assembly of Iron and Iron Oxide Janus Particles

Smoukov *et al.*[62] investigated the behavior of 8 and 34 nm iron-capped 4 µm polystyrene particles in the presence of a static magnetic field of 0.15 T and found that particles assemble into stable staggered (34 nm Fe cap) and metastable double chains (8 nm Fe cap). The top row of Figure 8.8 depicts optical micrographs of 4 µm iron-capped Janus particles (a) before exposure to the magnetic field, (b) during application of the 0.15 T magnetic field and (c) after removal of the magnetic field.[62] By introduction of a 75:25 Ar–O_2 mixture during the evaporation and variation of the iron evaporation rate, Ren *et al.*[63] were able to fabricate caps of ferromagnetic non-stoichiometric $Fe_{1-x}O$, ferromagnetic Fe_3O_4 and antiferromagnetic α-Fe_2O_3, which show staggered chain, double chain and no assembly behavior, respectively, in a 0.08 T magnetic field, as shown in Figure 8.8d–f. They further show that identical assembly behavior is observed for cap thicknesses ranging from 17 to 50 nm, with the assembly stability decreasing with decreasing cap thickness owing to reduced magnetic interactions, in good agreement with the predictions of the simulations discussed above. The two types of chains observed, staggered and double, show distinctly different mechanical properties. Whereas the staggered chains are highly flexible upon removal of the external magnetic field, double chains retain their straight configuration owing to their close-packed rigid structure. Interestingly, when longer chains are grown, the formation of slightly twisted double chains is observed, which could give some insight into the magnetic interactions between the particles within a chain. One puzzling difference between the two studies is the observation of double chains for 8 nm Fe-capped Janus particles and their low stability, *i.e.* the chains fall apart upon agitation in the absence of an external magnetic field. A potential explanation for the formation and low stability of the 8 nm Fe-capped Janus particle double chains is that their caps are oxidized when exposed to the aqueous assembly environment, resulting in weakening of their magnetic moments. This rationale would be in good agreement with the easy oxidation of iron in aqueous environments, the low amount of cap material and also the observed reduction of double chain stability in the presence of capillary forces during convective flow assembly when the Fe_3O_4 cap thickness is reduced from 50 to 17 nm.

8.3.4 Janus Particles in Biaxial Fields

The rich behavior found for Janus particles in electric and magnetic fields and the improved control over particle assembly demonstrated for isotropic, spherical particles (see Section 8.1.2), warrants the investigation of Janus particle assembly in biaxial fields. Figure 8.9 shows one result from such a study for the assembly behavior of non-magnetized 2.4 µm Fe_3O_4 capped-Janus particles when they are first exposed to a horizontal AC electric field

(V_{AC} = 187.5 V, f_{AC} = 75 kHz) followed by addition of a vertical magnetic field of 0.08 T.[64] Initially, the particles assemble into the typical staggered chain configuration (L1 = 43.8 μm), in which the Fe_3O_4 caps point alternatingly left and right such that their caps touch at the center of the chain (Figure 8.9a). Upon application of the vertical magnetic field of 0.08 T, the caps are magnetized by the external field and the chain rotates along the chain axis in order to align the long axis of the Fe_3O_4 caps with the magnetic field direction (Figure 8.9b). As a result of the magnetization and chain rotation, all caps are now aligned such that their magnetic dipoles are oriented parallel to each other in the energetically most unfavorable θ = 90° configuration. The highly unstable configuration in Figure 8.9b is rendered metastable by the horizontally applied AC electric field. The instability of the chain is apparent because the chain axis is no longer oriented perfectly parallel to the horizontal AC electric field. The presence of thermal fluctuations provides the particles in the chain with sufficient energy to overcome the AC field barrier and reorganize *via* a stepping process (Figure 8.9c–e) into the energetically more favorable θ = 0° configuration, in which the magnetic moments of the caps are aligned head-to-tail in the direction of the magnetic field. Simultaneously with the directional rearrangement, the chain also

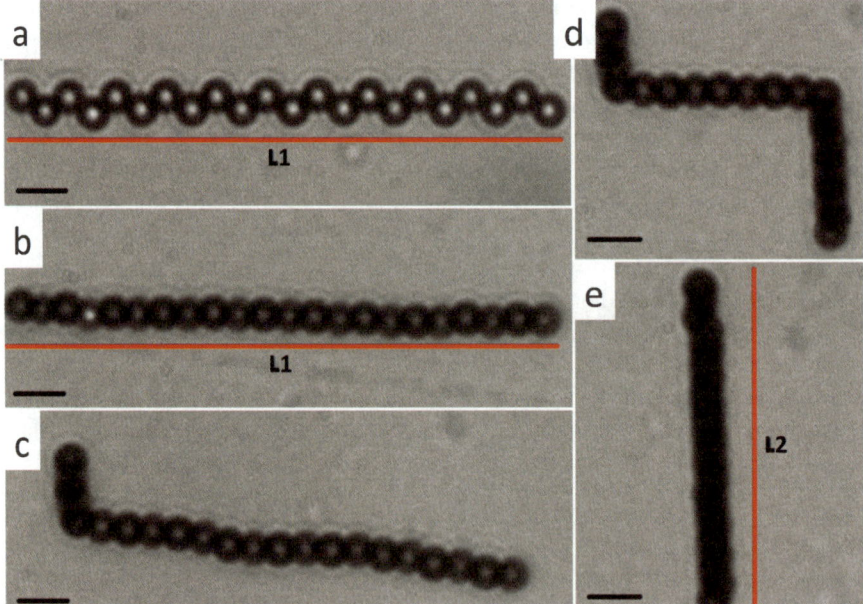

Figure 8.9 Behavior of Fe_3O_4-capped Janus particles in perpendicularly applied AC electric/magnetic biaxial fields. (a) Initial chain configuration with L1 = 43.8 μm in a horizontal AC electric field. (b) Immediate response to vertical magnetic field. (c)–(e) The chain orientation changes from horizontal to vertical in t = 24 s and the chain configuration changes from staggered to double with L2 = 28.4 μm.[64]

collapses from a staggered configuration into the shorter double-chain configuration (L2 = 28.4 μm) typical for Fe_3O_4-capped particles in a magnetic field. The assembly behavior observed can potentially be used for the specific placement of magnetic chains in designer materials. In addition, by careful choice of cap thickness/material and AC electric and magnetic field strength, this process can be rendered reversible.

8.4 Future Outlook

Janus particles represent a new class of materials with interesting properties that make them field addressable. There is no doubt that we will see more and more applications that will make use of the programmability of the Janus particle interactions. This final section reviews a couple of envisioned, conceptualized and realized applications of Janus particles. The earliest application of field-addressable Janus particles, so-called *Gyricon balls*, is in electronic paper. A Gyricon ball is a micrometer-sized sphere comprised of two polymer halves, where each half has either 20% of black or white pigments in it. A small amount of charge additive (0.1%) is added to each half to enhance the formation of charge across the particle volume.[112] The particle will rotate, showing its black or white half depending on the direction of the applied voltage, thereby allowing the display of black-on-white text. Embedment of electroresponsive photonic Janus particles into a Gyricon display[113–115] (*e.g.* Gyricon particles in a polydimethylsiloxane film swollen with hexadecane) was demonstrated by Kim *et al.*[116] and shows the potential for direction-independent reflection color and color-electronic ink displays. Figure 8.10 summarizes four selected applications for metal-capped Janus particles and Janus particles capped with magnetic material. Figure 8.10a illustrates the use of Janus particles by application of magnetic fields for a magnetoresponsive fluorescent switch allowing free writing on a well-defined Janus bead panel in analogy to the electronic paper described above.[98] The particles are fabricated using a microfluidic method that allows loading of one half with magnetic material and the other half with cadmium selenide quantum dots. A simulation of a microfluidic valve using the electroosmotic motion of a metallodielectric Janus particle is depicted in Figure 8.10b.[117] The basic function to be performed by the Janus particle is to block one of two outlet streams. The Janus particle moves to block the right channel upon application of a horizontal electric field because of vortices formed by the induced-charge electrokinetic flow around the conducting side of the Janus particle (top row). Application of a second, dominant vertical electric field results in the dislodging of the Janus particle (middle row) and its alignment with the dominant field followed by motion of the Janus particle to block the top channel (bottom row). Removal of the vertical field and re-application of the horizontal field to the system reverse the process.

Figure 8.10c provides the proof-of-concept for a third application that uses magnetic Janus particles for magnetolytic cancer therapy.[100] Janus particles

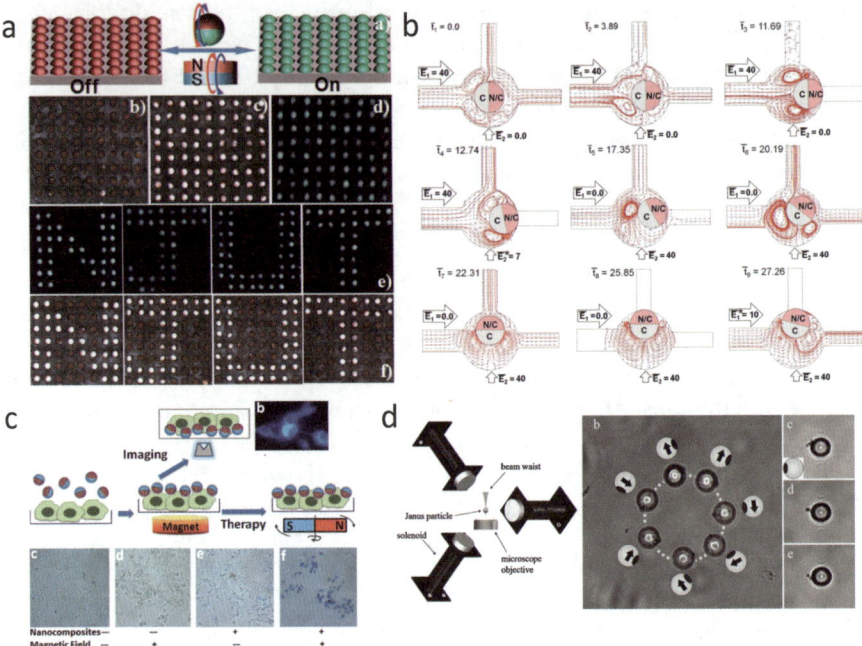

Figure 8.10   Applications of metal-capped and magnetic Janus particles. (a) Responsive Janus supraballs in a flexible bead display. (b) Simulation of a Janus particle-driven microfluidic valve. (c) Envisioned use of magnetic Janus particles in the destruction of cancerous cells. (d) Holonomic control of a Janus particle in an optomagnetic trap. Parts (a) and (d) reprinted with permission from references 98 and 106, respectively Copyright 2011 and 2009 Wiley-VCH Verlag GmbH. Part (b) reproduced from reference 117. Part (c) reprinted with permission from reference 100. Copyright 2010 American Chemical Society.

are prepared *via* ultrasonic emulsification of chloroform containing a mixture of an amphiphilic diblock copolymer and hydrophobic magnetic nanoparticles and an aqueous phase containing poly(vinyl alcohol) (PVA) followed by slow evaporation of the chloroform resulting in Janus particles. The PVA phase is labeled with a fluorescent dye. The Janus particles are settled on a cell layer by exposure to a magnetic field (Figure 8.10c, top panel). Subsequently, the cell layer can either be fluorescently imaged or exposed to a spinning magnetic field (50 rpm), which results in destruction of the cells.[100] The bottom panel of Figure 8.10c shows four cell cultures that have been exposed to various combinations of Janus particles and magnetic field. Only in the case where both Janus particles and magnetic field are applied is the number of live cells reduced by 77%. The fourth application is the use of magnetic Janus particles as sensors.[106,107] Figure 8.10d shows the set-up (left panel) required for the direct control of five degrees of freedom of a Janus dot particle, *i.e.* a Janus particle with a cap smaller than 50%. The cobalt-capped Janus dot particle is prepared *via* a templating method. The optical trap is used to control the three translational degrees of freedom, whereas the solenoid creates a rotating horizontal magnetic field that orients the Janus dot particle with respect to the surface normal of the imaging plane (middle panel). The rotating magnetic field can also be applied vertically, leading to the rotation of the particle around its axis in the imaging plane (right panel). The control over five degrees of freedom will permit the measurement of mechanical properties of large biomolecules and also the bottom-up assembly of very well-defined designer materials. In addition to the control of as many as five degrees of freedom, Janus particles in general also provide interesting opportunities to observe the Brownian rotation of a spherical object, which can then in turn be used to obtained physical information about the surrounding fluid.[103,107]

In conclusion, we have discussed the unusual properties of Janus particles in various external fields. The research field has revealed a rich variety of effects, the uncovering of which has greatly stimulated the recent developments in the theory of the polarization and interactions of such particles and the simulation of structures resulting from field-driven assembly. As the area is recent and still in turbulent development, many new effects and structures are likely to be reported in the coming years. As we have exemplified above, field-driven Janus particles are already finding applications in multiple new materials and devices; however, this is another field where we have so far only scratched the surface of the potential applications. The name of the two-faced Roman god has remained permanently embedded in present-day science.

Acknowledgements

I.K. acknowledges support by the National Science Foundation under a Career Award (No. CBET-06-44789) co-funded by the Division of Chemistry and the Office of Multidisciplinary Activities. O.D.V. acknowledges support from the NSF Research Triangle MRSEC (DMR-1121107).

References

1. S. C. Glotzer and M. J. Solomon, *Nat. Mater.*, 2007, **6**, 557.
2. M. Grzelczak, J. Vermant, E. M. Furst and L. M. Liz-Marzan, *ACS Nano*, 2010, **4**, 3591.
3. Special Issue, *Macromol. Rapid Commun.*, 2010, **31**, 99.
4. Special Issue, *Curr. Opin. Colloid Interface Sci.*, 2011, **16**, 81.
5. S. Jiang, Q. Chen, M. Tripathy, E. Luijten, K. S. Schweizer and S. Granick, *Adv. Mater.*, 2010, **22**, 1060.
6. I. Kretzschmar and J. H. Song, *Curr. Opin. Colloid Interface Sci.*, 2011, **16**, 84.
7. M. Lattuada and T. A. Hatton, *Nano Today*, 2011, **6**, 286.
8. F. Li, D. P. Josephson and A. Stein, *Angew. Chem. Int. Ed.*, 2011, **50**, 360.

9. E. Duguet, A. Désert, A. Perro and S. Ravaine, *Chem. Soc. Rev.*, 2011, **40**, 941.

10. H. Löwen, *J. Phys.: Condens. Matter*, 2001, **13**, R415.

11. P. J. Rankin, J. M. Ginder and D. J. Klingenberg, *Curr. Opin. Colloid Interface Sci.*, 1998, **3**, 373.

12. O. D. Velev and K. H. Bhatt, *Soft Matter*, 2006, **2**, 738.

13. O. D. Velev and S. Gupta, *Adv. Mater.*, 2009, **21**, 1897.

14. M. Conradi, M. Ravnik, M. Bele, M. Zorko, S. Zumer and I. Musevic, *Soft Matter*, 2009, **5**, 3905.

15. M. Conradi, M. Zorko and I. Musevic, *Opt. Express*, 2010, **18**, 500.

16. H. R. Jiang, N. Yoshinaga and M. Sano, *Phys. Rev. Lett.*, 2010, **105**, 268302.

17. I. Buttinoni, G. Volpe, F. Kümmel, G. Volpe and C. Bechinger, *arXiv:1110.2202v3 [cond-mat.soft]*, 2011.

18. R. Golestanian, T. B. Liverpool and A. Ajdari, *New J. Phys.*, 2007, **9**, 126.

19. J. R. Howse, R. A. L. Jones, A. J. Ryan, T. Gough, R. Vafabakhsh and R. Golestanian, *Phys. Rev. Lett.*, 2007, **99**, 048102.

20. S. J. Ebbens and J. R. Howse, *Langmuir*, 2011, **27**, 12293.

21. A. B. Pawar, Doctoral Thesis, City College of New York, 2009.

22. R. D. Deegan, O. Bakajin, T. F. Dupont, G. Huber, S. R. Nagel and T. A. Witten, *Nature*, 1997, **389**, 827.

23. R. D. Deegan, O. Bakajin, T. F. Dupont, G. Huber, S. R. Nagel and T. A. Witten, *Phys. Rev. E*, 2000, **62**, 756.

24. Y. H. Wang and W. D. Zhou, *J. Nanosci. Nanotechnol.*, 2010, **10**, 1563.

25. B. G. Prevo, D. M. Kuncicky and O. D. Velev, *Colloids Surf. A*, 2007, **311**, 2.

26. O. D. Velev, S. Gangwal and D. N. Petsev, *Annu. Rep. Prog. Chem.*, 2009, **105**, 213.

27. M. Trau, D. A. Saville and I. A. Aksay, *Langmuir*, 1997, **13**, 6375.

28. R. J. Hunter, *Foundations of Colloid Science*, 2nd edn, Oxford University Press, Oxford, 2001.

29. M. Bohmer, *Langmuir*, 1996, **12**, 5747.

30. D. C. Prieve, P. J. Sides and C. L. Wirth, *Curr. Opin. Colloid Interface Sci.*, 2010, **15**, 160.

31. T. M. Squires and M. Z. Bazant, *J. Fluid Mech.*, 2004, **509**, 217.

32. K. D. Barbee, A. P. Hsiao, M. J. Heller and X. H. Huang, *Lab Chip*, 2009, **9**, 3268.

33. R. Pethig, *Biomicrofluidics*, 2010, **4**, 022811.

34. K. H. Bhatt and O. D. Velev, *Langmuir*, 2004, **20**, 467.

35. S. Kim, R. Asmatulu, H. L. Marcus and F. Papadimitrakopoulos, *J. Colloid Interface Sci.*, 2011, **354**, 448.

36. T. Honegger, K. Berton, E. Picard and D. Peyrade, *Appl. Phys. Lett.*, 2011, **98**, 181906.

37. S. O. Lumsdon, E. W. Kaler and O. D. Velev, *Langmuir*, 2004, **20**, 2108.

38. J. H. E. Promislow and A. P. Gast, *Phys. Rev. E*, 1997, **56**, 642.

39. D. Wirtz and M. Fermigier, *Langmuir*, 1995, **11**, 398.
40. J. de Vicente, D. J. Klingenberg and R. Hidalgo-Alvarez, *Soft Matter*, 2011, **7**, 3701.
41. M. Lattuada, M. Furlan, A. Butté and M. Morbidelli, *Chimia*, 2009, **63**, 78.
42. S. Singamaneni, V. N. Bliznyuk, C. Binek and E. Y. Tsymbal, *J. Mater. Chem.*, 2011, **21**, 16819.
43. S. A. Majetich and M. Sachan, *J. Phys. D: Appl. Phys.*, 2006, **39**, R407.
44. G. K. Auernhammer, D. Collin and P. Martinoty, *J. Chem. Phys.*, 2006, **124**, 204907.
45. J. P. Ge, L. He, Y. X. Hu and Y. D. Yin, *Nanoscale*, 2011, **3**, 177.
46. B. Huke and M. Lucke, *Rep. Prog. Phys.*, 2004, **67**, 1731.
47. A. Snezhko, *J. Phys.: Condens. Matter*, 2011, **23**, 153101.
48. R. M. Erb, H. S. Son, B. Samanta, V. M. Rotello and B. B. Yellen, *Nature*, 2009, **457**, 999.
49. M. E. Leunissen, H. R. Vutukuri and A. van Blaaderen, *Adv. Mater.*, 2009, **21**, 3116.
50. Y. Nagaoka, H. Morimoto and T. Maekawa, *Langmuir*, 2011, **27**, 9160.
51. F. Smallenburg and M. Dijkstra, *J. Chem. Phys.*, 2010, **132**, 204508.
52. W. D. Ristenpart, I. A. Aksay and D. A. Saville, *Phys. Rev. Lett.*, 2003, **90**, 128303.
53. H. J. Schöpe, *J. Phys.: Condens. Matter*, 2003, **15**, L533.
54. J. Kleinert, S. Kim and O. D. Velev, *Langmuir*, 2012, **28**, 3037.
55. A. B. Pawar and I. Kretzschmar, *Macromol. Rapid Commun.*, 2010, **31**, 150.
56. A. Perro, S. Reculusa, S. Ravaine, E. B. Bourgeat-Lami and E. Duguet, *J. Mater. Chem.*, 2005, **15**, 3745.
57. K. J. Lee, J. Yoon and J. Lahann, *Curr. Opin. Colloid Interface Sci.*, 2011, **16**, 195.
58. A. Walther and A. H. E. Muller, *Soft Matter*, 2008, **4**, 663.
59. B. G. Prevo and O. D. Velev, *Langmuir*, 2004, **20**, 2099.
60. S. Gangwal, O. J. Cayre, M. Z. Bazant and O. D. Velev, *Phys. Rev. Lett.*, 2008, **100**, 058302.
61. S. Gangwal, O. J. Cayre and O. D. Velev, *Langmuir*, 2008, **24**, 13312.
62. S. K. Smoukov, S. Gangwal, M. Marquez and O. D. Velev, *Soft Matter*, 2009, **5**, 1285.
63. B. Ren, A. Ruditskiy, J. H. Song and I. Kretzschmar, *Langmuir*, 2012, **28**, 1149.
64. A. Ruditskiy, B. Ren and I. Kretzschmar, *Soft Matter*, 2012, under review.
65. F. Guzman, E. Cranston, M. Rutland and I. Kretzschmar, in preparation.
66. J. Vermant, *Nature*, 2011, **476**, 286.
67. P. J. Yunker, T. Still, M. A. Lohr and A. G. Yodh, *Nature*, 2011, **476**, 308.
68. Y. O. Popov, *Phys. Rev. E*, 2005, **71**, 036313.
69. T. Okuzono, M. Kobayashi and M. Doi, *Phys. Rev. E*, 2009, **80**, 021603.
70. A. M. Cazabat and G. Guena, *Soft Matter*, 2010, **6**, 2591.
71. K. L. Maki and S. Kumar, *Langmuir*, 2011, **27**, 11347.

72. W. Chen, J. Koplik and I. Kretzschmar, *Phys. Rev. Lett.*, 2012, under review.

73. R. D. Deegan, *Phys. Rev. E*, 2000, **61**, 475.

74. T. P. Bigioni, X. M. Lin, T. T. Nguyen, E. I. Corwin, T. A. Witten and H. M. Jaeger, *Nat. Mater.*, 2006, **5**, 265.

75. H. Bodiguel and J. Leng, *Soft Matter*, 2010, **6**, 5451.

76. R. Bhardwaj, X. H. Fang, P. Somasundaran and D. Attinger, *Langmuir*, 2010, **26**, 7833.

77. M. P. Allen and D. J. Tildesley, *Computer Simulations of Liquids*, 1st edn, Oxford University Press, New York, 1987.

78. D. Frenkel and B. Smit, *Understanding Molecular Simulation*, Elsevier, Amsterdam, 2002.

79. D. C. Rapaport, *The Art of Molecular Dynamics Simulation*, 2nd edn, Cambridge University Press, Cambridge, 1995.

80. A. S. Dimitrov and K. Nagayama, *Chem. Phys. Lett.*, 1995, **243**, 462.

81. S. Gangwal, A. B. Pawar, I. Kretzschmar and O. D. Velev, *Soft Matter*, 2010, **6**, 1413.

82. M. Z. Bazant and T. M. Squires, *Curr. Opin. Colloid Interface Sci.*, 2010, **15**, 203.

83. Y. Daghighi and D. Q. Li, *Microfluid. Nanofluid.*, 2010, **9**, 593.

84. T. Honegger, O. Lecarme, K. Berton and D. Peyrade, *Microelectron. Eng.*, 2010, **87**, 756.

85. Y. Daghighi, Y. D. Gao and D. Q. Li, *Electrochim. Acta*, 2011, **56**, 4254.

86. N. P. Pardhy and B. M. Budhlall, *Langmuir*, 2010, **26**, 13130.

87. D. Long and A. Ajdari, *Phys. Rev. Lett.*, 1998, **81**, 1529.

88. T. M. Squires and M. Z. Bazant, *J. Fluid Mech.*, 2006, **560**, 65.

89. M. S. Kilic and M. Z. Bazant, *Electrophoresis*, 2011, **32**, 614.

90. L. Zhang and Y. X. Zhu, *Appl. Phys. Lett.*, 2010, **96**, 141902.

91. A. M. Boymelgreen and T. Miloh, *Phys. Fluids*, 2011, **23**, 072007.

92. H. Morgan and N. G. Green, *AC Electrokinetics: Colloids and Nanoparticles*, Research Studies Press, Baldock, 2003.

93. K. D. Hermanson, S. O. Lumsdon, J. P. Williams, E. W. Kaler and O. D. Velev, *Science*, 2001, **294**, 1082.

94. A. P. Berke, L. Turner, H. C. Berg and E. Lauga, *Phys. Rev. Lett.*, 2008, **101**, 038102.

95. A. Ulman, *Chem. Rev.*, 1996, **96**, 1533.

96. A. B. Pawar and I. Kretzschmar, *Langmuir*, 2008, **24**, 355.

97. A. B. Pawar and I. Kretzschmar, *Langmuir*, 2009, **25**, 9057.

98. S. N. Yin, C. F. Wang, Z. Y. Yu, J. Wang, S. S. Liu and S. Chen, *Adv. Mater.*, 2011, **23**, 2915.

99. L. B. Zhao, L. Pan, K. Zhang, S. S. Guo, W. Liu, Y. Wang, Y. Chen, X. Z. Zhao and H. L. W. Chan, *Lab Chip*, 2009, **9**, 2981.

100. S. H. Hu and X. H. Gao, *J. Am. Chem. Soc.*, 2010, **132**, 7234.

101. A. K. F. Dyab, M. Ozmen, M. Ersoz and V. N. Paunov, *J. Mater. Chem.*, 2009, **19**, 3475.

102. I. Sinn, P. Kinnunen, S. N. Pei, R. Clarke, B. H. McNaughton and R. Kopelman, *Appl. Phys. Lett.*, 2011, **98**, 024101.

103. B. H. McNaughton, P. Kinnunen, M. Shlomi, C. Cionca, S. N. Pei, R. Clarke, P. Argyrakis and R. Kopelman, *J. Phys. Chem. B*, 2011, **115**, 5212.

104. B. H. McNaughton, M. Shlomi, P. Kinnunen, C. Cionca, S. N. Pei, R. Clarke, P. Argyrakis and R. Kopelman, *Appl. Phys. Lett.*, 2010, **97**, 144103.

105. L. Baraban, A. Erbe and P. Leiderer, *Eur. Phys. J. E*, 2007, **23**, 129.

106. R. M. Erb, N. J. Jenness, R. L. Clark and B. B. Yellen, *Adv. Mater.*, 2009, **21**, 4825.

107. B. H. McNaughton, R. R. Agayan, J. X. Wang and R. Kopelman, *Sens. Actuators B: Chem.*, 2007, **121**, 330.

108. M. Albrecht, G. H. Hu, I. L. Guhr, T. C. Ulbrich, J. Boneberg, P. Leiderer and G. Schatz, *Nat. Mater.*, 2005, **4**, 203.

109. L. Baraban, D. Makarov, M. Albrecht, N. Rivier, P. Leiderer and A. Erbe, *Phys. Rev. E*, 2008, **77**, 031407.

110. S. Kantorovich, R. Weeber, J. J. Cerda and C. Holm, *Soft Matter*, 2011, **7**, 5217.

111. C. J. Behrend, J. N. Anker and R. Kopelman, *Appl. Phys. Lett.*, 2004, **84**, 154.

112. J. M. Crowley, N. K.Sheridon and L. Romano, *J. Electrostat.*, 2002, **55**, 247.

113. T. Nisisako, T. Torii, T. Takahashi and Y. Takizawa, *Adv. Mater.*, 2006, **18**, 1152.

114. R. Ishikawa, M. Omodani and S. Maeda, *J. Imaging Sci. Technol.*, 2006, **50**, 168.

115. N. K. Sheridon, E. A. Richley, J. C. Mikkelsen, D. Tsuda, J. M. Crowley, K. A. Oraha, M. E. Howard, M. A. Rodkin, R. Swidler and R. Sprague, *J. Soc. Inf. Display*, 1999, **7**, 141.

116. S. H. Kim, S. J. Jeon, W. C. Jeong, H. S. Park and S. M. Yang, *Adv. Mater.*, 2008, **20**, 4129.

117. Y. Daghighi and D. Q. Li, *Lab Chip*, 2011, **11**, 2929.

CHAPTER 9

DNA Self-assembly: from Nanostructures to Macro-engineering

YI CHEN*[a], ABIGAIL K. R. LYTTON-JEAN*[a] AND HYUKJIN LEE*[b]

[a] The Koch Institute for Integrative Cancer Research at MIT, Department of Chemical Engineering, Massachusetts Institute of Technology, 77 Massachusetts Avenue, Cambridge, MA 02139, USA; [b] College of Pharmacy, Department of Chemistry and Nanoscience, Ewha Womans University, Pharmaceutical Science Building A-306, 11-1 Deaheyn-Dong, Seodaemun-Gu, Seoul 120-750, Korea
*E-mail: yic@mit.edu or aklytton@mit.edu or hyukjin@ewha.ac.kr

9.1   Introduction

Nanotechnology has developed rapidly for applications in engineering, medicine and materials.[1-3] There are two approaches for the construction of nanomaterials: 'top-down' and 'bottom-up.'[4-6] The 'top-down' systems encompass the microscopic manipulations of small numbers of atoms or molecules to form elegant patterns on the nanoscale. On the other hand, the 'bottom-up' approach entails many molecules self-assembling in parallel steps, using their molecular recognition properties, to form nanomaterials. DNA molecules have appealing features for use in nanotechnology.[7] DNA is composed of two complementary single strands which hybridize to form a DNA double helix, often referred to as a duplex. The duplex has a diameter of

RSC Smart Materials No. 1
Janus Particle Synthesis, Self-Assembly and Applications
Edited by Shan Jiang and Steve Granick
© The Royal Society of Chemistry 2012
Published by the Royal Society of Chemistry, www.rsc.org

about 2 nm and is composed of short structural repeating units spaced ~3.4 Å apart. Owing to its size and sequence-specific recognition properties between complementary single strands, DNA has become a key player in many bottom-up approaches in nanotechnology.

The most amazing feature of DNA is the complementary relationships that form between single strands. The single strands are composed of four different bases [adenine (A), thymine (T), guanine (G) and cytosine (C)], which will hydrogen bond together to form A–T and G–C interactions called Watson–Crick base pairs.[8] Owing to the sequence-specific complementarity between single strands, DNA can be used as 'molecular Velcro' to program the assembly of itself or nanomaterials to which the DNA is attached. Based on the powerful recognition properties of DNA, the construction of more complicated, well-defined structures becomes possible.

A limitation associated with the use of DNA as an assembly tool is the unbranched nature of the DNA molecule. Because DNA is linear, the structures composed of DNA are confined to varying lengths of DNA molecules or circular DNA molecules. In order to overcome this limitation, more diverse constructs such as branching DNA units have been developed.[9–15] The basis of this approach depends on the combination of two concepts: the Holliday junction and 'sticky ends.' The Holliday junction is a mobile junction between four strands of DNA.[16] A sticky end is a short. single-stranded overhang protruding from the end of a duplex DNA molecule. Sticky-ended cohesion is a good example of programmable molecular recognition: the diversity of possible sticky ends is significantly large and the product formed at the site of this cohesion is the classic DNA duplex. Hence sticky ends offer both predictable control of intermolecular associations and predictable geometry at the point of cohesion. Because of the uniqueness of a DNA stand, it provides a

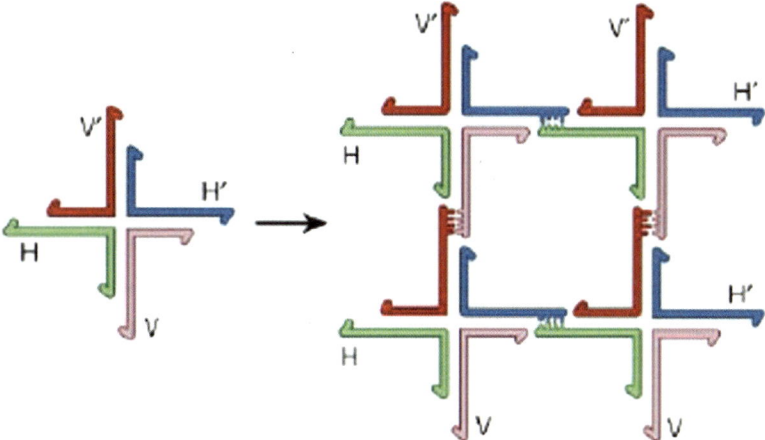

Figure 9.1 Assembly of branch DNA. Self-assembly of branched DNA molecules into a two-dimensional crystal. Adapted from N. C. Seeman, *Nature*, 2003, **421**, 427–431.

tractable, diverse and programmable system with remarkable control over intermolecular interactions, coupled with known structures for their complexes. A DNA branched junction forms from four DNA strands (Figure 9.1); those strands colored green and blue have complementary sticky-end overhangs labeled H and H', respectively, whereas those colored pink and red have complementary overhangs V and V', respectively. A number of DNA branched junctions bind together based on the orientation of their complementary sticky ends, forming a square-like unit with unpaired sticky ends on the outside, so that more units can be added to produce a two-dimensional crystal. This chapter describes how to use this principle in fabricating DNA nanomachines, DNA-enabled self-assembly of inorganic/organic nanoparticles and micro- to macro-engineering.

9.2   DNA Nanomachines

Nanoscale mobile devices have been developed in recent years for applications such as information processing, nanoelectronic devices, biosensors and regulation of chemical reactions and molecular assembly.[17–20] In addition to construction of DNA arrays, extensive efforts have also been devoted to the development of DNA nanomachines because of their potential applications in nanoelectronics, smart materials and biosensors.[11,13,21,22] DNA nanomachines can work *via* different mechanisms: (1) conformational changes induced by environmental changes, (2) motions fueled by strand displacement and (3) autonomous motion powered by enzymatic activity.

9.2.1   Conformational Changes Induced by Environmental Changes

Seeman's group reported the first DNA nanomachine based on the B–Z transition of DNA.[11] The nanomachine was composed of two stiff double-crossover (DX) structures connected by a double-helix fragment (Figure 9.2). The central DNA double helix has the base-paired sequence d(CG),[11] a 'proto-Z' sequence that can be converted to left-handed Z-DNA at high ionic strength. When the buffer contained only 10 mM $MgCl_2$ and 100 mM NaCl, the central DNA adopted the B conformation. However, when 0.25 mM $Co(NH_3)_6Cl_3$ was added, the central DNA duplex changed to a left-handed Z conformation. Since the structure of the DNA nanomachine was fairly rigid, the twisting of the central DNA duplex can produce well-defined, repeatable molecular motions of the order of several nanometers.

There are other DNA devices fueled by protons rather than by metal ions.[13,22] This nanomachine is based on an i-motif DNA structure, which is formed from sequences containing stretches of cytosine (C) residues. At low pH conditions, cytosine can be protonated at the N3 position. As a result, non-Watson–Crick base pairs (*e.g.* C–C+) can form. When a C-rich DNA strand is presented at a pH lower than 6.5, it will self-fold into an intramolecular quadruplex, called an i-motif. However, this motif will disassemble in a high

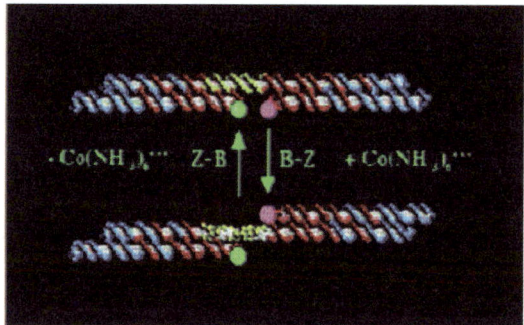

Figure 9.2 At the top, the connecting bridge is in the B conformation, so the two helices of the double-crossover segments are on the same side of the helix containing the bridge. At the bottom, the bridge is in the Z conformation, so the helices are on opposite sides of the helix. The change from B- to Z-DNA is caused by the addition of hexaamminecobalt(III) chloride to the solution. The change back is caused by removal of this reagent. The two large stippled circles represent fluorescent dyes attached to the device. The change of conformation from B- to Z-DNA is monitored by fluorescence resonance energy transfer (FRET) spectroscopy involving these two dyes. Adapted from C. Mao, W. Sun, Z. Shen and N. C. Seeman, *Nature*, 1999, **397**, 144–146.

pH environment. Therefore, the C-rich DNA strand can hybridize with its complementary strand, forming a stable Watson–Crick DNA duplex, when the pH has increased sufficiently. This process can be recycled again and again with the pH change, leading to a reliable conformational change between closed and open states. As shown in Figure 9.3, this nanomachine contains three DNA strands (F, L and S) and operates through the reversible formation–dissociation of a DNA triplex (the interaction of three DNA strands involving non-Watson–Crick base pairing).[22] At pH 8.0, the three strands associate to form an open complex (open state) consisting of three 15-base-pair duplexes and a single-stranded region. This single-stranded region adopts a random coil conformation in the presence of divalent cations such as Mg^{2+}, but its sequence is designed to allow it to bind to one of the duplexes to form a triplex. The formation of C+G−C triplets depends on the protonation of cytosine residues, which requires an acidic solution. Upon triplex formation, the complex as a whole becomes fairly compact (closed state). When the pH value increases to 8.0, the C+G−C triplets become unstable and the DNA triplex dissociates into a duplex and a single-stranded region; the DNA complex returns to its open state. The DNA machine continuously cycles between the closed and the open states when the solution pH oscillates between 5.0 and 8.0.

The conformational change was further demonstrated by the results of a fluorescence experiment. Strand F was labeled with Rhodamine Green and BHQ-1 quencher molecules at its 5- and 3-ends, respectively. In the open state of the DNA machine (pH 8.0), the fluorophore is well separated from the

(a)

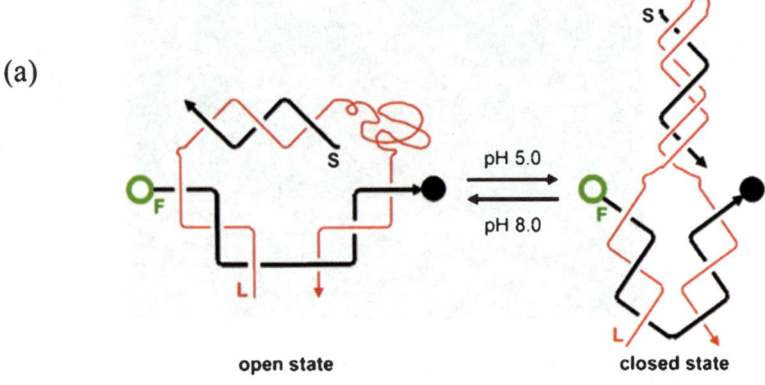

open state   closed state

(b)

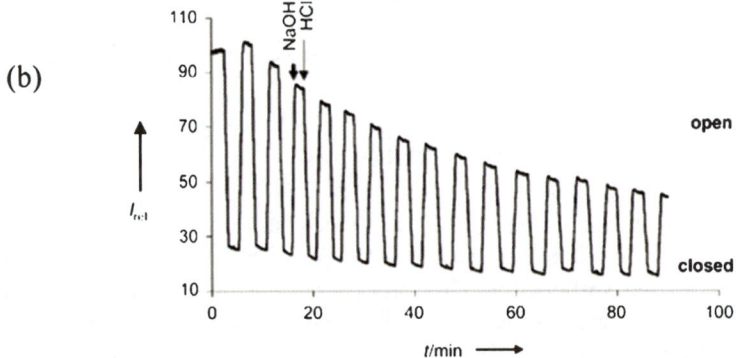

Figure 9.3   (a) The machine consists of three DNA strands: a strand with a fluorescent label (F), a long strand (L) and a short strand (S). The open and closed circles represent Rhodamine Green and black hole quencher-1 (BHQ-1), respectively. A DNA triplex involving the S and L strands forms and dissociates reversibly. (b) Fluorescence intensity was monitored at 530 nm (λ_{em} of Rhodamine Green) while the solution pH value oscillated between 5.0 and 8.0. Adapted from Y. Chen, S. H. Lee and C. Mao, *Angew. Chem. Int. Ed.*, 2004, **43**, 5335–5338.

quencher and a strong fluorescence signal was observed. In the closed state (pH 5.0), the fluorophore is near the quencher and the fluorescence signal decreased in intensity.

9.2.2   Motions Fueled by Strand Displacement

This approach was introduced by Yurke *et al.*[13] in 2000. They designed a DNA 'nanotweezer' which contains three DNA single strands (Figure 9.4). These three DNA strands can assemble into a nicked duplex in the middle, together with two single-stranded tails at both ends. The nick separated the duplex into two rigid arms of the tweezers. In its relaxed state, the two arms are open and apart from each other. When a new DNA strand, F, was introduced, it would hybridize with the two 'tails,' thus closing the tweezers. The F DNA contained three segments. Two of them are complementary to the tails of the tweezers, and the third region is an unpaired single-stranded overhang. To re-open the tweezers, a new DNA which is fully complementary to F was added. As a result, the new DNA would

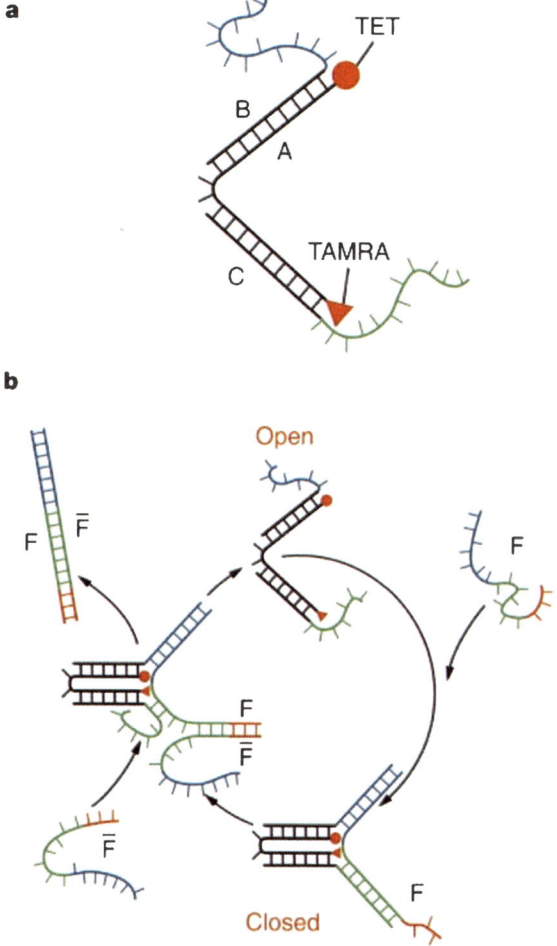

Figure 9.4   Construction and operation of the molecular tweezers. (a) Molecular tweezer structure formed by hybridization of oligonucleotide strands A, B and C. (b) Closing and opening the molecular tweezers. Closing strand F hybridizes with the dangling ends of strands B and C (shown in blue and green) to pull the tweezers closed. Hybridization with the overhang section of F (red) allows F to be removed from the tweezers, forming a double-stranded waste product F and allowing the tweezers to open. Adapted from B. Yurke, A. J. Turberfield, A. P. Mills, F. C. Simmel and J. L. Neumann, *Nature*, 2000, **406**, 605.

first bind to the single-stranded overhang of F and then gradually extract the whole F DNA from the tweezers, forming a DNA duplex as waste.

This process is called strand displacement, which is driven by forming a more stable complex (fully complementary DNA duplex in this case). Consequently, the tweezers were re-opened. The open and close motions of the DNA nanotweezers can cycle again and again. Each single motion can be precisely triggered by addition of a DNA strand. Strand displacement serves as a highly controllable means to control the DNA nanomachine.

9.2.3   Autonomous Motion Powered by Enzymatic Activity

Natural protein motors, including myosin, kinesin and F0F1-adenosine triphosphate synthase, can continuously work when there is energy.[23] Their movements are powered by the chemical process of ATP hydrolysis. To mimic this process, Chen *et al.* developed an autonomous DNA motor called a DNA enzyme.[21] The DNA enzyme binds to its RNA substrate through Watson–Crick base pairing and cleaves the RNA into two short fragments. This is accomplished as shown in Figure 9.5. It contains two 15-base-pair helical

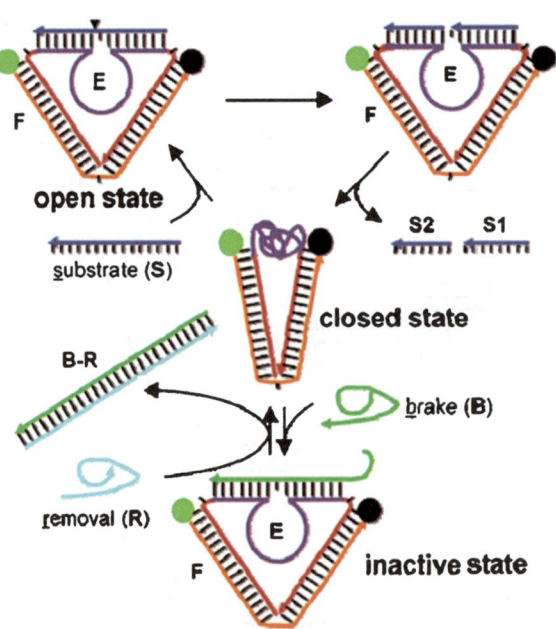

Figure 9.5   The DNA motor consists of two single strands, E and F. The E strand contains a 10–23 DNA enzyme domain, which is colored purple. The F strand has a Rhodamine Green fluorophore at the 5′-end (labeled as a solid green circle) and a black hole quencher-1 (BHQ-1) at the 3′-end (labeled as a solid black circle). Adapted from Y. Chen and C. Mao, *J. Am. Chem. Soc.*, 2004, **126**, 8626–8627.

domains, which are joined by a single-stranded 10–23 DNA enzyme at one end and a single-base hinge at the other. This 10–23 DNA enzyme consists of a 15-base catalytic core and two seven-base-long substrate-recognition arms. When there is no substrate, the single-stranded DNA enzyme collapses into a closed coil and the overall conformation of the DNA motor is fairly compact (closed state). When the DNA enzyme binds to its substrate, the single-stranded DNA enzyme forms a bulged duplex with the substrate. This bulged duplex pushes the two helical domains of the DNA motor apart and leads the motor to adopt an open state. After binding, the DNA enzyme cleaves its substrate into two short fragments (S1 and S2). The resulting fragments have a lower affinity for the DNA enzyme than the intact substrate and, therefore, dissociate from the DNA motor. Consequently, the DNA motor returns to the closed state and can undergo the next cycle of substrate binding (open state), cleavage and dissociation (closed state). The motor autonomously cycles between the open and closed states until the fuel (the enzyme substrate) runs out.

This DNA autonomous motor can be stopped and continue its movement by using strand displacement. The brake strand (B) is a DNA analog of strand S, which cannot be cleaved by the enzyme. Like strand S, strand B can bind to the recognition arms of the enzyme. However, strand B slightly extends complementary segments into the catalytic core by two bases and has a 10 base long tail. Strand B can form a longer bulged duplex with the enzyme than strand S does. Hence the enzyme will bind to strand B instead of strand S because of differing affinity. A brake removal strand (R) is fully complementary to strand B and they form a long duplex (R–B). Strand R first base pairs with strand B at the tail region and then pulls strand B out of the motor through branch migration. When strand B is removed from a motor, the enzyme becomes active again and the motor resumes motion. The brake can be reversibly added to and removed from the motor.

9.3 DNA-enabled Self-assembly of Inorganic/Organic Nanoparticles

Self-assembly is the process in which discrete components such as molecules and nanoparticles spontaneously organize themselves by specific interactions.[25] In the field of nanotechnology, directed self-assembly of nanoparticles has attracted much attention and is regarded as one of the most prominent nanofabrication processes for various technological applications.[26–29] Although directed self-assembly still employs the basic principles of self-assembly by modulating the building blocks, it facilitates the process with thermodynamic forces without going into intricate techniques. The invention and development of DNA-immobilized AuNPs (DNA-AuNPs) has led to the concept of synthetically directed inorganic/organic materials assembly.[30,31] Using simple base-pair hybridization of single-stranded DNA to its complementary strand, it is possible to control the assembly of materials on the nano-scale, which would have otherwise been nearly impossible. The majority of the work carried out thus far has shown that DNA can be used to control the placement, periodicity and distance between particles within the assembly.

9.3.1 Properties of DNA-modified Gold Nanoparticles

Gold nanoparticles (AuNPs), typically 1–100 nm in diameter, are a class of inorganic nanomaterials which have become widely investigated because they exhibit intense size-dependent optical properties.[32–34] Generally small AuNPs (~10 nm) exhibit a deep red color when in solution, which shifts to a light purple/pink as the diameter of the AuNPs increases (250 nm). For 13 nm AuNPs, the plasmon band is around 520 nm and the surface plasmon band red shifts as the diameter of the AuNPs increases, thus resulting in a color change from red to blue. Similarly, when AuNPs are aggregated into clusters, the surface plasmon band red shifts and a red-to-blue color change is observed.[34,35]

Surface immobilization of DNA on to the AuNPs fulfills two main purposes. First, it provides functionality and recognition properties to the AuNPs, and second, immobilized DNA strands stabilize AuNPs so they do not irreversibly aggregate in the presences of salt. In the absence of stabilizing molecules such as DNA, AuNPs will irreversibly aggregate *via* charge interaction in the presence of salt. DNA-AuNPs are synthesized by simple mixing of thiol-modified oligonucleotides, usually 20–30 bases, to the AuNP solution followed by sulfur–gold interaction on the surface of the AuNPs. A single thiol modification at either the 3′- or 5′-end of the oligonucleotide is sufficient for stable surface immobilization. In addition, a salt aging process is typically employed to achieve a greater DNA density on the AuNP surface by slowly increasing the salt concentration, thereby decreasing the shielding between the negatively charged DNA strands.

Generally, synthetic DNA is prepared using standard phosphoramidite DNA synthesis techniques in the solid phase and these techniques have difficulty preparing a long DNA strands of >100 bases. Hence it is challenging to prepare AuNPs with longer DNA sequences and there have been efforts to develop alternative methods to achieve this goal. These methods utilize standard DNA-AuNPs with short DNA sequences and then extend the length of the covalently immobilized DNA using the polymerase chain reaction (PCR).[36,37] The nanoparticle-bound DNA acts as the primer and the immobilized DNA strand is extended based on a circular DNA template. In addition to the DNA length that can be immobilized on AuNPs, the number of DNA strands on a single AuNP can be carefully modulated by various factors such as the salt concentration and AuNP size. For example, a 15 nm AuNP can have 20–250 strands of DNA covalently immobilized on the surface.[38,39] In addition, the AuNP size can be varied from a few nanometers to 250 nm in diameter, leading to a variety of nanoparticle sizes and DNA densities.[40]

By simple base-pair hybridization, DNA-AuNPs can be self-assembled into different structures by designing linker DNA with a complementary sequence.

AuNPs, for example, can be modified with either DNA-A or DNA-B and then assembled in solution by addition of a third DNA sequence (A'B') that is complementary to A and B. The assembly process can be monitored by observing a red-to-blue color change and a broaden and red shifting of the surface plasmon (Figure 9.6). Similarly to DNA duplex formation, the assembly of DNA-AuNP is thermodynamic and reversible. By denaturing the linking DNA duplexes in the AuNP assembly, the DNA-AuNPs can be dispersed at elevated temperature and a return of the red color is seen. This denaturing process, often referred to as 'DNA melting,' can also be induced by a decrease in salt concentration.

The melting transition exhibited by the deconstruction of DNA-AuNP aggregates is substantially different from that of unmodified DNA duplexes in two distinct ways. First, the melting transition of the AuNP assembly is significantly sharper than that of an unmodified DNA duplex. When the first derivative of the melting transition is analyzed, the full width at half-maximum (FWHM) is about 2–4 °C compared with an FWHM of about 12 °C for a DNA duplex. Second, the melting temperature (T_m) is significantly increased compared with that of an analogous DNA duplex under identical buffer conditions (Figure 9.7).[41] Both of these properties are functions of the DNA-

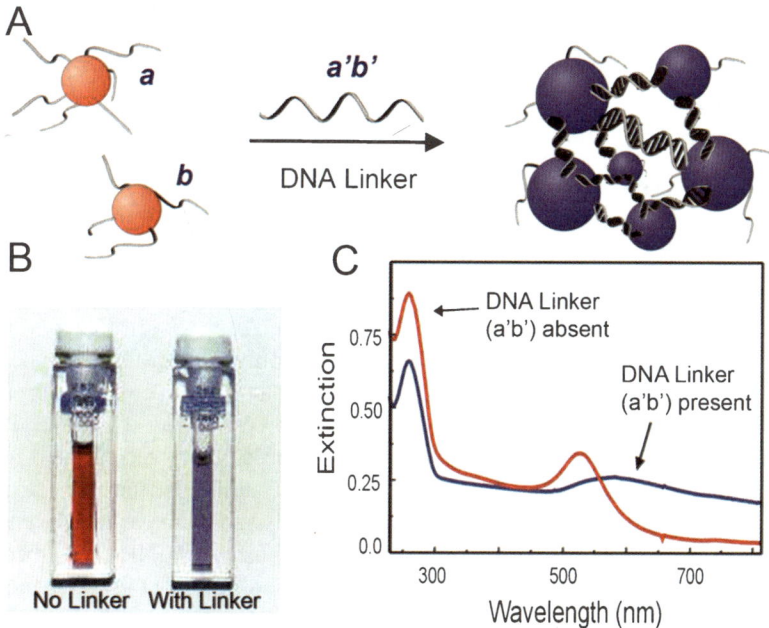

Figure 9.6   In addition to gold nanoparticles, DNA has also been used to assemble organic particles. Organic virus-like protein nanoparticles (VLPs) and inorganic AuNPs were assembled using DNA directed hybridization to create a binary system which resembles the non-compact diamond lattice.[48]

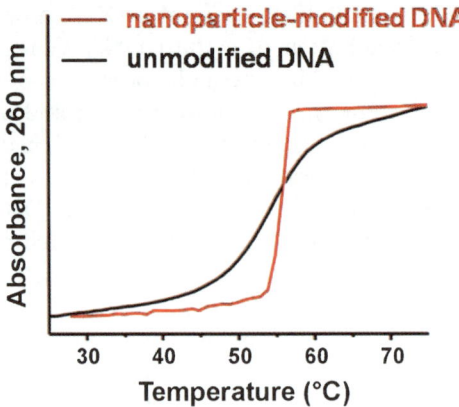

Figure 9.7   Melting transition of unmodified duplex DNA (black) and DNA linked
              DNA-AuNPs (red). Adapted from N. L. Rosi and C. A. Mirkin,
              Nanostructures in biodiagnostics. *Chem. Rev.*, 2005, **105**, 1547–1562.

AuNP complex and are important for many of the applications in which these
materials are used, such as biodetection and nano-therapeutics. Despite these
two major differences in the melting properties, DNA-AuNPs do show some
similar behavior to duplex DNA. For example, the stability of the DNA-
AuNP assemblies and the T_m increase as a function of increase in the DNA
length, DNA concentration and salt concentration are similar to those of
unmodified duplex DNA.

9.3.2   Directed Self-assembly of DNA-modified Gold Nanoparticles

9.3.2.1   *Colloidal crystallization*

The self-assembly of DNA-AuNPs described above exhibits a fractal-like
structure, with little internal ordering between the AuNPs.[42] In the past, there
have been various approaches to the assembly of different types of
nanoparticles into highly ordered crystalline structures or colloidal crystals,
by using electrostatic forces and van der Waals forces.[43] It has been speculated
that the sequence-specific interactions of DNA could be applied to the directed
assembly of nanoparticles into highly ordered crystalline structures.[44] As the
synthesis techniques for preparing monodisperse AuNPs have improved, the
ability to assemble highly ordered structures using AuNPs has become
possible. Therefore, scientists have begun to utilize different DNA sequences to
direct the assembly of DNA-AuNPs into highly ordered materials, by utilizing
programmable DNA base pairing.

  The strategy is similar to that of the organizational packing of atoms in a
crystalline lattice, where particles can be designated A, B, C, *etc.*, based on the
DNA sequence. From a thermodynamic standpoint, DNA-mediated nanoparticle self-assembly will preferentially generate the structure that is the most

energetically favored, yielding the largest number of DNA hybridization events. For example, if a single-component system is employed, where all particles are identical and self-complementary, all particles can bind to all other particles with equal affinity and the most particle-to-particle interactions can be achieved by a close-packed crystalline structure. In this type of structure, each particle meets 12 nearest neighbors and has a face-centered cubic (FCC) unit cell as shown in Figure 9.8. However, when a binary system is employed, where there are two different types of particles with either A or B DNA and only A particles can bind to B particles, the structure with the most particle-to-particle interactions is a non-close-packed structure with eight nearest neighbors. In this structure the unit cell is body-centered cubic (BCC).

It has been shown that DNA can be used to assemble nanoparticles into highly ordered crystals such as FCC and BCC structures.[45,46] To characterize the crystal structure in solution, small-angle X-ray scattering (SAXS) techniques were employed in which the electron density of the gold particles scatters and diffracts the X-rays. Depending on the crystal structure of the particles, the X-rays are diffracted into different patterns based on Bragg diffraction theory. DNA is a particularly powerful tool since it can direct nanoparticle assembly into multiple structures guided solely by DNA sequence. By adjusting the length of immobilized DNA strands, the overall crystal lattice spacing can also be tuned without changing the size of the inorganic core.

In addition to gold nanoparticles, DNA has also been used to assemble organic particles. Organic virus-like protein nanoparticles (VLPs) and inorganic AuNPs were assembled using DNA-directed hybridization to create a binary system which resembles the non-compact diamond lattice. By

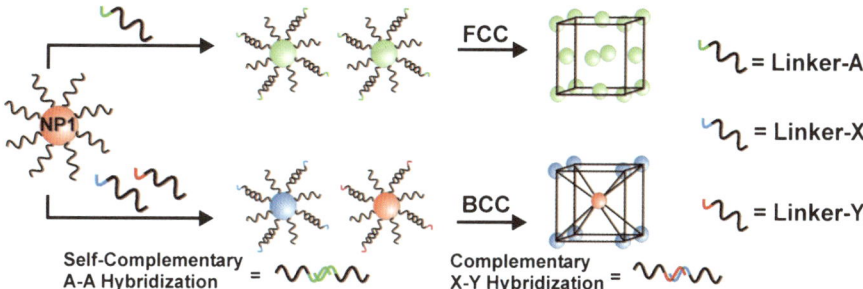

Figure 9.8 The green overhang of linker-A is self-complementary and therefore DNA-AuNPs hybridized to linker-A are self-complementary and behave as a single-component system. Maximizing the number of particle-particle interactions results in a close-packed structure depicted by an FCC unit cell. The red and blue overhang of linkers-X and -Y are complementary and DNA-AuNPs hybridized to linkers-X and -Y are complementary and behave as a binary system. Maximizing the number of particle–particle interactions results in a non-close-packed structure depicted by a BCC unit cell. Adapted from, S. Y. Park *et al.*, DNA-programmable nanoparticle crystallization. *Nature*, 2008, **451**, 553.

combining with the proper connecting oligonucleotides, these components form NaTl-type colloidal crystalline structures containing interpenetrating organic and inorganic diamond lattices, as confirmed by SAXS. DNA-mediated particle assembly is therefore shown to be compatible with particles having very different properties, provided that they are amenable to surface modification. The diamond crystal structure (non-compact) is particularly important in colloidal crystallization because the lattice spacing is amenable to use as photonic crystals in the visible range. Therefore, further work in the area has important potential applications.

9.3.2.2 Self-assembly of Asymmetrically Modified Gold Nanoparticles

The AuNPs described above were isotropically modified with DNA to create uniform surface coverage. To impart directionality to the DNA binding and obtain greater control over the AuNP assemblies, strategies have been employed to create asymmetrically modified particles. Asymmetrically modified particles have been designed where a particular DNA sequences is located on only one side of the AuNP.[50,51] These strategies require the hybridization of DNA-AuNPs to DNA that is immobilized on a flat surface. The hybridization of the particles to the surface blocks one face of the AuNPs and prevents the hybridization of other DNAs. The remaining DNA on the AuNP which is not blocked by the flat surface is then hybridized to a different DNA sequence. Using thermal control, the DNA-AuNPs can then be released

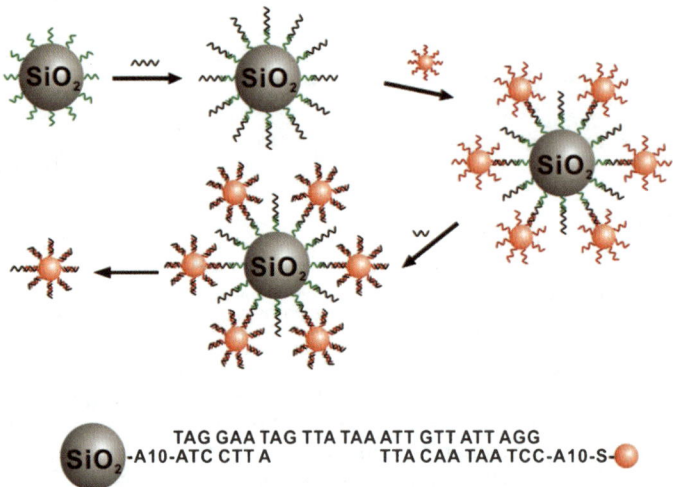

SiO₂-A10-ATC CTT A   TAG GAA TAG TTA TAA ATT GTT ATT AGG
                        TTA CAA TAA TCC-A10-S-

Figure 9.9   Synthetic scheme for the asymmetric functionalization of nanoparticles with DNA. Adapted from F. Huo, A. K. R. Lytton-Jean and C. A. Mirkin, Asymmetric functionalization of nanoparticles based on thermally addressable DNA interconnects. *Adv. Mater.*, 2006, **18**, 2304–2306.

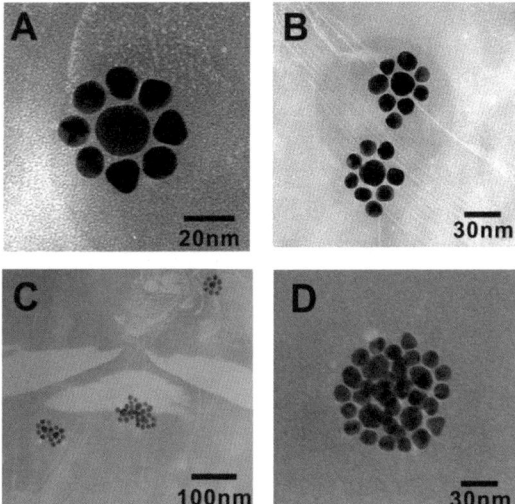

Figure 9.10   TEM images of satellite structures composed of 13 nm asymmetrically
functionalized nanoparticles and 20 nm symmetrically functionalized
gold nanoparticles. Images (a) and (b) show the separate satellite
structures and (c) and (d) show some clumped satellite structures.
Reprinted from F. Huo, A. K. R. Lytton-Jean and C. A. Mirkin,
Asymmetric functionalization of nanoparticles based on thermally
addressable DNA interconnects. *Adv. Mater.*, 2006, **18**, 2304–2306.

from the flat surface while keeping the other DNA sequences hybridized to the
rest of the AuNP. When the flat surface is removed, it leaves an 'empty' face
which can be hybridized to a different DNA sequence, leading to a DNA-
coated Janus AuNP. An example of this strategy is described in Figure 9.9.
The hybridization of the different DNA sequences to the DNA-AuNP can
then be made permanent by ligating the DNA sequences using DNA ligase, if
desired. These particles have been assembled into arrangements such as
satellite structures which have been verified by transmission electron
microscopy (TEM) (Figure 9.10).

9.4   Micro- to Macro-engineering by Self-assembly

One of the most important goals of self-assembly is to achieve macro-sized
objects by self-assembly[17,24]. By doing so, it would help to bridge the molecular
scale and macro worlds and further the design of new materials with novel
characteristics. The simplest practical route for producing precisely designed
3D macroscopic objects is to form a crystalline arrangement by self-assembly.
This requires a motif that has a robust 3D structure, dominant affinity
interactions between parts of the motif when units self-associate and
predictable structures from these interactions.[24] Fulfilling all of these three
criteria is challenging, but DNA offers the opportunity to achieve this goal.

The first successful example of a pure DNA crystal was the construction of a tensegrity triangle motif.[52] This motif is constructed by three four-arm junctions fused together to form a triangle shape. Each side of the triangle is a DNA duplex. At each vertex, the two double-helix domains of the four-arm junction point in different directions. The overall three-helix domains of the triangle have threefold symmetry. Therefore, they do not share the same plane and produce an alternating over-and-under motif. In detail, there are three sets of single strands in the triangle. The three identical parts combine with three blue strands to form the arm. The three green strands complete the double helices in the arm to comprise the four-arm junctions at each vertex. The sticky

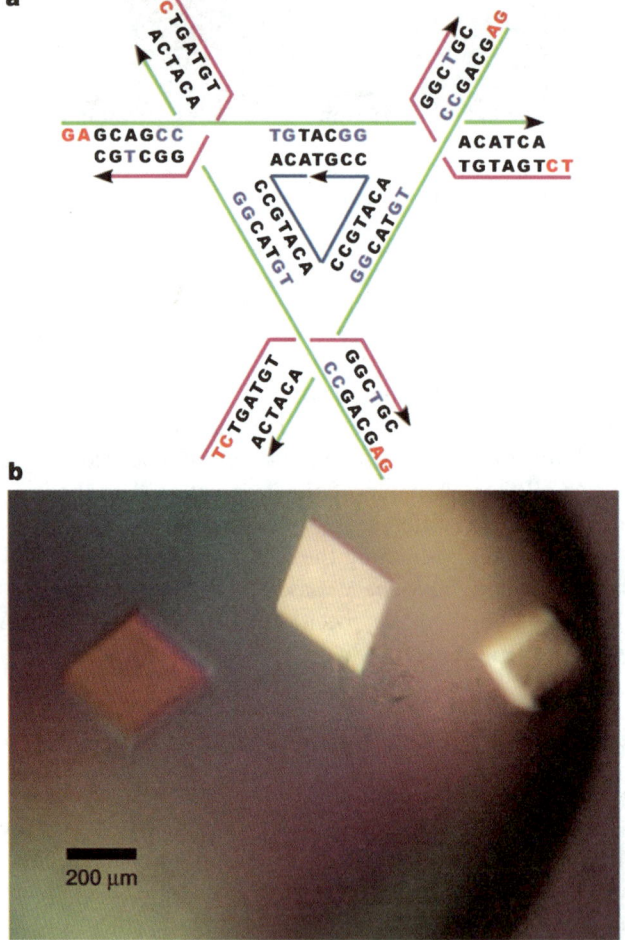

Figure 9.11 Design, sequence and crystal pictures. (a) Schematic of the tensegrity triangle. (b) An optical image of the crystals of the tensegrity triangle. The rhombohedral shape of the crystals and the scale are visible. Adapted from J. Zheng *et al.*, *Nature*, 2009, **461**, 74–77.

ends at each arm are used to link with other triangles to form 3D arrays. This tensegrity triangle motif will form rhombohedral structures after they link to each other through sticky-end hybridization. In this design, the rhombohedral edge can be turned from around two double-helical turns in length (21 bases) to four double-helical turns (41 bases). Hence it can give out enough cavities to accommodate different nanocomponents with different size.

The pictures of these rhombohedral crystals are shown in Figure 9.11. Unlike self-assembled two-dimensional crystals, typically a few micrometers in scale or 3D DNA nanoparticle crystals, these self-assembled 3D crystals are macroscopic objects, exceeding 250 mm in dimension. Furthermore, these macroscopic objects have predictable atomic structures based on rational design. The arrangement of the molecules in this crystal is the result of sticky-end cohesion. X-ray diffraction experiments show that the real structure of this crystal is consistent with the design. As a further demonstration of the ability to program DNA crystalline arrangements in 3D using sticky-ended cohesion,

Table 9.1 The cross-sectional area and cavity size are derived from the lattice parameters. Cross-sections and cavity sizes are estimated by subtracting two radii of the double helix (~ 10 Å) from the unit cell dimensions. The space group indicates whether deliberate three-fold rotational averaging has been performed; it has for those in $R3$, not for those in $P1$. Edge lengths and inter-junction distances (within triangles) are given in nucleotide pairs. Crystal 1 is the work reported here. The structures of crystals 3 and 7 have been determined by molecular replacement; others are in progress. Adapted from J. Zheng *et al.*, *Nature*, 2009, **461**, 74–77.

Crystal number	Edge length (nucleotide pairs)	Space group	Inter-junction pairs	Rhombohedral cell dimensions	Resolution (Å)	Cross-section (nm²)	Cavity size (nm³)
1	21	$R3$	7	$a = 69.2$ Å $\alpha = 101.4°$	4.0	23	103
2	21	$P1$	7	$a = 68.0$ Å $\alpha = 102.6°$	5.0	23	101
3	31	$R3$	17	$a = 102.0$ Å $\alpha = 112.7°$	6.1	62	366
4	31	$P1$	17	$a = 100.9$ Å $\alpha = 111.6°$	6.3	61	373
5	32	$R3$	18	$a = 103.6$ Å $\alpha = 113.6°$	6.5	64	367
6	32	$P1$	18	$a = 103.3$ Å $\alpha = 112.2°$	6.5	64	395
7	42	$R3$	17	$a = 134.9$ Å $\alpha = 110.9°$	11.0	123	1104
8	42	$P1$	17	$a = 133.7$ Å $\alpha = 111.3°$	14.0	120	1048
9	42	$R3$	28	$\alpha = 134.9$ Å $\alpha = 117.3°$	10.0	117	643

eight other rhombohedral lattices from related tensegrity triangles were constructed. The crystals and their structures are summarized in Table 9.1. It not only indicates that are the unit cells, lengths and angles close to the predicted values, but also shows that these crystals contain cavities that can exceed 1000 nm³, which can be used to host other components.

9.5 Conclusion

DNA, in addition to containing genetic information, has also been recognized as a useful material in the field of nanotechnology. This chapter outlines the basic principles of DNA self-assembly and several applications for advancing the field of structural DNA nanotechnology. Indeed, based on self-assembly, structural DNA nanotechnology has already become an interdisciplinary research field including chemistry, materials science, computer science, biology and physics. Because of this powerful method of precise control from the nanometer scale to macro-engineering, it is believed that this field will progress rapidly and new, exciting directions will emerge in the near future.

References

1. H. G. Craighead, *Science*, 2000, **290**, 1532–1535.
2. N. C. Seeman, *Nature*, 2003, **421**, 427–431.
3. A. P. Alivisatos, K. P. Johnsson, X. G. Peng, T. E. Wilson, C. J. Loweth, M. P. Bruchez and P. G. Schultz, *Nature*, 1996, **382**, 609–611.
4. R. P. Feynman, in *Miniaturization*, ed. H. D. Gilbert, Reinhold, New York, 1961.
5. T. A. Jung, R. R. Schlittler, J. K. Gimzewski, H. Tang and C. Joachim, *Science*, 1996, **271**, 181–184.
6. M. D. Struthers, R. P. Cheng and B. Imperiali, *Science*, 1996, **271**, 342–345.
7. C. P. Collier, G. Mattersteig, E. W. Wong, Y. Luo, K. Beverly, J. Sampaio, F. M. Raymo, J. F. Stoddart and J. R. Heath, *Science*, 2000, **289**, 1172–1175.
8. J. D. Watson and F. H. C. Crick, *Nature*, 1953, **171**, 737–738.
9. N. C. Seeman, *J. Biomol. Struct. Dyn.*, 1985, **3**, 11–34.
10. R. Ma, N. R. Kallenbach, R. D. Sheardy, M. L. Petrillo and N. C. Seeman, *Nucleic Acids Res.*, 1986, **14**, 9745.
11. C. Mao, W. Sun, Z. Shen and N. C. Seeman, *Nature*, 1999, **397**, 144–146.
12. J. Zheng, P. E. Constantinou, C. Micheel, A. P. Alivisatos, R. A. Kiehl and N. C. Seeman, *Nano Lett.*, 2006, **6**, 1502.
13. B. Yurke, A. J. Turberfield, A. P. Mills, F. C. Simmel and J. L. Neumann, *Nature*, 2000, **406**, 605.
14. E. Winfree, F. Liu, L. A. Wenzler and N. C. Seeman, *Nature*, 1998, **394**, 539–544.
15. D. Liu and S. Balasubramanian, *Angew. Chem. Int. Ed.*, 2003, **42**, 5734.
16. Y. Liu and S. West, *Nat. Rev. Mol. Cell Biol.*, 2004, **11**, 937–944.

17. G. M. Whitesides, J. P. Mathias and C. T. Seto, *Science*, 1991, **254**, 1312–1319.

18. S. W. Santoro and G. F. Joyce, *Proc. Natl. Acad. Sci. U. S. A.*, 1997, **94**, 4262.

19. R. D. Vale and R. A. Milligan, *Science*, 2000, **288**, 88.

20. C. P. Collier, G. Mattersteig, E. W. Wong, Y. Luo, K. Beverly, J. Sampaio, F. M. Raymo, J. F. Stoddart and J. R. Heath, *Science*, 2000, **289**, 1172–1175.

21. Y. Chen, M. Wang and C. Mao, *Angew. Chem. Int. Ed.*, 2004, **43**, 3554–3557.

22. Y. Chen, S. H. Lee and C. Mao, *Angew. Chem. Int. Ed.*, 2004, **43**, 5335–5338.

23. F. J. Nedelec, T. Surrey, A. C. Maggs and S. Leibler, *Nature*, 1997, **389**, 305–308.

24. Y. Chen, C. Mao, *J. Am. Chem. Soc.*, 2004, **126**, 8626–8627.

25. B. A. Grzybowski, C. E. Wilmer, J. Kim, K. P. Browne and K. J. M. Bishop, *Soft Matter*, 2009, **5**, 1110–1128.

26. C. B. Murray, C. R. Kagan and M. G. Bawendi, *Annu. Rev. Mater. Sci.*, 2000, **30**, 545–610.

27. S. A. Claridge, A. W. Castleman, S. N. Khanna, C. B. Murray, A. Sen and P. A. Weiss, *ACS Nano*, 2009, **3**, 244–255.

28. D. V. Talapin, J. Lee, M. V. Kovalenko and E. V. Shevchenko, *Chem. Rev.*, 2010, **110**, 389–458.

29. S. C. Glotzer and M. J. Solomon, *Nat. Mater.*, 2007, **6**, 557–562.

30. C. A. Mirkin, R. L. Letsinger, R. C. Mucic and J. J. Storhoff, *Nature*, 1996, **382**, 607–609.

31. A. P. Alivisatos, K. P. Johnsson, X. G. Peng, T. E. Wilson, C. J. Loweth, M. P. Bruchez Jr and P. G. Schultz, *Nature*, 1996, **382**, 609–611.

32. H. C. van de Hulst, *Light Scattering by Small Particles*, Dover, New York) 1981, pp. 397–400.

33. C. Burda, X. Chen, R. Narayanan and M. A. El-Sayed, *Chem. Rev.*, 2005, **105**, 1025–1102.

34. J. Yguerabide and E. E. Yguerabide, *Anal. Biochem.*, 1998, **262**, 137–156.

35. B. L. V. Prasad, S. I. Stoeva, C. M. Sorensen and K. J. Klabunde, *Langmuir*, 2002, **18**, 7515–7520.

36. W. Zhao, Y. Gao, S. A. Kandadai, M. A. Brook and Y. Li, *Angew. Chem. Int. Ed.*, 2006, **45**, 2409–2413.

37. N. S. R. Pena, S. Raina, G. P. Goodrich, N. V. Fedoroff and C. D. Keating, *J. Am. Chem. Soc.*, 2002, **124**, 7314–7323.

38. S. J. Hurst, A. K. R. Lytton-Jean and C. A. Mirkin, *Anal. Chem.*, 2006, **78**, 8313–8318.

39. L. M. Demers, C. A. Mirkin, R. C. Mucic, R. A. Reynolds, R. L. Letsinger, R. Elghanian and G. Viswanadham, *Anal. Chem.*, 2000, **72**, 5535–5541.

40. J. S. Lee, D. S. Seferos, D. A. Giljohann and C. A. Mirkin, *J. Am. Chem. Soc.*, 2008, **130**, 5430.

41. J. J. Storhoff, A. A. Lazarides, R. C. Mucic, C. A. Mirkin, R. L. Letsinger and G. C. Schatz, *J. Am. Chem. Soc.*, 2000, **122**, 4640–4650.

42. N. L. Rosi, C. A. Mirkin, Nanostructures in biodiagnostics *Chem. Rev.*, 2005, **105**, 1547–1562.

43. S. Y. Park, J. S. Lee and D. G. Georganopoulou, *J. Phys. Chem. B*, 2006, **110**, 12673–12681.
44. C. B. Murray, C. R. Kagan and M. G. Bawendi, *Annu. Rev. Mater. Sci*, 2000, **30**, 545–610.
45. O. D. Velev, *Science*, 2006, **312**, 376–377.
46. S. Y. Park, A. K. R. Lytton-Jean, B. Lee, S. Weigand, G. C. Schatz and C. A. Mirkin, *Nature*, 2008, **451**, 553.
47. D. Nykypanchuk, M. M. Maye, D. van der Lelie and O. Gang, *Nature*, 2008, **451**, 549–552.
48. P. Cigler, A. K. R. Lytton-Jean, D. G. Anderson, M. G. Finn and S. Y. Park, *Nat. Mater.*, 2010, **9**, 918–922.
49. F. Huo, A. K. R. Lytton-Jean and C. A. Mirkin, *Adv. Mater.*, 2006, **18**, 2304–2306.
50. X.-Y. Xu, N. L. Rosi, Y. Wang, F. Huo and C. A. Mirkin, *J. Am. Chem. Soc.*, 2006, **128**, 9286–9287.
51. J. Zheng, J. Birktoft, Y. Chen, R. Sha, T. Wang, P. Constantinou, C. Mao, S. Ginell and N. Seeman, *Nature*, 2009, **461**, 74–77.
52. D. Liu, W. Wang, Z. Deng, R. Walulu and C. Mao, *J. Am. Chem. Soc.*, 2004, **126**, 2324–2325.

Janus Particle Localization and Tracking for Studies of Particle Dynamics

STEPHEN M. ANTHONY*[a] AND MINSU KIM[b]

[a] Department of Immunology, University of Texas Southwestern Medical Center, 6000 Harry Hines Boulevard, Dallas, TX 75390, USA; [b] Department of Physics, University of California San Diego, 9500 Gilman Drive, La Jolla, CA 92093, USA
*E-mail: santhony@mailaps.org

10.1 Introduction

The heterogeneity of the two faces of Janus particles (their defining characteristic) presents opportunities for new experimental measurements, at the same time posing some unusual challenges for particle detection and monitoring. In particular, Janus particles open a new avenue for the exploration of rotational dynamics at the single particle level, since not only the position of the individual Janus particles, but also the orientation of their heterogeneous surfaces can be monitored, introducing monitoring of two rotational dimensions. Alternative methods to study the rotational diffusion of colloids are fairly limited. Although time-resolved fluorescence anisotropy is widely used to study the rotation of molecules,[1] the fluorescence lifetime of the dye employed (typically on the order of nanoseconds) prevents exploration of the rotation of materials which have longer rotational times, limiting the utility of this technique to the molecular scale. Similarly, dynamic light scattering and fluorescence recovery after photobleaching have allowed studies of the ensemble-average

RSC Smart Materials No. 1
Janus Particle Synthesis, Self-Assembly and Applications
Edited by Shan Jiang and Steve Granick
© The Royal Society of Chemistry 2012
Published by the Royal Society of Chemistry, www.rsc.org

rotational diffusion of colloids,[2,3] but such techniques are limited to the properties of the ensemble and cannot dissect the constituent elements.

Therefore, localization and tracking of Janus particles allow otherwise unobtainable glimpses into the rotational and translational dynamics of colloidal particles. Such measurements are most conveniently accomplished by ensuring that the Janus particles are optically heterogeneous, with hemispherical Janus particles presenting the simplest case. Simultaneously, however, the existence of optical heterogeneity makes the many techniques routinely used to localize particles, such as those reviewed by Cheezum *et al.*,[4] inapplicable. Whereas those techniques were primarily concerned with signal-to-noise limitations, localizing Janus particles presents an entirely different challenge, one suggested by the name applied to some of the initially studied Janus particles,[5,6] modulated optical nanoprobes or MOONs, namely that as a Janus particle rotates, its image corresponds to the various phases of the moon.

Therefore, whereas for existing localization techniques the major concern was typically signal-to-noise ratio, resolving the Janus particles signal from the background generally does not pose a problem. Typically, Janus particles are fabricated such that they are ~ 1–2 μm in size, like those discussed in this chapter, and as such are readily fabricated containing many dyes or with an opaque hemisphere which generates a readily observable shadow under direct illumination. Hence approximately localizing Janus particles rarely poses significant difficulty. Similarly, as noted by Anker and co-workers,[5,6] the total observed intensity from Janus particles varies sinusoidally with the zenith angle, allowing that coordinate to be readily measured. Instead, the initial question for Janus particles concerns how one locates the center of a Janus sphere with precision since as the Janus sphere rotates, appearing like the different phases of the moon (Figure 10.1), the particle's center is not the center of brightness. (A limiting case is that, when observing the crescent phase, the center of the sphere will display no fluorescence at all.)

While approximate localization and determination of the zenith angle from the total intensity can be accomplished with minimal magnification, when magnified and viewed through an appropriate microscope the lateral, zenith and azimuthal coordinates can be determined much more precisely. The simplest case corresponds to Janus particles consisting of fluorescent spheres with opaque hemispherical caps, which mimic the moon in all its phases. Another common geometry uses silica (or otherwise transparent) spheres with an opaque hemispherical cap. When such particles are observed under white light illumination, the phenomenon is similar, except that there exists symmetry between images observed with zenith angles above and below $\pi/2$, as the shadow must always occupy at least half the circle. Further, although hemispherical Janus faces represent the simplest convenient geometry, particles with differing Janus balance can also be fabricated, in addition to particles with more than two faces, providing too many geometries to cover all possibilities. As such, localization techniques described here will be limited to those necessary for Janus particles with approximately hemispherical faces, although many of the principles can readily be extended to more complex geometries.

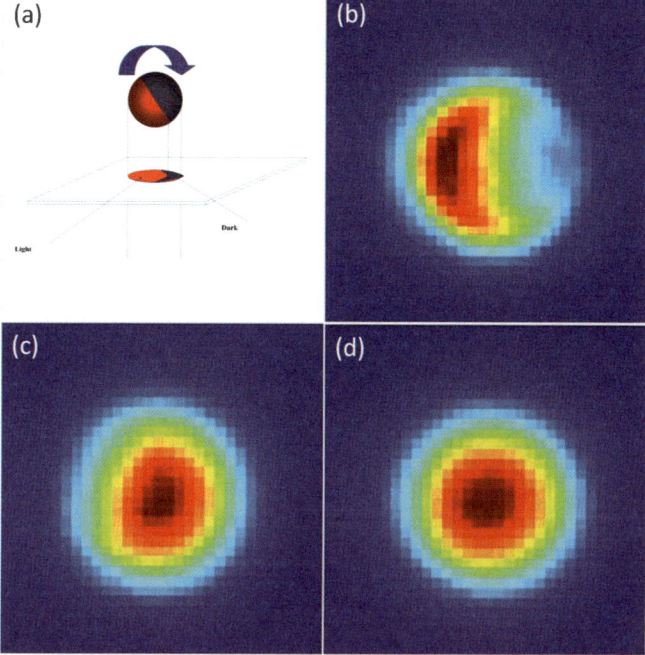

Figure 10.1 Illustrative images displaying the moon-like appearance of a spherical fluorescent particle with a hemispherical opaque Janus coating. (a) Schematic illustration of this idea. (b)–(d) Images of a Janus particle, 2 μm diameter, for three orientations ranging from crescent to full moon. The color denotes the varying intensity of the image. Adapted with permission from S. M. Anthony, L. Hong, M. Kim and S. Granick, *Langmuir*, 2006, **22**, 9812. Copyright 2006 American Chemical Society.

Here, we begin by demonstrating methods to locate the center of hemispherical Janus particles precisely, regardless of their orientation. However, in order to measure rotational dynamics, the rotational orientation, which is only accessible for Janus spheres, not homogeneous spheres, must also be determined. As noted above, measuring the zenith angle can be relatively straightforward, but determination of the azimuthal angle requires more effort. Further, the utility of Janus particles as a probe is significantly extended if the Janus particles need not be observed in strict isolation, so techniques are explored that allow spatial and orientational localization of the particles when their respective images overlap.

10.2 Isolated Particle Localization

10.2.1 Spatial Localization

Before applying any subsequent analysis, as a first step the spatial position of each Janus particle must be determined. While numerous particle localization methods exist, as reviewed by Cheezum *et al.*,[4] those techniques have primarily focused on overcoming a different experimental limitation, namely limited signal-to-noise ratio. Fortunately, the typically high contrast achievable between the observable portion of a Janus particle and its background allows the spatial position to be located by an alternative method, provided that the particle is divided into hemispheres (which, for the examples shown here, was confirmed by scanning electron microscopy). Extension to Janus geometries

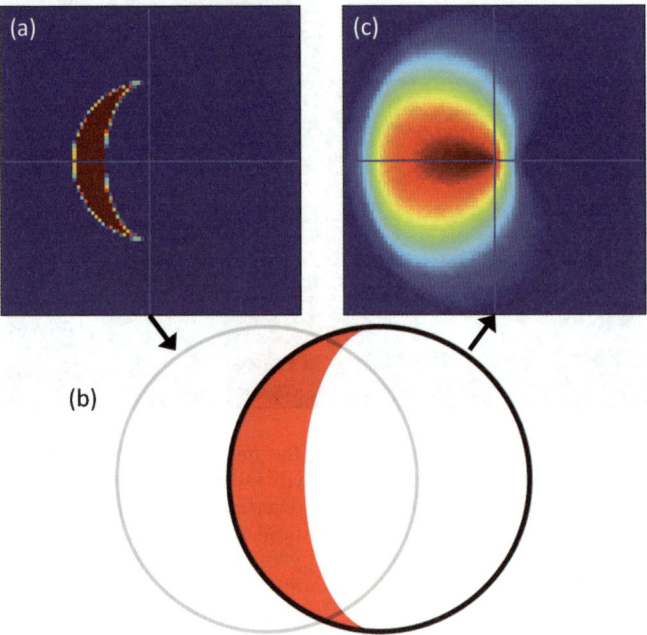

Figure 10.2   How to locate the center of mass of Janus spheres, whose fluorescence image is inherently anisotropic depending on how the particle is oriented relative to the observer. (a) Simulated image produced by a Janus sphere in the crescent phase, its center located at the indicated cross-hairs. Note that, regardless of orientation relative to the observer, the bright region will always have two points separated by the diameter of the Janus sphere. Note also that diffraction blurs the experimental images, but that, for clarity, diffraction blurring was not calculated for the simulated images. (b) A sole circle whose diameter matches the Janus sphere's diameter contains the fluorescent image completely. (c) Image produced by convolving the original image with a circle of this same diameter. The maximum in this image corresponds to the deduced center of the Janus sphere. Uncertainty in its location is greater parallel to the tilt of the Janus sphere than perpendicular to it; however, error in this direction has no effect upon calculating the azimuthal angle. The color at each point indicates the relative quality of the overlap between a circle centered at this point and the crescent at left, with the best overlap corresponding to the centered circle. Adapted with permission from S. M. Anthony, L. Hong, M. Kim and S. Granick, *Langmuir*, 2006, **22**, 9812. Copyright 2006 American Chemical Society.

other than equal hemispheres is also possible, provided that a signal is observable from at least an entire hemisphere (opaque cap limited to a portion of a hemisphere). The key factor is that when applying the above constraints, regardless of the orientation of the Janus particle, the observed signal (neglecting diffraction) would always contain two points separated by the diameter of the Janus sphere. Therefore, if the image is convolved with a circular disk having the same diameter as the Janus sphere, the location of the brightest point corresponds to the center of the Janus sphere, since centering the disk at any other point would not completely contain the bright region. The idea is illustrated in Figure 10.2.

10.2.2 Angular Localization

Once the spatial coordinates of the Janus sphere have been determined, the next challenge is to determine the angular orientation relative to the observer (the microscope). Specifying the coordinate system of the Janus sphere, the convention applied here was to utilize the coordinates specifying the center of the bright hemisphere of the Janus sphere, with $\theta = 0°$ corresponding to a full-moon orientation (Figure 10.3). Therefore, for the case of spheres with hemispherical Janus geometry, the total brightness observed from the entire sphere after background subtraction would be approximately co-sinusoidal:

$$\frac{A}{2} \times (1 + \cos\theta) \tag{10.1}$$

where A corresponds to the maximum observable brightness. Conveniently, when determining θ, the consequences of both limited depth of focus of the microscope objective and improper focus are generally negligible, since the primary effect of shifting the focal plane is to blur the image. Therefore, provided that the region of interest utilized for determining the total brightness is sufficiently large to contain the entire blurred image of the Janus sphere, the total brightness observed will be unaffected.

Equation (10.1) provides a convenient guideline, but was developed before multiple varieties of Janus spheres were in use.[5,6] Depending on the specific nature and geometry of the Janus sphere being investigated, the exact formula will vary. When considering signal due to the shadow cast by a Janus sphere with a hemispherical opaque cap under trans-illumination, eqn (10.1) would be exact, except that for trans-illumination, no geometry results in a full-moon appearance, with the maximum intensity instead observed when $\theta = 90°$. As such, the proper formula is instead

$$A \times (1 - |\cos\theta|) \tag{10.2}$$

where A once again corresponds to the maximum observable brightness. Note that the two formulas are exactly equivalent for $90° \leq \theta \leq 180°$ once

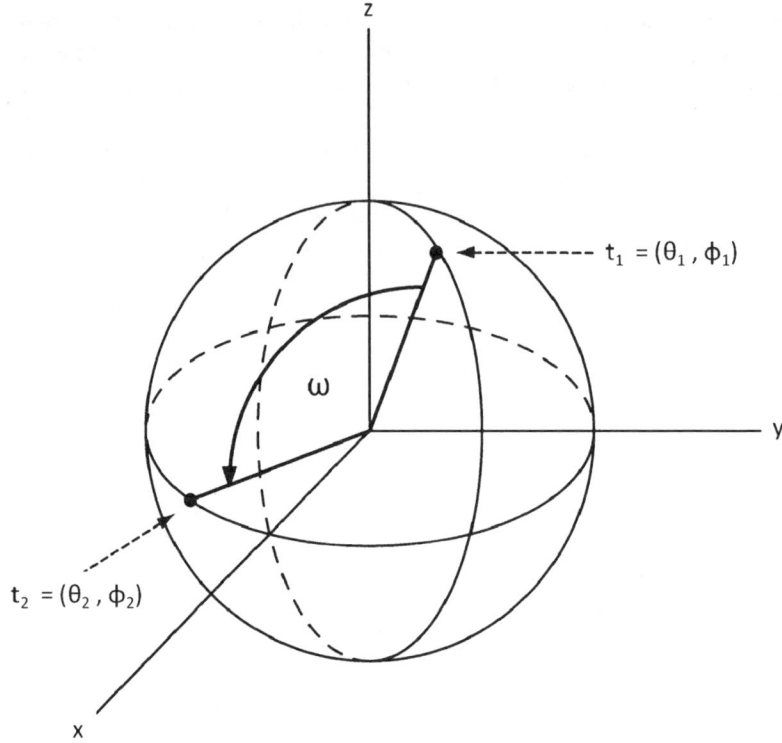

Figure 10.3   The measured quantities, θ and φ, completely specify the position of the bright hemisphere of the Janus sphere at any given time. For many purposes, such as determining rotational diffusion, rather than the specific orientation of the Janus sphere at a given point in time, the quantity of interest is how much the Janus sphere has rotated from its previous position. For such cases, the angle of variation, ω, is computed over the given time interval. Adapted with permission from S. M. Anthony, L. Hong, M. Kim and S. Granick, *Langmuir*, 2006, **22**, 9812. Copyright 2006 American Chemical Society.

consideration is given that, for this geometry, maximum intensity occurs at $\theta = 90°$ and the intensity is symmetric around $\theta = 90°$. When instead Janus spheres based upon fluorescent particles under epi-illumination are considered, the formula depends upon whether the component fluorescent particles have fluorophores embedded throughout or are surface labeled. For the former case, eqn (10.1) represents merely a convenient approximation, while the exact formula can be determined by calculating the fraction of the volume of the sphere not obstructed from view by the metal coating, yielding as an exact solution

$$A \times \left(1 - \frac{\theta}{\pi} + \frac{9}{64} \cdot sin(2 \times \theta)\right) \tag{10.3}$$

On the other hand, when the Janus sphere is derived from a surface-labeled fluorescent sphere, the observed intensity dependence upon angle differs substantially and is readily determined to be

$$A \times \frac{\pi - \theta}{\pi} \tag{10.4}$$

Regardless of the geometry, for sufficiently long consecutive time series of images, all orientations will be sampled and the brightest value of the Janus sphere reveals the value of A after background correction. Alternatively, for shorter series, it is possible to select images with easily determined θ, such as half-moon orientation and also calculate A. It is worth noting that the determination of θ for Janus spheres does not require their imaging, but can be accomplished using the intensity only measured with spot detectors such as photodiodes.

When, however, images of Janus spheres are acquired, it becomes possible to calculate the azimuthal angle, φ, a colloid measurement previously unavailable and requiring only a single image.[8] For any orientation other than $\theta = 0°$ or $180°$ (in which case φ is irrelevant), within the region corresponding to the Janus sphere, both light and dark portions are observed. As such, the azimuthal angle φ can be determined from the orientation of the line connecting the center of the Janus sphere, located previously, with the center of the bright pixels of the image, mathematically specified by the centroid of that portion of the image. Most uncertainty in locating the center of the Janus sphere is collinear with this line and therefore this error does not propagate into calculating the angle. In principle, the zenith angle, θ, can also be determined from the distance separating the two points that define the line. However, for typical image sizes, this method is less accurate than the one just described, especially since here the error in locating the center of the Janus sphere becomes relevant. However, unlike the previous method of calculating θ, this method requires only a single image rather than a time series.

Although the combination of θ and φ completely specifies the orientation of one axis of the sphere, for many cases, such as measuring rotational diffusion, the quantity of interest is the amount that the Janus sphere has rotated. This quantity is specified by the angle of variation, ω, as shown in Figure 10.3. Using the angle of variation, the rotational diffusion constant, D_r, of a Janus sphere can be computed using eqautions developed long ago by Perrin.[9] For small angular displacements, the approximation

$$\langle \omega^2 \rangle = 4 D_r \Delta t \tag{10.5}$$

can be employed. However, the rigorous equation incorporating the effects of the bounded nature of angles is

$$\langle \sin^2 \omega \rangle = \frac{2}{3} \left(1 - e^{-6D_r \Delta t} \right) \tag{10.6}$$

which can be rewritten as

$$D_{\mathrm{r}}\Delta t = \frac{-\ln\left(1 - \frac{3}{2}\langle\sin^2\omega\rangle\right)}{6} \qquad (10.7)$$

In both cases, the uncertainty in ω can be determined from the intercept at $\Delta t = 0$.

10.2.3 Experimental Validation

The experimental validity of these methods was evaluated in a model system,[7] spherical particles at dilute concentrations, allowing comparison with established theoretical models. Images in aqueous solution of hemispherically coated 2 μm diameter Janus spheres having an opaque gold hemisphere and a fluorescent core were acquired via epi-fluorescence microscopy with a total magnification of 157.5× (for detailed experimental conditions and analysis, see Anthony et al.[7]). Analysis of the trajectories of six Janus spheres over a 75 s interval revealed close experimental correspondence with theory, with accurate results obtained even from analysis of the single particle. To determine the translational diffusion, the mean square displacement (MSD) of the particle positions was computed and its slope determined. In order to obtain an analogous figure for rotational diffusion, eqn (10.7) was employed. In Figure 10.4b, the rotational dynamics are plotted according to eqn (10.7). From the slope of the linear fit, the rotational diffusion constant was found to be 0.178 rad^2 s^{-1}, that is, 585 $\pm$ 35 deg^2 s^{-1}. Since the reported error was based on the variation in the computed rotational diffusion constant throughout the time sequence, this error represents the error in measurement for this sample, but does not include the error introduced by the polydispersity. This agrees well with the Debye–Stokes–Einstein rotational diffusion constant of 540 deg^2 s^{-1} for a 2 μm sphere.

On the basis of the y-intercept, the average uncertainty in orientation was 14°. This suggests an average uncertainty of $\sim 10°$ in the determination of the zenith and azimuthal angles for this method under these conditions. However, it is important to note that the uncertainty in these measurements itself depends on the zenith angle. With respect to the zenith angle, the accuracy decreases when the zenith angle is near one of its extreme bounds. Since there exists an approximately co-sinusoidal relationship between the zenith angle and the brightness, smaller uncertainties in brightness produce correspondingly larger uncertainties in the zenith angle when it is near its extreme values. Similarly, determining the angle of the azimuth depends on the zenith angle. For a zenith angle of 90° (half moon), the azimuthal orientation is clear; however, as the zenith approaches its limits, discerning the azimuthal orientation becomes increasingly difficult.

Simultaneously, the translational dynamics were determined (Figure 10.4a). From the y-intercept of 0.0033 $\mu\mathrm{m}^2$, the uncertainty in determining the

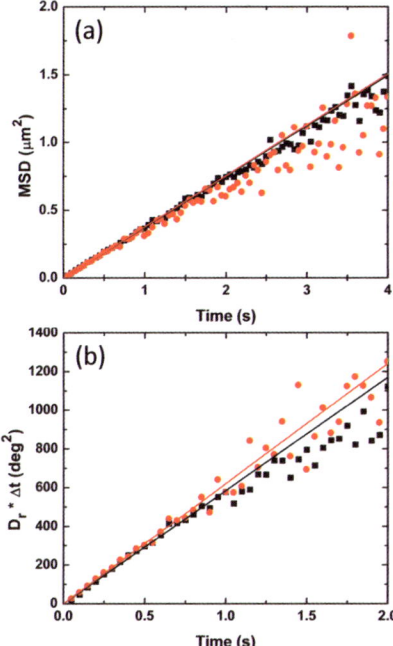

Figure 10.4   The efficacy to quantify the rotational and translational diffusion of Janus particles is illustrated. (a) The translational data are plotted both for the average of six Janus spheres (black squares) and for a single Janus sphere (red circles). Fitting yielded an ensemble-average translational diffusion constant for the 2 μm Janus spheres of 0.094 ± 0.003 μm^2 s^{-1} (black line). Based on the y-intercept of 0.0033 μm^2, the uncertainty in position for the Janus spheres is 0.04 μm. As can be seen from the virtual overlap of the two fits, while the ensemble data are obviously less noisy, information from a single trajectory of a single particle can be analyzed quantitatively with the same conclusion. (b) As a result of the boundary conditions on angular displacements, the mean squared angular displacement is an approximation limited in utility to small angles. Here, the more rigorous equation [eqn (10.7)], which accounts for these limitations, is employed. [The right-hand side of eqn (10.7), equivalent to $D_r \times \Delta t$, is plotted here.] The resulting ensemble average rotational data for the 2 μm Janus spheres (black squares) is nicely linear. A vertical offset of 82 deg^2, resulting from the angular uncertainty of 14°, was subtracted. The slope of this plot yields a rotational diffusion coefficient of 585 ± 35 deg^2 s^{-1} (black line). For comparison, similar data for a single Janus sphere are also shown (red circles). From even this single sphere, it is possible to determine the rotational diffusion constant of 620 ± 100 deg^2 s^{-1}, albeit with lower accuracy (red line). The deviations from the linear fits after ~ 1 s are not significant, as the uncertainty of those points is much greater. Additionally, the error for those points is not normally distributed, such that, on average, one should expect to observe a downward curvature. Adapted with permission from S. M. Anthony, L. Hong, M. Kim and S. Granick, *Langmuir*, 2006, **22**, 9812. Copyright 2006 American Chemical Society.

position of the Janus spheres is 0.04 μm or approximately ½ pixel. The ½ pixel accuracy is likely not the fundamental limit of the available accuracy, but rather reflects that in the interests of computational efficiency, the Janus spheres were localized only to the nearest pixel in that work.[7]

On the basis of the slope, the translational diffusion constant was determined as 0.094 ± 0.003 μm^2 s^{-1}. As pointed out in the original article,[7] this value is less than would be anticipated from the Stokes–Einstein equation for dilute particles of this diameter in solution (0.22 μm^2 s^{-1}), reflecting the hydrodynamic influence of the nearby wall, since the particles tracked were those which had sedimented to near the bottom of the sample cell. The expected ratio of diffusion parallel to this wall, relative to bulk diffusion, is given by

$$\frac{D}{D_0} = 1 - \frac{9}{16}\frac{r}{z} + \frac{1}{8}\left(\frac{r}{z}\right)^3 - \frac{45}{256}\left(\frac{r}{z}\right)^4 - \frac{1}{16}\left(\frac{r}{z}\right)^5 \qquad (10.8)$$

where r is the radius of the particle and z is the separation of the center of the particle from the wall.[10] Assuming a Boltzmann distribution based upon gravitational potential energy for the separation of the particles from the wall, and given the calculated density of the Janus spheres, 1.45 g cm^{-3}, the mean distance between the bottom of the particle and the wall is 0.22 μm. Then, considering the Boltzmann distribution of in-plane diffusion coefficients of particles located at each expected distance from the wall, the expected ratio of these quantities is 0.47, which should be compared with the observed ratio of 0.45. This is excellent agreement, especially considering that this estimate does not even consider the second-order effect that particles also diffuse normal to the solid surface.

10.3   Optically Overlapping Particle Localization

Although the above methods work well for isolated Janus spheres, different techniques[8] must be employed when the concentration increases and the particles are not clearly separated. In general, concentration has not been as significant a concern for colloidal tracking, since colloids are frequently large enough that the perturbations at the edges can be neglected, using only the intensities near the centers to localize the particles with sufficient accuracy for many purposes. As such, traditional methods of colloidal localization and tracking are commonly applied for concentrations ranging from dilute solutions to glasses.[11] When considering Janus particles, on the other hand, the edges cannot simply be ignored, as in some orientations the only observations stem from the edges. Typical methods designed to deal with particles in close proximity generally require the images of the particles to be rotationally symmetrical, a condition which Janus spheres fail to satisfy.

The problem presented by Janus spheres in close proximity, therefore, is that once diffraction, lensing due to refractive index mismatches, and particle excursions from the image plane are considered, the images of the Janus spheres will overlap, taking diverse forms depending on the relative orientations of the adjacent Janus spheres. The need for algorithms capable of analyzing overlapped Janus sphere images is motivated not simply by concentrated particle suspensions, as the same problem arises for lower particle concentrations due to Janus spheres coming into transient close contact in the course of Brownian motion. These transient interactions are of particular interest as they provide a convenient medium to explore the complex hydrodynamic forces such that solvent displacement and particle collisions mediate between particles,[12,13] for which the rotational information available from Janus spheres has great utility. Whereas the translational hydrodynamic cross-correlation of particles has been studied,[12,14,15] comparable studies of rotational hydrodynamic cross-correlation have so far been limited to theory and computation of limited order,[16,17] and have scarcely been explored experimentally at the single-particle level owing to the image analysis difficulties resolved by recent work.[7,8]

In contrast to traditional methods of colloidal tracking,[11] the method[8] developed for analyzing overlapping images of Janus spheres employs and extends overlapping object recognition algorithms.[18,19] On the experimental side, analysis of Janus spheres in close proximity is facilitated by using trans-illumination of Janus spheres with one opaque hemisphere. The algorithm benefits substantially if, for the corresponding sphere lacking the Janus opaque coating, the perimeter of such a sphere can be distinguished from the background as shown in Figure 10.5a. Such contrast can be induced either if the substrate sphere absorbs some fraction of the light or through a slight mismatch in refractive index between the transparent substrate of the Janus sphere and the surrounding medium, and may also be enhanced through various phase-contrast imaging techniques. As such, in practice for trans-illumination, substantial effort would need to be made to avoid having any such signal.

Figure 10.5 shows representative images both of an uncoated sphere and spheres with opaque hemispherical Janus coatings. As can be seen, for the uncoated sphere, its center is darker than the background and a bright halo surrounds the image, an indication of the substantial lensing due to the refractive image mismatch of the sphere. Now contrast what happens when imaging Janus spheres: the coated region is effectively opaque and casts a shadow, but the entire perimeter of the sphere's outline remains discernable in all orientations (Figure 10.5b–d).

Similarly to the fluorescent Janus sphere, the point on the sphere used to specify its angular orientation is the point at the center of the hemisphere which produces the dominant signal. Defining the azimuthal angle, therefore, poses no additional problems compared with fluorescent Janus spheres, with a full range of observable angles $0 \leq \varphi \leq 2\pi$. However, working with the shadows cast by Janus spheres with opaque hemispheres under trans-illumination introduces an angular ambiguity not present with fluorescent Janus spheres with epi-illumination, namely that a mirror symmetry exists in

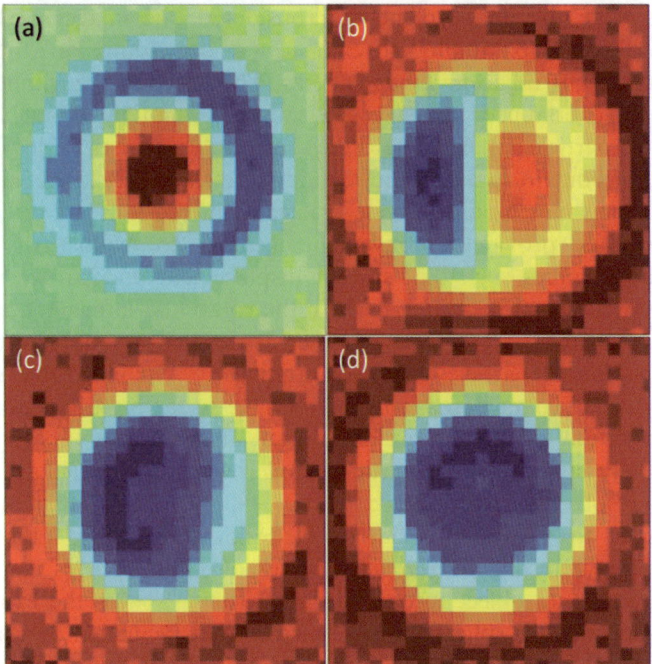

Figure 10.5   Sample trans-illumination images showing an uncoated silica sphere (a) and Janus spheres in half-moon (b), gibbous-moon (c) and full-moon (d) orientations. Images are pseudo-colored with their minimum value in blue and their maximum value in red, with absolute ranges differing between figures. Adapted with permission from S. M. Anthony, M. Kim and S. Granick, *Langmuir*, 2008, **24**, 6557. Copyright 2008 American Chemical Society.

the observations as the zenith angle is varied around $\theta = \pi/2$; this is because, regardless of whether the hemispherical metal cap is oriented towards or away from the observer, the opaque region is effectively the same. Therefore, while the zenith angle θ should be the polar angle from the z-axis, with $0 \leq \theta \leq \pi$ (Figure 10.3), due to the ambiguity the observed zenith angle θ ranges from 0 to $\pi/2$ back to 0. Although representing a drawback in comparison with fluorescent Janus spheres, when working with Janus spheres in close proximity the drawback is more than offset by the benefits of resolving the entire perimeter of the Janus sphere from the background made possible by trans-illumination. The presence of the complete perimeter greatly facilitates locating the particle center, otherwise a challenge when particles are in close proximity as the method previously described for isolated particles is inapplicable. Additionally, greater accuracy in determining the center also improves the determination of the azimuthal angle φ.

The algorithmic process of tracking Janus spheres under trans-illumination consists of four major consecutive steps:

1. Extracting the image information into a more concise form.
2. Approximately locating the position of each Janus sphere.
3. Separating the overlapping Janus sphere images.
4. Extracting the more precise position and orientation from the now-isolated images.

10.3.1 Image Preprocessing

Owing to the nature of the later steps, suppressing the high spatial frequency noise in the image constitutes an important first step. To this end, first a low-

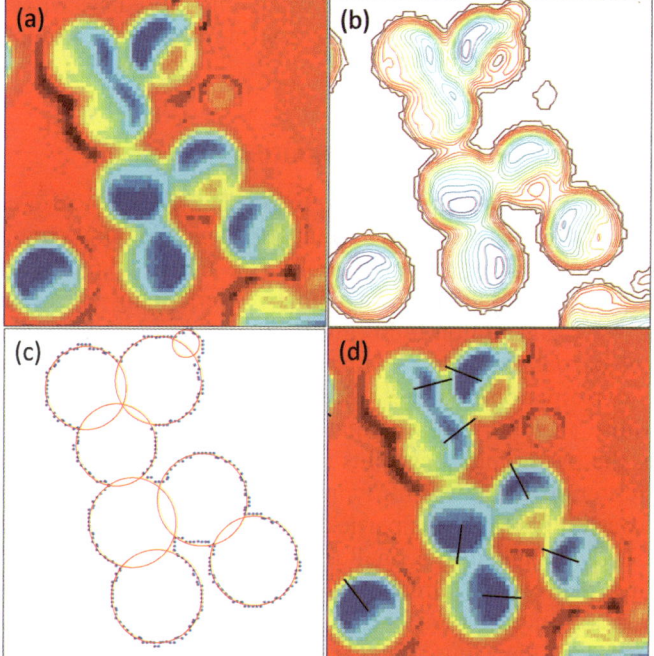

Figure 10.6 Demonstration of the general process to track Janus spheres whose images overlap. (a) Original image as recorded by the CCD camera. (b) The same image converted to contour plot form. For clarity, fewer contours are shown than are typically used in actual calculations. (c) Vertices of the exterior contour (blue dots) are fitted using an overlapping object recognition algorithm. This determines the circles of which the contour is composed (red circles). This procedure is robust, able to fit circles with half of their perimeter obscured. It recognizes particles of different sizes, including a piece of dust on the microscope objective (the small circle to the upper right). (d) Original image with overlaid black lines depicting the final fit. The lines run from the center of each particle outward, in this way indicating the angle (φ) in the plane of the image. Adapted with permission from S. M. Anthony, M. Kim and S. Granick, *Langmuir*, 2008, **24**, 6557. Copyright 2008 American Chemical Society.

pass filter (standard boxcar filter, three pixels in each direction) is convolved with the image. Provided that the apparent size of the Janus sphere is several pixels in diameter, the filter is effective in smoothing the image without significant loss in clarity of the moon-like images of the Janus spheres. To place this in perspective, an apparent diameter of around 20 pixels is readily obtained for even 1 μm diameter Janus spheres using 100× magnification and commonly used CCD cameras (detector pixel size 6 μm). Next, a simple background subtraction is applied. Given the generally strong signal of Janus spheres, the intensity of the background pixels can readily be discerned on an intensity histogram of the image. Provided that the background intensity is relatively uniform throughout the image, the mean value of the background can simply be subtracted from the entire image. As a final preprocessing step, the image is converted to a contour plot using only those lines whose intensity is above the background level, resulting in contour plots similar to those shown in Figure 10.6b.

10.3.2   Overlapping Object Recognition

The second step constitutes approximately locating the position of each Janus sphere. For any singular or cluster of Janus spheres, the outer contour should consist of one or more overlapping circles (see Figure 10.6). Discriminating circles is a process frequently dealt with by overlapping object recognition algorithms.[18,19] Segmentation following Honkanen et al.[18] generally results in a number of partial perimeters of the spheres. Those segments can then be grouped by corresponding spheres using clustering algorithms related by Shen et al.,[20] whose method assumes that if two circles overlap by a certain percentage, the overlapping segments should be considered part of the same circle. This criterion works well for typical Janus particle experiments, since the overlap between different Janus particles arises primarily from diffraction and hence inherently is limited to a small fraction of the entire circle.[8] Finally, all segments are fitted with a direct least-squares circle-fitting method. Further analysis is made only of those circles whose fitted diameter is close to the known diameter of the Janus spheres.

10.3.3   Separation of Overlapping Janus Spheres

Up to this point of the image analysis, the images of Janus spheres in close proximity to one another still overlap owing to diffraction, introducing significant amounts of distortion. Even for simple silica spheres, it is known that traditional centroid-based tracking algorithms can overestimate the particle separation by up to tens or even hundreds of nanometers for this reason. A common method to correct it treats image overlap as a linear superposition and then exploits the symmetry of spheres by using a simple reflection algorithm to subtract contributions from adjacent particles.[21,22] This works well for pairs of particles but not for larger groupings of particles; the method is restricted to relatively low particle concentrations. Unfortunately, such symmetry-based methods are inapplicable to Janus spheres. However, assuming the same principle of linear superposition of image overlap, an iterative subtraction process can instead be employed to separate and resolve the particles (see Figure 10.7).[8] Although limited to discrete clusters (as opposed to near monolayers), the method can handle clusters of several particles (see Figure 10.6), allowing a wide range of concentrations to be explored.

The strategy proceeds as follows. Given a contour plot that consists of a linear superposition of the images of the two Janus spheres, the objective is to subtract the image of one of them, thus revealing the unperturbed image of the other. To accomplish this requires some bootstrapping because to implement it requires knowing the unperturbed image of one of the Janus spheres, exactly what this process is seeking to obtain. Progress, however, can be made incrementally. Although the entire unperturbed image for any overlapping Janus sphere is unavailable, for those contours closest to the background the contribution from each separate Janus sphere can be relatively easily distinguished, as the contour of each Janus sphere possesses a roughly circular region centered near the center of the Janus sphere. As contours are considered whose intensities are further from the background, they possess smaller circular regions, perhaps only a hemisphere or even less in the case of overlap between hemispheres on adjoining particles. Additionally, even without overlap from adjacent particles, these contours are often incomplete circles (except when the Janus sphere is oriented coaxially with the observer). By fitting the circular portion, the contribution of one Janus sphere to the image of adjacent Janus spheres can be determined (see Figure 10.7b).

In this iterative fashion, although no completely unperturbed image is ever available to subtract, unperturbed contours can be subtracted one level at a time. In the process, each subsequent layer becomes less perturbed, such that, by the subtraction of the final contour level, instead of the initially fairly perturbed contours, instead these contours are nearly unperturbed (see Figure 10.7d). The initial image is replicated once per particle involved, with the set of replicates eventually yielding a set of unperturbed images. For each iteration, one replicate is chosen and used to determine what to subtract from replicates representing adjacent particles, then the next replicate and the next successively until all replicates have been employed in that iteration. Subtractions can be performed rapidly using the publicly available General Polygon Clipping Library.[23]

Finally, only relatively unperturbed images of each Janus sphere remain, where as a verification, the portions subtracted can be seen to correspond well to the images of the adjacent particles. The accuracy of this process is unaffected by the relative orientation of the neighboring Janus spheres, placing no symmetry constraints. At the end, contour plots nearly equivalent to those that would have been obtained from isolated Janus spheres at the same orientation and at the same position are left.

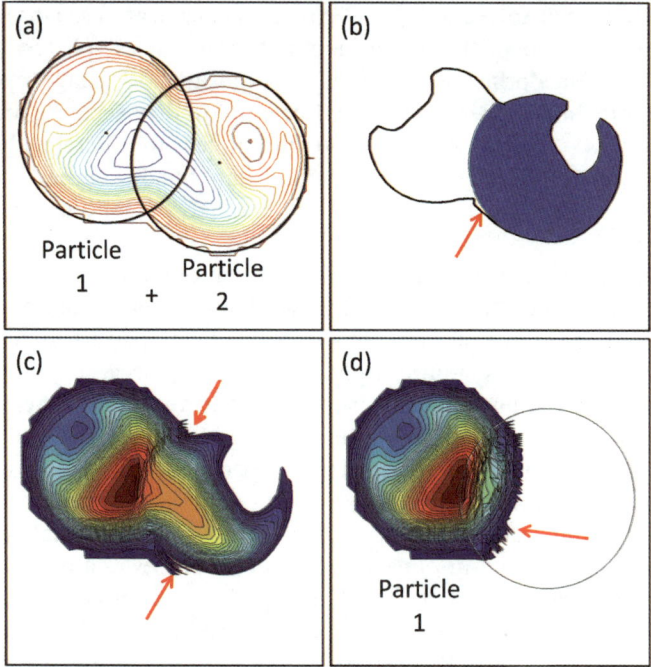

Figure 10.7   A technique[8] to separate and resolve the elements of overlapping circular images is illustrated for a pair of Janus spheres. (a) The starting point is a contour plot of the overlapping images. Note that from the overlapping object algorithm, the center (shown here by dots) and the diameter (shown here by black circles) of each particle are approximately known. To begin, two copies of this initial contour plot are considered: one for particle 1 (left) and one for particle 2 (right). (b), A contour from particle 2 (black line) has been selected and a trial attempt was made to determine the portion of this contour that is genuinely attributable to particle 2 (blue area), which is then subtracted from particle 1. Note that, because this approximation is inexact, slivers of particle 2 (red arrow) may not be subtracted. In subsequent iterations, the same procedure is applied to a contour from the new version of particle 1, iterating back and forth between the two particles. (c) Illustration of the imputed image of particle 1, part way through the iterative procedure. Note that while the subtraction continues to improve relative to (b), the slivers noted in (b) (red arrows) continue to accumulate. (d) After the iterations are completed, the image of particle 1 has been isolated. The slivers, whose presence in the earlier images was highlighted in the earlier panels of this figure, have been largely cropped off, using the criterion that they extend too far from the center of particle 1. Some image perturbation does remain from overlap with particle 2, whose position is indicated by the circle, but this is mostly roughness in the contour of the circles. This roughness is easy to notice visually but has minimal effect on subsequent calculations. Adapted with permission from S. M. Anthony, M. Kim and S. Granick, *Langmuir*, 2008, **24**, 6557. Copyright 2008 American Chemical Society.

10.3.4 Refining the Position and Extracting the Orientation

As a final step, since the image is now that of an isolated Janus particle, the techniques described previously for localizing isolated particles could be employed; similarly, the technique described in this step could be applied to localizing isolated particles if their images were converted to contour plots. However, for isolated particles which do not require this more extensive method, the previous method is much simpler to implement. Exploiting a property of trans-illuminated Janus spheres (see Figure 10.5), regardless of orientation, the contours close to the background form complete circles. As such, performing circle fitting on several such contours and averaging them allows accurate determinations of the center of the Janus sphere, with final precision of <10 nm.[8]

Next, the angles are determined. Calculation of the azimuth angle proceeds from the fact that the image of an isolated Janus sphere must be symmetrical across this angle, so the centroid of each contour of the image must lie along the azimuth angle. Once the centroids of the contours have been computed, a linear fit is then applied, which is trivially transformed from a line to a vector by determining which direction holds the highest intensity (see Figure 10.8). Calculation of the zenith angle exactly follows the technique detailed previously for isolated particles.

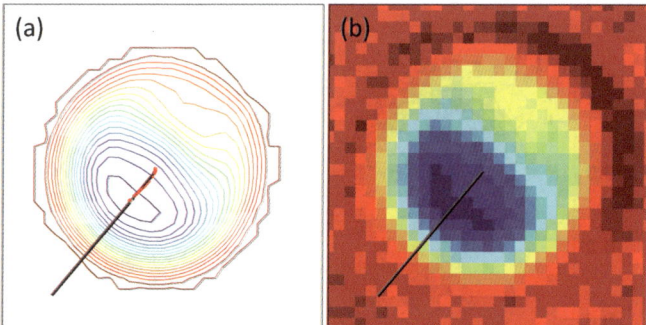

Figure 10.8 (a) Five most nearly circular contour rings are used to determine the center (terminus of the black line) of the particle. A linear fit to the centroids of each contour (red dots) yields the azimuth angle (black line) of the Janus particle. (b) Comparison of the raw image with the fit (black line). Adapted with permission from S. M. Anthony, M. Kim and S. Granick, *Langmuir*, 2008, **24**, 6557. Copyright 2008 American Chemical Society.

10.4 Probing Translational and Rotational Dynamics

The ability to localize (both spatial and angular orientation) Janus particles described above represents a powerful new tool, providing the capability to monitor translational and rotational motion simultaneously in a wide range of environments, a capability that is only beginning to be exploited. One example is a recent study[24] which employed Janus particles to unravel colloidal rotation near the colloidal glass transition. Monodisperse colloidal-sized spherical particles in liquid suspension are known to enter jammed glassy states as the volume fraction is raised.[25–27] Prior to the use of Janus particles as a probe, although numerous single particle studies of glassy systems had been undertaken, yielding a wealth of knowledge regarding the glassy state, understanding of the system was limited to translational dynamics.[28,29]

Comparison of the trajectories of angular and translational displacement showed qualitatively similar behavior: restricted motion followed by leaps (Figure 10.9). However, the locations of angular persistence decorrelate more rapidly than those for spatial position, leading to a mean-square angular displacement that shows no plateau regime, but sub-Fickian behavior (Figure 10.10). Angular and translational displacement also correlate and the degree of correlation changes for different concentrations of colloidal particles (Figure 10.11).

In the study by Kim *et al.*,[24] Janus spheres were introduced into the colloidal glassy system, high concentrations of poly(methyl methacrylate) (PMMA) spheres dispersed in a cyclohexylbromide–decalin mixture in a proportion adjusted empirically to match the refractive index and density of the particles. Janus particles were prepared by hemispherically coating the parent particles

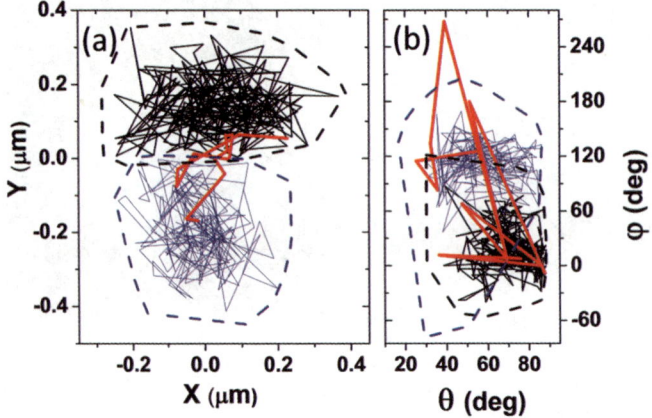

Figure 10.9 Typical positional (a) and angular (b) trajectories at a volume fraction of 0.51; the axes are spatial positions x and y (a) and angles θ and φ (b) following the conventions given in Figure 10.3. Regarding position, the particle jumps from the first cage (medium black line) to the second cage (thin blue line) rapidly (thick red line). This is accompanied by distinct leaps in angle, separating the angular trajectory into two regions. However, the rotation is not a caged motion and two regions easily overlap in a longer time. Solid lines, trajectories during 15 min; dashed lines, boundaries of trajectories during 40 min. Adapted with permission from M. Kim, S. M. Anthony, S. C. Bae and S. Granick, *J. Chem. Phys.*, 2011, **135**, 054905. Copyright 2011 American Institute of Physics.

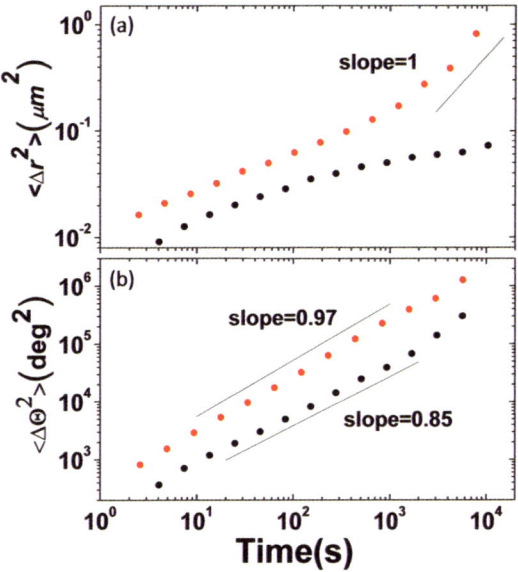

Figure 10.10 Positional mean squared displacement (a) and angular mean squared displacement (b) are plotted on log–log scales against time at volume fractions 0.49 and 0.59 (red and black circles, respectively). For clarity, only two volume fraction data are shown. Adapted with permission from M. Kim, S. M. Anthony, S. C. Bae and. S. Granick, *J. Chem. Phys.*, 2011, **135**, 054905. Copyright 2011 American Institute of Physics.

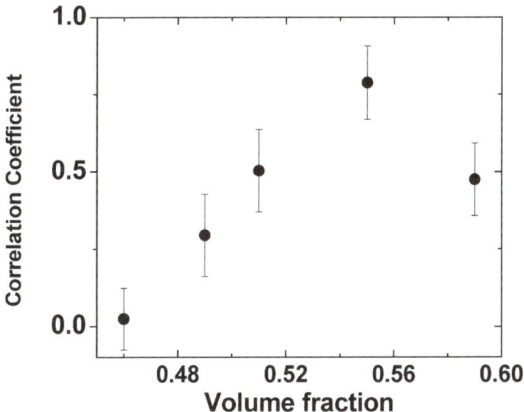

Figure 10.11 The correlation C between translation and rotation plotted versus volume fraction, C being defined as

$$C = \frac{E[(X_1 - \mu_1)(X_2 - \mu_2)]}{\sqrt{E\left[(X_1 - \mu_1)^2\right]} \times \sqrt{E\left[(X_2 - \mu_2)^2\right]}}$$

where X_1 is a rotational displacement, X_2 is a translational displacement and μ_1 and μ_2 are mean values of X_1 and X_2. Adapted with permission from M. Kim, S. M. Anthony, S. C. Bae and. S. Granick, *J. Chem. Phys.*, 2011, **135**, 054905. Copyright 2011 American Institute of Physics.

with 12 nm of aluminum and subsequently treating the metal layer so as to be chemically identical with the parent particles.

10.5   Conclusion

Janus particles represent a new class of colloidal probes, with the unique ability not only to monitor spatial dynamics, but also to monitor rotational dynamics simultaneously at the single-particle level. Techniques have been developed which allow the determination of two degrees of angular motion with high accuracy for a wide range of systems. The availability of such a probe opens up new opportunities for experiments, the most prominent among them being the study of the translational–rotational coupling of colloids in concentrated suspensions.

Acknowledgements

S.M.A. acknowledges the National Science Foundation for financial support in the form of a Graduate Research Fellowship. Support from the US Department of Energy, Division of Materials Science, is acknowledged under Award No. DE-FG02-07ER46471 through the Frederick Seitz Materials Research Laboratory at the University of Illinois at Urbana-Champaign. The National Science Foundation is acknowledged both for grant support under NSF-CBET-0853737 and for the use of facilities purchased with grants NSF-DMR-0605947 and NSF-CMS-0555820. The Donors of the Petroleum Research Fund are acknowledged for their support, administered by the American Chemical Society, #45523-AC7.

References

1. J. R. Lakowicz, *Principles of Fluorescence Spectroscopy*, Springer, New York, 2nd edn, 1999.
2. M. T. Cicerone, F. R. Blackburn and M. D. Ediger, *J. Chem. Phys.*, 1995, **102**, 471.
3. J. Kanetakis, A. Tolle and H. Sillescu, *Phys. Rev. E*, 1997, **55**, 3006.
4. M. K. Cheezum, W. F. Walker and W. H. Guilford, *Biophys. J.*, 2001, **81**, 2378.
5. J. N. Anker and R. Kopelman, *Appl. Phys. Lett.*, 2003, **82**, 1102.
6. J. N. Anker, C. Behrend and R. Kopelman, *J. Appl. Phys.*, 2003, **93**, 6698.
7. S. M. Anthony, L. Hong, M. Kim and S. Granick, *Langmuir*, 2006, **22**, 9812.
8. S. M. Anthony, M. Kim and S. Granick, *Langmuir*, 2008, **24**, 6557.
9. F. Perrin, *Ann. Sci. École Normale Supérerieure*, 1928, **45**, 1.
10. L. P. Faucheux and A. J. Libchaber, *Phys. Rev. E*, 1994, **49**, 5158.
11. J. C. Crocker and D. G. Grier, *J. Colloid Interface Sci.*, 1996, **179**, 298.

12. J. C. Meiners and S. R. Quake, *Phys. Rev. Lett.*, 1999, **82**, 2211.
13. M. Reichert and H. Stark, *Phys. Rev. E*, 2004, **69**, 031407.
14. B. X. Cui, H. Diamant, B. H. Lin and S. A. Rice, *Phys. Rev. Lett.*, 2004, **92**, 258301.
15. B. X. Cui, H. Diamant and B. H. Lin, *Phys. Rev. Lett.*, 2002, **89**, 188302.
16. H. G. Hermanns and B. U. Felderhof, *J. Chem. Phys.*, 2007, **126**, 044902.
17. B. Cichocki, M. L. Ekiel-Jezewska and E. Wajnryb, *J. Chem. Phys.*, 1999, **111**, 3265.
18. M. Honkanen, P. Saarenrinne, T. Stoor and J. Niinimaki, *Meas. Sci. Technol.*, 2005, **16**, 1760.
19. F. Pla, *Comput. Vision Image Understanding*, 1996, **63**, 334.
20. L. P. Shen, X. Q. Song, M. Iguchi and F. Yamamoto, *Pattern Recognit. Lett.*, 2000, **21**, 21.
21. R. Verma, J. C. Crocker, T. C. Lubensky and A. G. Yodh, *Macromolecules*, 2000, **33**, 177.
22. J. C. Crocker, J. A. Matteo, A. D. Dinsmore and A. G. Yodh, *Phys. Rev. Lett.*, 1999, **82**, 4352.
23. A. Murta, *General Polygon Clipper Library, Version 2.32*, http://www.cs.man.ac.uk/toby/alan/software/gpc.html (last accessed May 25, 2012).
24. M. Kim, S. M. Anthony, S. C. Bae and S. Granick, *J. Chem. Phys.*, 2011, **135**, 054905.
25. P. N. Pusey and W. van Megen, *Nature*, 1986, **320**, 340.
26. E. Bartsch, V. Frenz, S. Moller and H. Silescu, *Physica A*, 1993, **201**, 363.
27. W. van Megen and S. M. Underwood, *Phys. Rev. E*, 1994, **49**, 4206.
28. E. R. Weeks, J. C. Crocker, A. C. Levitt, A. Schofield and D. A. Weitz, *Science*, 2000, **287**, 627.
29. W. K. Kegel and A. van Blaaderen, *Science*, 2000, **287**, 290.

Janus Balance and Emulsions Stabilized by Janus Particles

SHAN JIANG[a] AND STEVE GRANICK*[b]

[a] The Koch Institute for Integrative Cancer Research at MIT, Department of Chemical Engineering, Massachusetts Institute of Technology, 77 Massachusetts Avenue, Cambridge, MA 02139, USA; [b] Departments of Materials Science and Engineering, Chemistry and Physics, University of Illinois, Urbana, IL 61801 USA
*E-mail: sgranick@illinois.edu

11.1 Introduction

Janus particle-stabilized emulsions share the characteristics of both emulsions stabilized by surfactant molecules and emulsions stabilized by homogeneous particles. On the one hand, Janus amphiphilic particles can be viewed as the colloidal version of small amphiphilic molecules, and much can be learned from the theory and application developed on small surfactants, where emulsion type is usually determined by the packing of surfactant molecules at interface.[1] Surfactants with small tails are known to stabilize oil-in-water emulsions, whereas those with large tails stabilize water-in-oil emulsions. On the other hand, homogeneous particles can also stabilize emulsions (Pickering emulsions), where colloidal particles spontaneously adsorb at the emulsion interface due to the intermediate hydrophobicity of particle surfaces.[2] For example, in a typical water/oil emulsion, the particle surface is more hydrophobic than the water phase, but less than the oil phase. In order to minimize the surface energy, colloidal particles find the interface the most favorable place to stay, with a certain percentage of the particle surface

RSC Smart Materials No. 1
Janus Particle Synthesis, Self-Assembly and Applications
Edited by Shan Jiang and Steve Granick

immersing in water or in oil. Completely removing particles from interface requires extra energy. These particle-decorated emulsion droplets are stable against coalesce. It has been demonstrated experimentally that the wettability of the particle surface is directly correlated with the emulsion type (oil-in-water or water-in-oil) and the stability of emulsions.

It was de Gennes's original idea that Janus particles would adsorb at a liquid/liquid or liquid/air interface to form a monolayer,[3] though it is true that homogeneous particles with proper surface chemistry will also adsorb. However, when the particle size is less than 10 nm, the adsorption energy at a liquid/liquid interface for homogeneous particles becomes comparable to the thermal fluctuation, which implies that particles can be easily displaced from the interface. Under these conditions, Janus geometry will adsorb much more strongly at the interface and offer higher emulsion stability, which can be critical for emulsions stabilized by small particles of ~ 10 nm. The first step to understanding emulsions stabilized by Janus particles is to understand the principles of Janus particles adsorbed at a planar interface, since many basic aspects of emulsions can be understood by simply calculating the free energy of the system without considering the curvature of the interface. It is expected that Janus particles will adsorb at an interface with their hydrophobic side facing the more hydrophobic (oil) phase and the hydrophilic side facing the other way. The adsorption obviously depends on the geometry of the Janus particles or *Janus balance*.

We quantify the notion of 'Janus balance,' defined as the dimensionless ratio of work to transfer an amphiphilic colloidal particle (a 'Janus particle') from the oil/water interface into the oil phase, normalized by the work needed to move it into the water phase.

The concept of Janus balance is similar to the HLB (hydrophilic–lipophilic balance) value of surfactant molecules,[4,5] which is the single most important parameter that characterizes the surfactant molecules. This value is a measure of the degree to which a surfactant molecule is hydrophilic or lipophilic. It provides an easy way to categorize and choose surfactants for different purposes. For example, water-in-oil emulsions usually require low HLB surfactants, whereas oil-in-water emulsions often require higher HLB surfactants. Inspired by the HLB concept from molecular surfactants, the concept of 'Janus balance' can also be used to characterize Janus particle-stabilized emulsions.

Intuitively, Janus balance is simply the balance between the hydrophobic and hydrophilic portions of the Janus particles, which is partially determined by the geometry of the Janus particles. Although 50:50 is the most studied geometry for Janus particles, it may not be optimal for emulsions with various oil and water phases. In order to find the best geometry for stabilizing emulsions, a rigorous quantification of the Janus balance concept is derived from thermodynamics by considering the surface energy. As has been the case for the HLB of surfactant molecules, the Janus balance concept may be useful for predicting the stabilizing efficiency and functions of the Janus particles.

Based upon the model and calculation, Janus particles show stronger adsorption at an emulsion interface than conventional homogeneous particles. However, very few experiments have been attempted to test the theory fully. It is still challenging to fabricate Janus particles in large quantities with well-controlled geometry. We present here a few experimental examples of how Janus particles can stabilize emulsions. On the other hand, more sophisticated simulation and calculation demonstrate that other factors may also contribute significantly to the emulsions stability, such as curvature, capillary wave and rotation of the particles at interface.[6,7]

11.2   Janus Particles at a Planar Interface

Janus particles were synthesized by directional coating of Au on a monolayer of silica particles, creating perfect 50% coverage. In order to generate different geometry, Au coatings were etched to different extents. Furthermore, the Au-coated surface was modified by octadecanethiol (ODT) applied to render them hydrophobic, whereas the bare silica surface was hydrophilic. Janus particles were dispersed in 2-propanol and then gently placed at the water/toluene interface. The water phase was gelled at a lower temperature with a gelling agent (1% agarose) after trapping particles at the interface. Using the gel trapping technique, the toluene phase was replaced with polydimethylsiloxane (PDMS), which was slowly

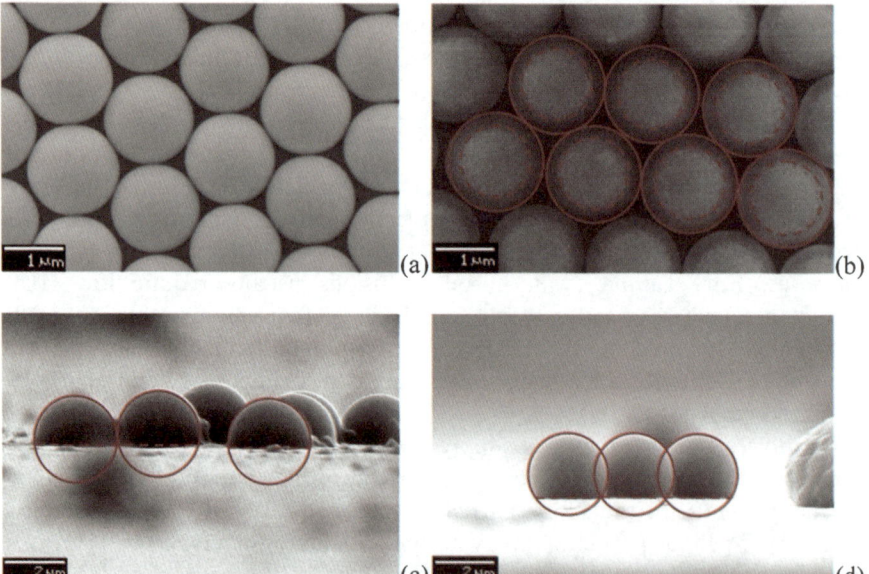

Figure 11.1   SEM images of Janus particles and PDMS surface. (a) Janus particles before etching; (b) Janus particles after etching for a certain period of time; (c) Janus particles (before etching) adhered at the PDMS surface; (d) etched Janus particles adhered at the PDMS surface.

cross-linked at room temperature overnight. Janus particles adsorbed at the interface surface were then adhered to the PDMS surface, which was available for further imaging by electron microscopy. The geometry of Janus particles at the interface could be simply visualized by scanning electron microscopy (SEM)and the contact angle could be quantified by analyzing the images.

Silica particles 2 μm in diameter were used in this study. Figure 11.1a and b show the geometry of Janus particles before and after etching. The Au coating corona was gradually etched away from the boundary, leaving behind less and less coverage as the etching time progressed. Figure 11.1c and d show the side view of the Janus particles adsorbed at the PDMS interface.

Figure 11.2 shows the contact angle at the toluene/water interface *versus* the geometry of Janus particles. Obviously, the embedding of Janus particles fits very well with the geometry of the particles. This strongly suggests that Janus particles were pinned at the boundary dividing the hydrophobic and hydrophilic areas.

In a separate study, similar behavior was observed with much larger Janus bubbles. For these large bubbles, the Au coating can be easily distinguished with strong contrast by electron microscopy. It was further confirmed that the hydrophobic Au coating was facing the oil phase, whereas the hydrophilic side faced water.

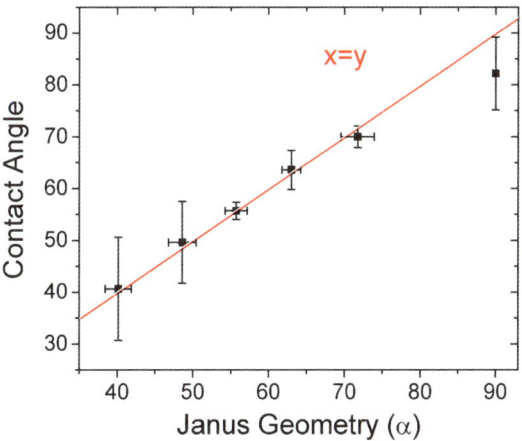

Figure 11.2   Geometry of Janus particles versus the contact angle (°) of particles adhered at the PDMS surface. The red line is where the geometry of Janus particles fits perfectly with the contact angle, which corresponds to the case where Janus particles are pinned at the hydrophilic–hydrophobic boundary.

11.3   Janus Balance

Based on the preliminary experiment on Janus particles adsorbed at the oil/water interface, we can set up a simple model to calculate the adsorption energy.[8] The calculations presented below can be simply extended to Janus particles whose shape is not spherical, but for simplicity (and without loss of generality) we consider a single Janus sphere of radius R sitting at a flat oil/water interface.

The interfacial energies (γ), referring to interface of the polar side (P), apolar side (A) against the oil side (O) and the water side (W), are γ(PO), γ(PW), γ(AO) and γ(AW), respectively. Figure 11.3 shows that the geometry of the Janus particle is quantified by the angle α, which determines the position of the boundary dividing the apolar (hydrophobic) and polar (hydrophilic) regions on the particle. Hydrophobicity is characterized by the angles θ_p (hydrophilic side) and θ_a (hydrophobic side), corresponding to the three-phase contact angle at the oil/water interface of a homogeneous particle consisting of this same hydrophilic or hydrophobic surface chemical makeup. For consistency, this angle is always measured starting from the center line pointing towards the water phase. In this way, θ_p is necessarily larger than 90°, and θ_a is necessarily smaller than 90°. For perfect hydrophilic and hydrophobic surfaces, $\theta_p = 180°$ and $\theta_a = 0°$.

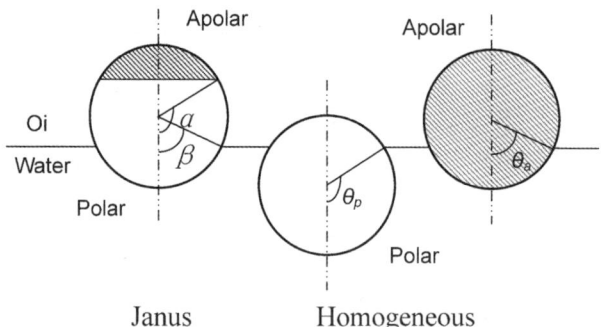

Figure 11.3   The geometry of a Janus particle at the oil/water interface and the contact angle at its hydrophilic and hydrophobic sides. The relative areas of the polar and apolar particle surface regions are parameterized by angle α and the contact angle of the Janus particle at interface β. The contact angle of the hydrophilic side is θ_p and the contact angle of the hydrophobic side is θ_a. Reproduced with permission from reference 8.

11.3.1   Contact Angle of Janus Particles at an Interface

The first step in the following argument is to find the contact angle β of the Janus particle, determined by minimizing the free energy of the Janus particle at the oil/water interface. As shown in Figure 11.4, when $\alpha <\theta_a< 90° <\theta_p$, the contact angle of the Janus particle equals the contact angle of a hypothetical homogeneous particle that possesses the same surface chemical makeup as the hydrophobic side of the Janus particle, that is, $\beta = \theta_a$. The physical meaning is evident when one considers the free energy change. Whenever the particle is moved up or down to lower the free energy, the change in free energy is the same provided that the entire hydrophilic moiety is immersed in the water phase – no matter what the size of the hydrophilic moiety is. In fact, this statement holds even when the surface area of the hydrophilic moiety is zero, *i.e.* if the particle is homogeneously hydrophobic. This means that the contact angle of this Janus particle is the same as that of a hypothetical homogeneous particle with the same hydrophobic surface.

However, when $\theta_a < \alpha < \theta_p$, the result is different: $\beta = \alpha$. The physical meaning is that moving the particle up or down necessarily increases the free energy, either by increasing the amount of polar surface area in contact with oil or the amount of apolar surface area in contact with water. In the same spirit, in the case where $\theta_a < 90° < \theta_p < \alpha$, it follows that $\beta = \theta_p$. The physical meaning is that moving the particle up or down necessarily increases the free energy by the same reasoning.

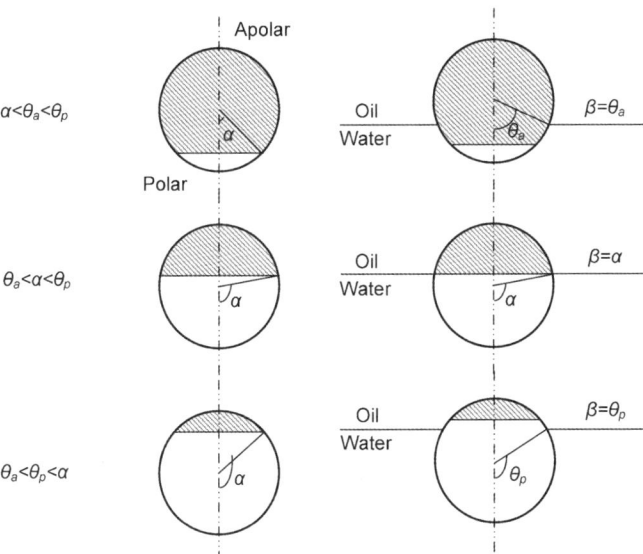

Figure 11.4 The contact angle for a Janus particle at the oil/water interface, with parameters defined in the caption of Figure 11.3. Reproduced with permission from reference 8.

11.3.2 Adsorption Energy

For simplicity, we first discuss the case when $\theta_a < \alpha < \theta_p$. It is worth pointing out that when the particle is moved away from the interface, the water–oil surface area increases. The energy to desorb this particle from equilibrium into the oil (water) phase follows considering the surface areas induced or eliminated by this process, as quantified below:

$$\frac{E(\alpha)_{\text{oil}}}{2\pi R^2} = \frac{1}{2}\gamma(\text{OW})\sin^2\alpha + [\gamma(\text{PO}) - \gamma(\text{PW})] \times (1 - \cos\alpha) \qquad (11.1)$$

$$\frac{E(\alpha)_{\text{water}}}{2\pi R^2} = \frac{1}{2}\gamma(\text{OW})\sin^2\alpha + [\gamma(\text{AW}) - \gamma(\text{AO})] \times (1 + \cos\alpha) \qquad (11.2)$$

where $E(a)_{\text{oil}}$ is the energy to desorb a particle from equilibrium into the oil phase and $E(a)_{\text{water}}$ is the energy to desorb a particle from equilibrium into the water phase. Substituting the contact angles q_a and q_p specified by Young's equation, eqns (11.3) and (11.4), then eqns (11.5) and (11.6) result:

$$\gamma(\text{OW})\cos\theta_p = \gamma(\text{PW}) - \gamma(\text{PO}) \qquad (11.3)$$

$$\gamma(\text{OW})\cos\theta_a = \gamma(\text{AW}) - \gamma(\text{AO}) \qquad (11.4)$$

$$\frac{E(\alpha)_{\text{oil}}}{2\pi R^2} = \gamma(\text{OW})\left[\frac{\sin^2\alpha}{2} + \cos\theta_p(\cos\alpha - 1)\right] \qquad (11.5)$$

$$\frac{E(\alpha)_{\text{water}}}{2\pi R^2} = \gamma(\text{OW})\left[\frac{\sin^2\alpha}{2} + \cos\theta_a(\cos\alpha + 1)\right] \qquad (11.6)$$

11.3.3 Quantification of Janus Balance

This gives the Janus balance (J), defined as the energy needed to desorb the particle from equilibrium into the oil phase, normalized by the energy needed to desorb it from equilibrium into the water phase:

$$J = \frac{\sin^2\alpha + 2\cos\theta_p(\cos\alpha - 1)}{\sin^2\alpha + 2\cos\theta_a(\cos\alpha + 1)} \qquad (11.7)$$

Equation (11.7) shows that the Janus balance depends not just on the respective areas of hydrophilic and hydrophobic chemical makeup, quantified by α, but also on the hydrophobicity of the two sides, quantified by θ_a and θ_p. Janus balance defined in this way considers all factors that affect the thermodynamics of the particle's adsorption.

If θ_a and θ_p are fixed, J increases as α increases (since $\cos\theta_p < 0$), which corresponds to a larger hydrophilic area. If α is fixed, J increases when θ_a or θ_p

increases, which corresponds to the hydrophilic part becoming more hydrophilic or the hydrophobic part becoming less hydrophobic. The larger the magnitude of J, the more hydrophilic is the solid surfactant. The same trend holds for the HLB value of surfactant molecules: surfactants with larger HLB have more affinity for water.

What about the extreme limits of these parameters? If a particle has a homogeneous chemical makeup ($\alpha = \theta_p = \theta_a$, where α refers simply to the contact angle of the homogeneous particle at the oil/water interface), the following equation results; for clarity, β substitutes α by symmetry:

$$J = \frac{(1 - \cos \beta)^2}{(1 + \cos \beta)^2} \tag{11.8}$$

where β is the contact angle of the homogeneous particle at the oil/water interface. Some years ago, in a visionary book, Kruglyakov[5] already defined the HLB value of a homogenous particle at the oil/water interface, deriving an expression equivalent to eqn (11.8).

Similarly, eqns (11.9) and (11.10) yield expressions for the Janus balance for the other two cases. When $\alpha < \theta_a < \theta_p$:

$$J = \frac{\sin^2 \theta_a + 2\cos \theta_p (\cos \alpha - 1) + 2\cos \theta_a (\cos \theta_a - \cos \alpha)}{\sin^2 \theta_a + 2\cos \theta_a (\cos \theta_a + 1)} \tag{11.9}$$

and when $\theta_a < \theta_p < \alpha$:

$$J = \frac{\sin^2 \theta_p + 2\cos \theta_p (\cos \theta_p - 1)}{\sin^2 \theta_p + 2\cos \theta_a (1 + \cos \alpha) + 2\cos \theta_p (\cos \theta_p - \cos \alpha)} \tag{11.10}$$

11.3.4 An Example

As silica is commonly used as a platform from which to construct Janus colloidal particles, the Janus balance concept is now illustrated with numbers that are realistic for these systems. As a demonstration of calculation, the values of θ_a and θ_p were set to 70° and 165°, respectively for (hydrophilic) silica and (hydrophobic) silica modified by silane agent. These values can vary with different oils used in the system.

Figure 11.5 shows the adsorption energy and the Janus balance value plotted as functions of the α parameter. If the adsorption energy is defined as the minimum of E_{oil} and E_{water}, then $J = 1$ corresponds to the maximum of the adsorption energy. Here $J = 1$ when $\alpha \approx 61°$. This is a striking counterexample to the naïve thought that a Janus particle, half hydrophilic and half hydrophobic in area, gives the highest adsorption energy. Instead, it is necessary to weight physical area by interfacial energy.

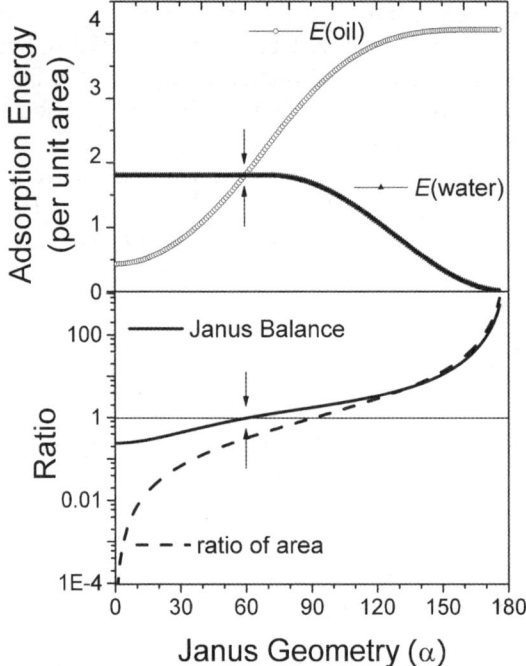

Figure 11.5 A typical example. The adsorption energy and Janus balance for a Janus particle with hydrophilic ($\theta_p = 165°$) and hydrophobic ($\theta_a = 70°$) areas. The Janus balance for this example equals 1 when $\alpha = 61°$ as indicated by the arrows, corresponding to maximum adsorption energy. The dashed line, drawn for reference, is the ratio of hydrophilic to hydrophobic surface area on the Janus particle. Reproduced with permission from reference 8.

11.3.5 Outlook and Potential Implications

It has been shown by both experiment and calculation that for the same surface, the contact angle depends on the oil involved. Therefore, even for the same Janus particle, the Janus balance can differ according to the oil. The same arises concerning the determination of the HLB value for small surfactant molecules: standard oil must be used for calibration otherwise the HLB value depends on the oil. It is necessary to calculate the Janus balance value for a specific system in order to design the best geometry of Janus particles to stabilize the emulsion. Then α should be chosen to ensure that J is as close to unity as possible.

Even for particles of homogeneous chemical makeup, the Janus balance concept can apply when considering the contact angle hysteresis to desorb the particle from equilibrium into the water or oil phase. In this case of homogeneous particle makeup, θ_a and θ_p become the advancing or receding angle.

Another system to which this concept may apply is that when particles coexist together with surfactants. Synergy between surfactants and particles to stabilize emulsions has been studied.[9] Since how much adsorbs depends on the phase from which it adsorbs, it is natural to expect that, according to the HLB, more surfactant will adsorb from the oil phase and less from the water phase or *vice versa*. However, as surfactant molecules adsorb less strongly than particles, their quantity adsorbed may change in response to the presence of adsorbed particles. This projected situation is complex and requires additional modeling and calculation to quantify the overall adsorption energy.

Finally, we note several factors, ignored here, that might potentially affect the emulsion stability and emulsion type under certain conditions. Particle–particle interaction and bending energy at a curved oil/water interface were not considered but the calculation presented here can be extended to the situation of having a curved interface.[10,11] Line tension was not considered, although it becomes increasingly important as the particle size becomes smaller.[7] Gravity and roughness of the particle were also ignored.[12] In analogy with the usefulness of the classical HLB of surfactant molecules, the concept of Janus balance may be useful in facilitating the design of Janus particles as emulsifiers, in predicting their self-assembly behavior[13] and in building a bridge between theory and practical applications.

11.4 Emulsions Stabilized by Janus Particles

There are a few synthetic methods now known to fabricate Janus particles in large quantities. With these methods,[14,15] it is now possible to explore the properties of emulsions stabilized by Janus particles.

11.4.1 An Example

We investigated the ability of Janus amphiphilic particles to stabilize emulsions. Janus particles 500 nm in diameter with different geometries were synthesized using an emulsion-based method. Janus particles (1.0 wt%) were initially dispersed in toluene and then mixed with water. Emulsification was achieved using a Turux 18 dispersing element running at 18 000 rpm for 1 min. The emulsion was then held at room temperature and immediately photographed to record the position of the phase separation. Emulsion stability was evaluated by monitoring the changes of the phases. For all the emulsions stabilized by the Janus particles studied, there were no obvious changes observed after 3 weeks. In a control experiment, the particles were harvested from the wax droplets before the silane vapor deposition. As demonstrated in Figure 11.6, particles without chemical modification could not stabilize these emulsions. This also indicates that surfactant added during the synthesis had been cleaned in the washing step. Figure 11.7 shows tracking of the phase change over longer than 3 weeks. The slight decrease in toluene and emulsion phase may be due to a small amount of evaporation. Hence it is

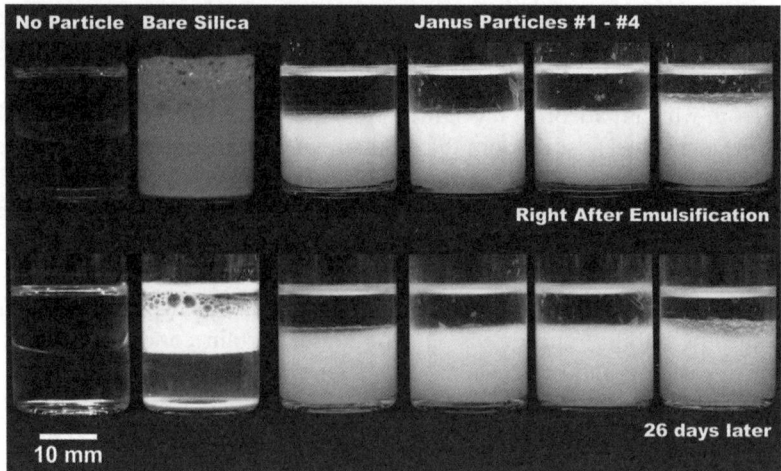

Figure 11.6 Photographs of emulsions. Emulsification is carried out by the dispersing element running at 18 000 rpm for 1 min. Particles are 500 nm in diameter. The relative portion of hydrophobic and charged regions is parameterized by an angle α, the inclination angle corresponding to the radian from the center of hydrophobic part to its edge. Janus particles #1–#4 have geometries (α) of 45, 47, 57 and 72°, respectively. The emulsions stabilized by Janus particles can be dispersed in toluene, but not water, thereby indicating that the emulsion type is water-in-oil. Reproduced with permission from reference 14.

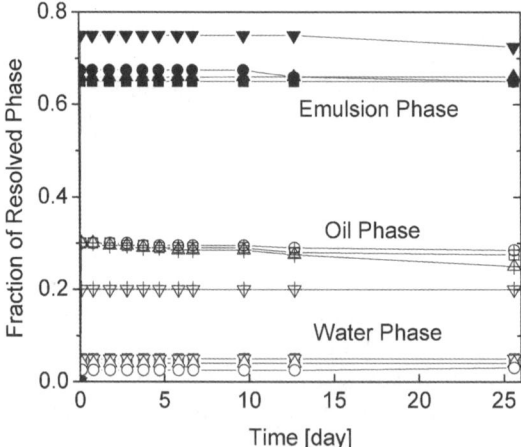

Figure 11.7 Long-term stability of the emulsions described in Figure 11.6: fraction of the resolved phase plotted against time. The slight decrease may be due to the evaporation of the oil (toluene). Reproduced with permission from reference 14.

clear that Janus amphiphilic particles could stabilize water–toluene emulsions. For all the particles investigated with different Janus balance, water-in-toluene emulsions were formed and stabilized for extended times.

11.4.2 Other Progress

Two parallel experimental directions are seen to highlight the seminal importance of particle size. Micron-sized Janus particles have been used to study the assembly behavior of Janus particles at the single particle level. For nanometer-sized Janus particles, emulsions have been characterized by measuring the emulsion droplet size and surface tension, because of the difficulty in direct imaging. The results from both directions confirm that Janus particles adsorb strongly at the interface. However, when particles approach a smaller size, additional factors come into play.

At micron sizes, Janus particles were found to adsorb very tightly at the interfaces of emulsions. Dumbbell amphiphilic micron particles were found to jam the emulsion interface and stabilize emulsion droplets of different shapes.[16]. Janus particles were found to form percolated network structures at a planar oil/water interface, whereas homogeneous particles formed ordered arrays.[17] The difference in assembly structures is due to the difference in the particle–particle interactions at the interface, which can have an important influence on emulsion stability. Lee and co-workers tracked particle diffusion at such interfaces and suggested that the attraction between Janus particles may be due to the quadrupolar capillary interactions induced by the undulation of the three-phase contact line around the Janus particles[18].

For nanometer-sized particles, it was found that the adsorption of Janus particles led to a decrease in surface tension.[19] Simulation showed that the rotation of Janus particles contributed more significantly when the particle size was below 10 nm and led to weaker adsorption at the interface. However, particle–particle interactions were not considered in those simulation and theoretical models.

To understand the phase diagram fully and set up truly realistic models for emulsions can be a challenging task. However, the complexity of the real world endows emulsion systems with the capacity for great practical versatility. Emulsions are widely used in many fields, from the food and oil industries to the biomedical field. For many applications, Janus particles may help prolong the shelf-life of emulsions. Emulsions can also be designed to be environmentally sensitive for drug delivery purposes. Janus particle-stabilized emulsions offer an interesting alternative for many difficult to tackle problems.

Acknowledgement

This work was supported at the University of Illinois by the US Department of Energy, Division of Materials Science, under Award DE-FG02-07ER46471 through the Frederick Seitz Materials Research Laboratory at the University of Illinois at Urbana-Champaign.

References

1. J. N. Israelachvili, *Intermolecular and Surface Forces*, Academic Press, New York, 2nd edn, 1991.
2. B. P. Binks and S. O. Lumsdon, *Langmuir*, 2000, **16**, 8622–8631.
3. P. G. de Gennes, *Rev. Mod. Phys.*, 1992, **64**, 645–648.
4. W. C. Griffin, *J. Soc. Cosmet. Chem.*, 1950, **1**, 16.
5. P. M. Kruglyakov, *Hydrophile–Lipophile Balance of Surfactants and Solid Particles*, Elsevier, Amsterdam, 2000.
6. K. D. Danov, B. Pouligny and P. A. Kralchevsky, *Langmuir*, 2001, **17**, 6599–6609.
7. R. Aveyard, J. H. Clint and T. S. Horozov, *Phys. Chem. Chem. Phys.*, 2003, **5**, 2398–2409.
8. S. Jiang and S. Granick, *J. Chem. Phys.*, 2007, 127.
9. B. P. Binks, J. A. Rodrigues and W. J. Frith, *Langmuir*, 2007, **23**, 3626–3636.
10. Y. Hirose, S. Komura and Y. Nonomura, *J. Chem. Phys.*, 2007, **127**, 054707.
11. P. A. Kralchevsky, I. B. Ivanov, K. P. Ananthapadmanabhan and A. Lips, *Langmuir*, 2005, **21**, 50–63.
12. Y. Nonomura, S. Komura and K. Tsujii, *J. Phys. Chem. B*, 2006, **110**, 13124–13129.
13. Q. Chen, J. K. Whitmer, S. Jiang, S. C. Bae, E. Luijten and S. Granick, *Science*, 2011, **331**, 199–202.
14. S. Jiang, Q. Chen, M. Tripathy, E. Luijten, K. S. Schweizer and S. Granick, *Adv. Mater.*, 2010, **22**, 1060–1071.
15. L. Hong, S. Jiang and S. Granick, *Langmuir*, 2006, **22**, 9495–9499.
16. J. W. Kim, D. Lee, H. C. Shum and D. A. Weitz, *Adv. Mater.*, 2008, **20**, 3239–+.
17. T. Brugarolas, B. J. Park, M. H. Lee and D. Lee, *Adv. Funct. Mater.*, 2011, **21**, 3924–3931.
18. B. J. Park, T. Brugarolas and D. Lee, *Soft Matter*, 2011, **7**, 6413–6417.
19. D. L. Cheung and S. A. F. Bon, *Soft Matter*, 2009, **5**, 3969–3976.

CHAPTER 12

Applications of Janus and Anisotropic Particles for Drug Delivery

ZHIYONG POON AND PAULA T. HAMMOND*

The Koch Institute for Integrative Cancer Research at MIT, Department of Chemical Engineering, Massachusetts Institute of Technology, 77 Massachusetts Avenue, Cambridge, MA 02139, USA
*E-mail: hammond@mit.edu

12.1 Overview

The interactions of micro- and nanoparticles with various cells inside the body determine their effectiveness as carriers for drug delivery.[1-6] These interactions are governed by the nanoparticle's inactive or active material properties,[2,7-13] which include surface chemistry,[14-16] size,[9,17,18] shape,[17,19-22] mechanical properties[23] and charge.[24,25] Accordingly, much work has been directed at examining these properties in an effort to build highly effective drug delivery particles, and new ways of manipulating a nanoparticle's surface properties, internal composition or dimensions can potentially offer the drug delivery community more sophisticated levels of control over their *in vivo* properties, which is much sought after in materials science applied to nanomedicine. The challenge is in defining an optimal set of material properties that can confer adequate systemic circulation, immune evasion, molecular targeting and drug release after administration into the body.[3,17,26-28] This is often difficult because of the interdependence of material parameters in nanoparticle design;

RSC Smart Materials No. 1
Janus Particle Synthesis, Self-Assembly and Applications
Edited by Shan Jiang and Steve Granick
Published by the Royal Society of Chemistry, www.rsc.org

the enhancement of one property for particle performance in a specific *in vivo* application can often adversely affect another desired property.

Anisotropic modification schemes for drug delivery particles allow compartmentalization of the particle volume into multiple domains, each imparting a different functionality to the resultant particle system.[29-31] Not only is this an elegant way to create multifunctional drug delivery particles, but also the anisotropy of the system can impart improved and synergistic functionalities not possible with isotropic materials.[32-36] These design capabilities are therefore particularly exciting for applications in systemic drug delivery due to nanoparticle design constraints for *in vivo* use. This chapter is focused on recent anisotropic particle designs for cancer drug delivery applications and also highlights some examples of their use in a broader biomedical setting. The systems discussed include Janus and patchy particles, systems with anisotropic interior compartmentalization and particles with different geometries.

12.2   Nanoparticle Design to Overcome Barriers to Drug Delivery

The capability of nanocarriers to improve or enable therapy lies in their careful design to confer additional properties to the resultant system so that they are better able to circumvent *in vivo* barriers to delivery.[10,37,38] There are three main levels of physiological barriers preventing the effective delivery of a blood-borne agent. The first level is systemic[1,39] in nature and includes the mechanisms of excretion, clearance and biodistribution. The second set of barriers relates to permeation of the tissue structure of tumors or targeted organs and the extracellular distribution of the agent within the tissue.[40] The final steps involve cell entry, the process of bringing the therapeutic from outside the cell to its specific site of action within the cell.[41]

Several design features have been identified to help nanoparticle systems overcome these barriers.[10] To get around systemic barriers, nanoparticles must first be stable enough to withstand the effects of changes in ionic strength, dilution and shear effects in the blood so that premature destabilization of the system is avoided. At the same time, strategies to avoid mechanisms of clearance by the reticuloendothelial system (RES) and the kidneys allows the nanoparticle to attain a long blood circulation profile,[7,42,43] which is required for effective nanoparticle targeting. As a general rule of thumb, nanoparticles are designed to be in the size range 5–200 nm to avoid clearance by filtration in the spleen and kidneys. Fenestrations in the spleen are typically 200–500 nm in size and will sieve out particles that are larger than ~ 200 nm.[1] The glomerulus inside the kidneys also filters out substances in blood smaller than ~ 5 nm. To overcome RES clearance, a method for attenuating the processes of opsonization – adsorption of plasma proteins on the nanoparticle surface – is necessary to prevent uptake of nanoparticles by macrophages and monocytes and to thus prolong circulation times. This can be achieved by employing a stealth coating around the nanoparticle that is highly hydrated and provides steric repulsion to resist protein adsorption. Stealth coatings used in nanocarriers are typically hydrophilic electron acceptors such as poly(-ethylene glycol) (PEG), which forms an effective water shell around each particle.[5,44–47] More recently, it has been found that other types of hydrogen bonding or charged polymers can also exhibit a stealth quality on surfaces, including negatively charged polymers such as some polysaccharides and certain zwitterionic or mixed-charge polymers.[48–50]

When a substantial circulation profile has been achieved, circulating nanoparticles will have a greater chance to escape from tumor vessels and, over time, accumulate within the tumor interstitials.[27,51] Aberrant blood vessels that supply solid tumors contribute to this process, as they are usually enlarged and lack a continuous layer of endothelium.[52] Poor lymphatic drainage of tumors also contributes to this process; therefore, the net accumulation of high molecular weight species within solid tumors is usually positive. This effect, termed enhanced permeation and retention (EPR), is the result of poor lymphatic drainage and aberrant blood vessels within the tumor, causing a buildup of high molecular weight species within the tumor microenvironment.[4]

Before the nanoparticle is presented to tumor cells for uptake, it must first effectively diffuse through the extracellular matrix of the tumor and allow the therapeutic to be accessible to cells further away from vessels. If anticancer medicines are unable to access all of the cells within a tumor that are capable of regenerating it, then, whatever their mode of action or potency, their effectiveness will be compromised.[53] Most of the cells in the human body are within a few cell diameters of a blood vessel. This not only facilitates the delivery of oxygen and nutrients to the cells that form the tissues of the body, but also permits the efficient delivery of most medicines. As tumor cells often have the potential for more rapid proliferation than the cells that form blood capillaries,[54] the proliferation of tumor cells forces vessels apart, reducing vascular density and creating a population of cells distant ($>$100 μm) from blood vessels,[55] a process that is exacerbated by a poorly organized vascular architecture,[56,57] irregular blood flow[58] and the compression of blood and lymphatic vessels by cancer cells.[59] In addition, the disorganized vascular network and the absence of functional lymphatics[60] cause increased interstitial fluid pressure (IFP).[61,62] Finally, the composition and structure of the extracellular matrix (ECM) can slow the movement of molecules within the tumor.[63,64] Overall, these characteristics of the tumor microenvironment limit the delivery of nanoparticles to cells that are situated far from functioning blood vessels. Although significant efforts have been expended on establishing the effects of microenvironment on drug delivery, overcoming this barrier effectively is an area that is least explored with nanoparticle systems. As the factors that can impede nanoparticle diffusion within the tumor will vary from tumor to tumor, it is difficult to provide a unifying strategy to address this obstacle. Typically, nanoparticles that are sub-100 nm in size provide an advantage in this area,[65] but this can be changed with different nanoparticle surface chemistries. Combinatorial delivery systems have also been explored to pretreat the microenvironment of the tumor to aid the diffusion of drug or nanoparticles, and this seems to be a promising approach to pursue.[61]

Once a drug carrier is able to overcome extracellular barriers to penetrate the tumor interstitium, its means of delivery depends upon uptake of the drug from the carrier to the tumor cells. Receptor-mediated drug delivery approaches allow for the direct intracellular delivery of a cancer drug to specifically targeted tumor cells via endocytosis.[27,66,67]The concepts of multi-valent cell targeting are important here to increase cell uptake. This is followed by the need for endosomal escape mechanisms, cytosolic trafficking and for gene delivery, finally penetrating the nuclear envelope, depending on where the specific site of action the therapeutic is.[10,37] The strategies for attaining these events are varied and include the use of polymers that promote the proton sponge effect,[68] cell-penetrating and endosomal escape peptides[69] and the attachment of nuclear localization sequences.[70]

12.3 Examples of Nanoparticle Systems: Liposomes, Micelles and Dendrimers

Circumventing the above-mentioned systemic barriers *via* controlling the pharmacologic properties of anti-cancer therapeutics will assist their more efficient delivery to targets within the tumor.[6,45,71,72] This can be achieved using nanoparticle-based medicines.[10,46,47,71] These forms of therapy are typically comprised of therapeutic entities, such as small-molecule drugs, nucleic acids, proteins or peptides that are assembled with non-therapeutic components, which affect the pharmacokinetics and pharmacodynamics of the therapeutics they carry. The end goal of all such systems is to increase the concentration of a therapeutic agent in its site of action while limiting systemic exposure. Numerous drug delivery vehicles[6,71] have been accomplished to achieve this goal, including liposomes, micelles and macromolecular drug carriers.

Figure 12.1 shows examples of conventional nanoparticle systems used in drug delivery. A number of successful drug delivery systems have been based on liposomes and commercialization of such systems has begun to take place.[6] For example, Doxil, a PEGylated liposomal doxorubicin formulation, has been approved for the treatment of Kaposi's sarcoma and recurrent ovarian cancer. Benefits of this treatment include decreased cardiac toxicity associated with free drug (doxorubicin) therapy and an overall greater anti-tumor effect. However, liposomal treatments are subject to limitations in drug type and capacity, and are prone to skin allergies and lower cell permeation rates due to their larger size.[73]

Macromolecular micelle systems based on self-assembling polymers offer another means of approach to drug delivery. A key advantage of these systems is the opportunity to tune the polymer synthetically as a means of optimizing

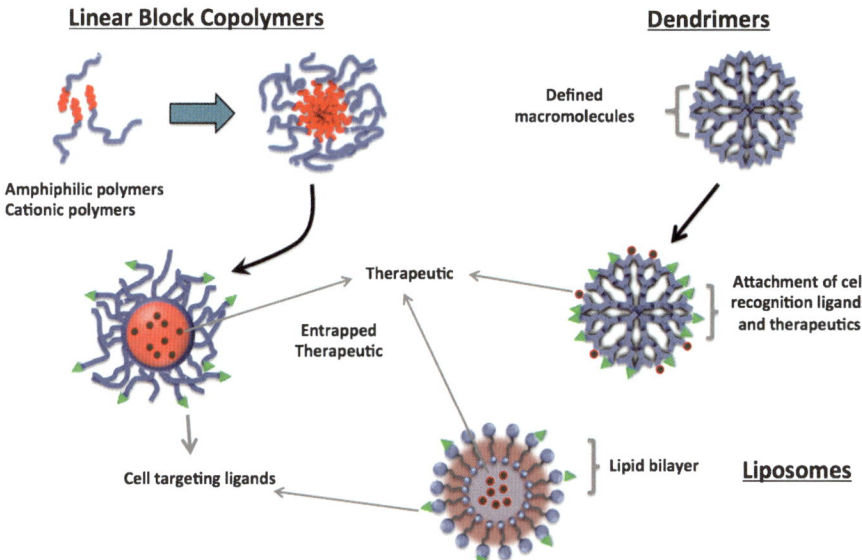

Figure 12.1 Schematic diagram of conventional drug delivery nanocarriers.

delivery.[47,74] These self-assembling systems are composed of hydrophobic and hydrophilic domains and, as a result, form a micellar structure under aqueous conditions. This structure is composed of an interior core region that is the more hydrophobic block and an outer shell composed of a water-soluble, hydrophilic block that exists in a highly hydrated state. The most promising systems in this area have been linear block copolymers with a biocompatible or biodegradable block, such as PEO-*b*-PLA [poly(ethylene oxide)-*b*-poly(lactic acid)] or PEO-*b*-PBLA [PEO-*b*-poly(benzyl L-aspartate)]. Several copolymer micellar formulations have also undergone Phase I or II clinical trials[75–80] or are currently involved in trials.[81,82] An example is Genexol, which is a PEO–PLA block copolymer that is the first approved copolymer nanoparticle formulation for paclitaxel delivery. PEO–PBLA copolymers functionalized with doxorubicin side groups and loaded with free doxorubicin drug (NK911) have also been investigated extensively and shown to yield highly significant improvements in doxorubicin blood circulation and higher antitumor activity than free doxorubicin in mouse tumors. Some promising success has been found in Phase I trials with NK105 paclitaxel-loaded copolymers based on synthetically modified versions of PEO–PBLA; these systems, which appear to exhibit increased dose efficacy and *in vivo* micellar stability, are now undergoing Phase II studies. However, key issues for polymeric micelle systems include stability of the carrier and encapsulated drug in the bloodstream and the need for the incorporation of highly selective targeting moieties that can lead to higher levels of intracellular uptake of particles in the tumor.

Dendrimers[83–86] are highly branched and symmetrical macromolecules ($\sim$5 nm in size). These polymeric scaffolds can be constructed from biocompatible and biodegradable elements with very densely packed exteriors and porous interior cavities to facilitate precise incorporation of drugs and other relevant molecules for drug delivery. Their unique branched topologies confer dendrimers with properties that differ substantially from those of linear polymers. These properties, which include high levels of molecular symmetry and terminal units, have been exploited as a means of creating targeted drug and gene delivery systems, including charged, hybrid dendritic vehicles and self-assembled linear-dendritic block copolymer systems, and newer hyper-branched polymer systems are also now being investigated for delivery or imaging applications.

12.4 Anisotropic, Patchy and Janus Particles in Systemic Drug Delivery

Methods for the construction of anisotropic particles are covered in a separate chapter and will not be discussed in depth in the following sections. Here we focus on applications of anisotropic particle in biomedicine. In this regard, three main features have been explored: non-uniform surface functionalizations, internal particle compartmentalization and anisotropic geometries.

12.4.1 Particles with Anisotropic, Janus or Patchy Surfaces

Anisotropic surface control of particles can offer interesting properties with importance to drug delivery because they are the dominant component that influences how nanoparticles interact with the *in vivo* environment. Accordingly, modifications of the surface chemical composition, charge and topology have profound effects on particle–particle and particle–cell interactions. Typically, a hydrated, antifouling material such as PEG is required as a coating on the surface to render the particle stable in aqueous environments and also 'stealth'-like to opsonins in the body that act to remove foreign bodies;[7,87] however, this stealth layer is non-specific to receptor targets and adversely affects processes of receptor binding for cellular uptake.[27,28,51,66] As a solution to this problem, ligands specific to receptors found on target cells may be used to functionalize PEG further and typically a high degree of ligand functionalization is sought to attain multivalency in cell targeting. In recent years, however, there has been growing appreciation that cell–ligand binding interactions are not influenced solely by multivalency, but also by the nature of the 3D presentation of the ligand.[88–92] The arrangement of the ligands on the nanoscale can significantly influence receptor binding events and ligand-mediated cellular processes; it has been demonstrated in several hallmark studies that *via* the rational design of ligand clusters, even small increases in valency can increase cell-binding affinities by factors of up to a 1000.[93–95] In contrast to bulk ligand functionalization strategies, the careful and rational design of targeting groups for drug delivery systems will also prevent unnecessary masking of PEG groups by excess ligands, which either does not confer any additional targeting benefit or causes undesirable loss of selectivity or levels of toxicity.[3,27,92,96–98] Within the context of targeted nanoparticle therapies, the intricate balance between these biophysicochemical parameters was investigated theoretically and experimentally. The arrangement of ligands in groupings or cluster arrangements were demonstrated to lead to unexpected biological responses[99–103] that can be translated and made beneficial to aspects of nanoparticle drug delivery.[92,104–108]

Binding affinity is an important aspect of nanoparticle targeting that is desired to achieve higher levels of therapeutic efficacy. Increasing the avidity of nanoparticles to cells allows a longer nanoparticle residence time on the cell surface to maximize the physiological response of therapy at any administered concentration.[26,109] Theoretical and experimental efforts to understand and improve avidity *via* manipulating the manner in which ligands are presented to the cell receptors have shown that the degree of functionalization and also the arrangement of particular ligands can greatly modulate the resulting avidity of the system to target cells. Changing the global ligand density uniformly on the nanoparticle surface improves binding to cells to a certain extent, but saturation of bound receptors and cellular uptake is typically observed at higher ligand densities.[97,98,105,106] This behavior is a result of the limitations on the size and positioning of surface receptors on the cell membrane that can co-localize and binding to a ligand-functionalized nanoparticle. On the other hand, there are also important limitations with regard to the ligand and linker molecule on the targeted vehicle. For example, long PEG linkers used to attach chemical ligands or peptides covalently to a nanoparticle surface provide a steric repulsion that can lower the ability of bound receptor regions to aggregate together on the cell membrane surface and the fact that the ligand has a tether that is relatively long indicates an ability of the ligands to explore larger regions of space, thus lowering the probability that specific binding will occur with ligands that are closely located. This point is a critical one because, along with increases in binding avidity with molecular binding, it is the localized clustering of these receptors, which are usually free to diffuse along the cell membrane surface, that leads to receptor 'cross-linking' processes and the initiation of receptor-mediated signaling and endocytosis or uptake of nanoparticles.

On nanoparticles with homogeneous ligand distributions, the bound fraction of ligands to cell receptors at equilibrium has an optimum and excess ligands at the point of contact can compete undesirably for binding. Anisotropic or patchy nanoparticle designs allow the engagement of cell surface receptors with a predetermined optimized number of targeting ligands, such that the bound ligand fraction becomes high. Computer simulations (Figure 12.2a) show that these arrangements give very high affinities compared with nanoparticles presenting the same amount of ligands that are homogeneously distributed.[104] Patchy ligand configurations can also give rise to interesting new binding kinetics. When closely packed, ligands can easily substitute one another during binding and their tethers are on average less stretched. These design

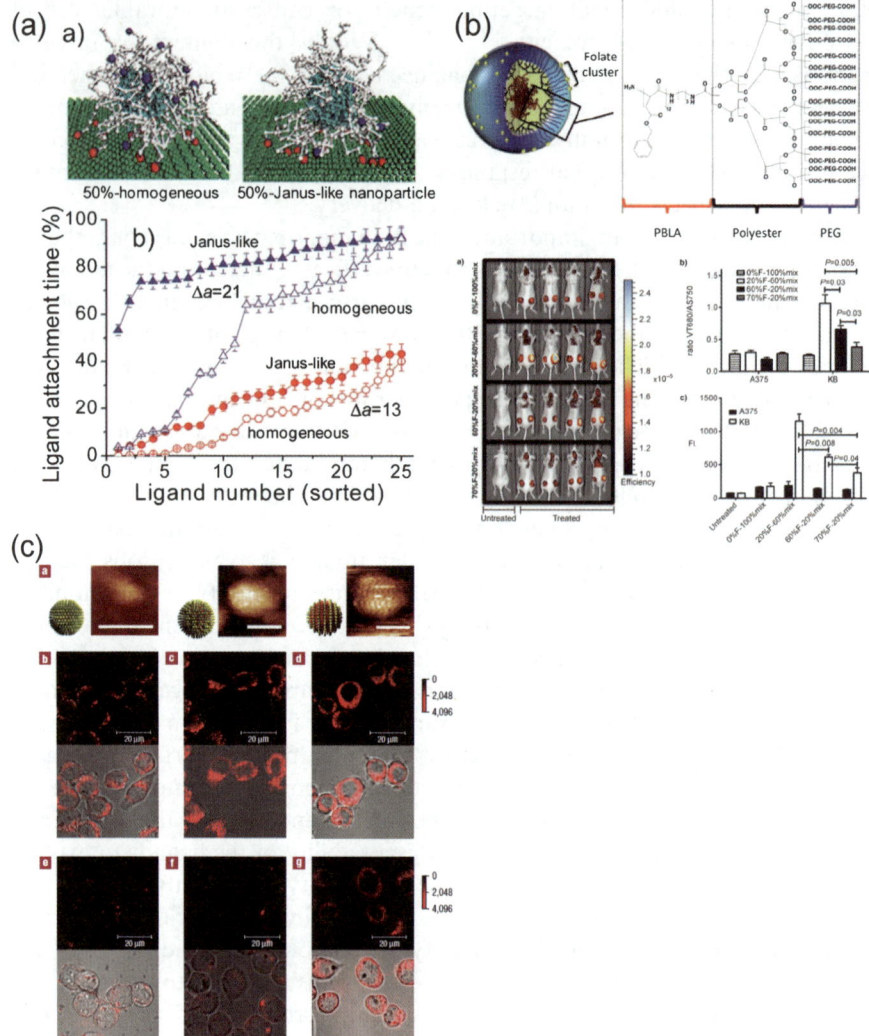

Figure 12.2 The arrangement of ligands on the surface of nanoparticles into Janus[104] (a), patchy[106] (b) or anisotropic[110] (c) formats has a profound impact on nanoparticle–cell interactions. Theoretical modeling of Janus particles (a) demonstrates that the bound fraction of ligands to cell receptors on particles with Janus-like ligand arrangements is higher compared with particles with homogeneously distributed ligands, leading to greater stability of the targeted system to cell receptors. Anisotropic or patchy nanoparticle designs permit the engagement of cell surface receptors with a predetermined optimized number of targeting ligands, such that the bound ligand fraction becomes high. This effect is observed experimentally in (b) with self-assembling polymers capable of forming nanoparticles with ligand clusters presented as patches on the surface of the nanoparticle. Optimum ligand sizes conferring an enhanced level of tumor cell targeting were observed for certain formulations [20% folate

considerations allow fast and strong binding to cell receptors, but their overall binding avidity is dependent on the optimal bound fraction at equilibrium.

Experimentally, polymers comprising a hydrophobic linear component and a hydrophilic dendritic component were used to investigate the effect ligand cluster size and arrangement, at constant multivalency or total number of ligand functional groups, have on cell targeting[106] (Figure 12.2b). The manner in which a model targeting ligand (folate) is presented to the cell on the nanoparticle surface – homogeneously or in specific spatially defined groupings on a nanoparticle surface – was controlled using the self-assembly of linear dendritic polymers with different folate functionalizations. A series of patchy micelles ($\sim$ 80 nm in diameter) were created using this approach, which had a statistically similar amount of folate presented globally but in different cluster arrangements. Micelles that present the same amount, but have different spatial arrangements of folate in variable sized clusters, allow meaningful comparisons to be made to determine the effect of folate spatial presentation on cellular internalization. KB tumor cells were used to evaluate the targeting and binding of the micelles to folate receptors that are overexpressed on the cell surface. There was a clear optimum in performance in folate cluster size; the highest cell associated fluorescence was observed for cells incubated with a patchy micelle formulation that presented 3–5 folates per cluster on the micelle surface. The measured EC_{50} of this particular formulation was also the lowest and the approximated dissociation rate constant (k_{off}) of the different micelles also shows that the optimal formulation had the longest dissociation time (2×10^{-5} s^{-1}) and a binding avidity that was two orders of magnitude lower than

(F)–60% mix], which were much higher than for targeted nanoparticles that presented the same amount of ligand content in a homogeneous manner. Intravital optical microscopy of mice induced with two different subcutaneous tumors (left flank, tumors without receptor of interest; right flank, tumors with receptor of interest) show enhanced retention and targeting of the optimized 20% F–60% mix formulation nanoparticle at the 48 h time point. Anisotropic arrangements of hydrophilic and hydrophobic surface functional groups on nanoparticles (c) also contribute significantly to their subcellular localization in cells. Gold nanoparticles prepared with different ratios of hydrophilic and hydrophobic functional groups gave rise to ordered or disordered amphiphilic ligand shells which have different degrees of penetration into dendritic cells. While both nanoparticles bearing a hydrophilic ligand shell (left nanoparticle) or a disordered amphiphilic ligand shell (middle nanoparticle) were taken up by cells, they remained in endosomes and were visualized by the punctate fluorescence in cells. In contrast, nanoparticles bearing an amphiphilic 'striped' ligand shell (right nanoparticle) were found to be present in the cytosol instead of the endosomes, indicated by their diffused fluorescence in the cell cytosol. These different effects were observed at 37 °C (middle panel) and, importantly, also at 4 °C (bottom panel), indicating that an energy-independent mechanism, resulting from the interaction of the ordered ligand shell with the cell membrane, is responsible for the endosomal escape into the cytosol.

the monofunctional folate systems (K_D on the order of 10 pM). Confocal analysis and competitive binding experiments confirmed that binding and targeting are folate receptor (FR) mediated and that the mechanism for internalization of targeted LDP is dependent on both energy-driven endocytosis and the presence of folate receptors. *In vivo*, drug delivery viable micelles presenting folate cluster sizes of the optimal range of 3−5 also targeted more significantly tumors expressing the corresponding receptor of interest. Compared with other micelle formulations, those presenting the optimal folate clusters showed 3−4-fold enhanced targeting to tumors. The cooperativity of multivalent ligand binding to receptors was used to interpret these results. The patchy micelle system is a negatively cooperative system; as $K_{\text{multivalent}} < (K_{\text{monovalent}})^{\alpha N}$ (where K is the dissociation constant, N is the total number of ligands and α is the degree of cooperativity), each successive ligand interaction with a receptor is less favorable than the previous interaction. The calculated degree of cooperativity shows that all formulations were negatively cooperative ($\alpha < 1$ for all formulations), but among the tested micelles, those presenting 3−5 folate clusters had the highest (least negative) α, indicating that it is possible to achieve a particular ligand configuration that optimizes the use of each ligand, making the system less negatively cooperative (α closer to 1). Similar enhancements to targeting by ligand clustering were also observed with other therapeutic systems;[111,112] these dramatic improvements to cell targeting clearly illustrate the importance of the ligand arrangement to targeted drug delivery.

The patterning of ligands in anisotropic arrangements on nanoparticle surfaces has not only been shown to modulate receptor-mediated targeting, it has also been explored as a means to gain entry into the cytosol *via* direct cell penetration. Amphiphilic gold nanoparticles (~ 6 nm in diameter) presenting mixed ligand shells of hydrophilic 1-mercapto-1-undecanesulfonate (MUS) and hydrophobic 1-octanethiol (OT) were also used to investigate the interactions of anisotropic ligand shells with cell membranes for the creation of cell-penetrating nanoparticles[110] (Figure 12.2c). Simple functionalization reactions allowed the synthesis of these particles with single- or multicomponent ligand shells and both the composition and morphology of multicomponent ligand particles can be easily and precisely controlled to create ligand shells with random or ordered arrangements. Using a set of gold nanoparticles that present the same hydrophobic/hydrophilic ligand, regularly arranged ribbon-like domains on the surface enabled nanoparticles to penetrate the cell membrane in a non-endocytic manner, resulting in their accumulation in the cytosol rather than in an endosomal compartment, as is typical with most nanoparticle systems. This property is unique to specific surface functional group arrangements on the surface, as particles with identical surface group content but lacking structural order accumulated within endosomes instead. Furthermore, under conditions that inhibit cellular endocytosis, similar observations were made, indicating that the mechanism of cellular entry for nanoparticles is not energy or cell dependent. Nanoparticle penetration was also found to be unlike the mechanisms reported for cationic polymers or nanoparticles, which can also enter cells in a non-endocytotic manner, but create substantial membrane damage during the process. These cell-penetrating gold nanoparticles traverse the cell membrane without damaging the cell membranes. A possible explanation for this unique property is that the nanometer scale of the hydrophobic/hydrophilic ordered arrangement of ligands allows the nanoparticles to fuse and cross the membrane without causing membrane poration, which can lead to cytotoxicity. This is exciting to the drug delivery community because a significant endosomal escape barrier prevents RNAi therapeutics from reaching their intended targets in the cytosol and methods for overcoming this handicap are highly desirable.

12.4.2 Interior Particle Compartmentalization

The ability to compartmentalize specific substances into different regions within a particle using self-assembly approaches, microfluidics or electrohydrodynamic co-jetting processes presents opportunities for creating a high level of internal organization in micro- and nanoparticles.[29–31,34] Individual compartments within the particle can be designed for selective degradation, contraction/swelling or with different drug release kinetics (Figure 12.3). These features offer a simple solution for drug delivery applications where more than a single therapeutic has to be administered in a sequential manner.

Anisotropic internal structures also have the potential to be exploited for use as particle reactors with several internal microenvironments that work in concert to maintain biochemical reaction networks.[115–120] Simple metabolic processes have been performed with enzymes confined within a particle compartment made accessible only to specific substrates. These nano-components serve as reactors that house enzymes specific to a certain class of molecules.[114,121–123] Entrapment of sets of different nano-compartments within a larger particle compartment then permits biochemical interaction between these individual compartments. A key issue in manipulating these network systems is the control of compartment permeability to different molecules. Various material systems, such as polymersomes, dendrimers and layer-by-layer capsules, have been used to create these semipermeable compartments.[118,119,124] The polymeric vesicles that confine enzymatic reactions have been explored for use as synthetic organelles for cellular enzyme therapy.[125] For diseases that originate from a dysfunction or absence of enzymes, this delivery strategy has been demonstrated to be an effective way for incorporating missing enzymes into cells; compared with free enzyme, these synthetic organelles maintain their activity for longer periods.

12.4.3 Anisotropic Geometries

Under physiological shear conditions, erythrocytes distribute towards the center of blood vessels, whereas platelets move laterally near vessel walls. This unique rheological phenomenon, termed margination, facilitates clotting for a variety of

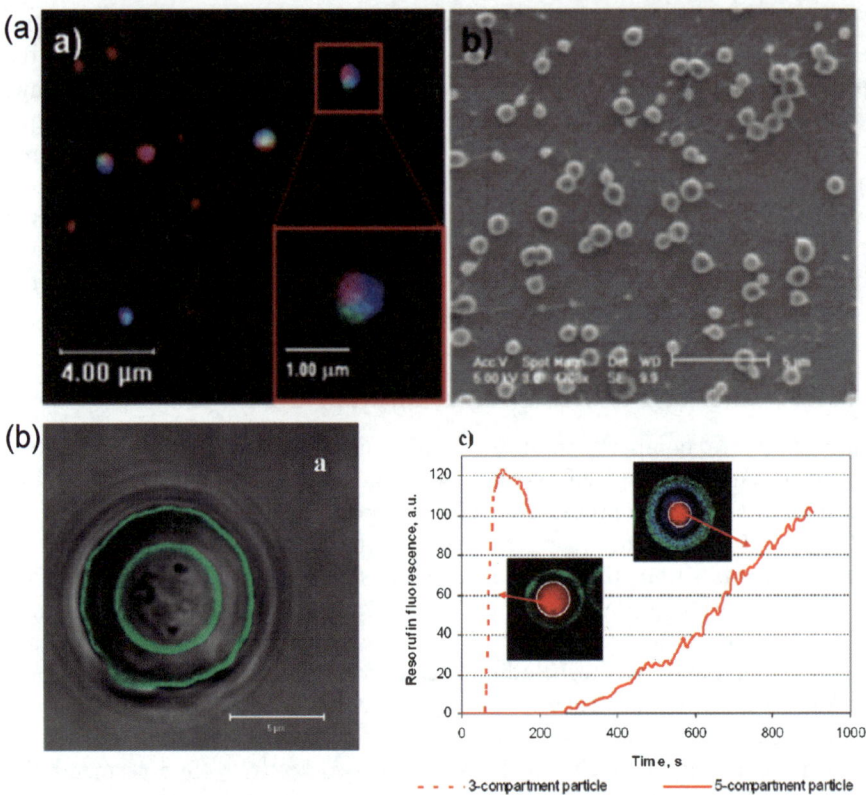

Figure 12.3   Two examples of anisotropic particle compartmentalization schemes that can be designed for selective degradation, contraction/swelling or with different drug release kinetics. Anisotropic compartmentalization particles offer an elegant and simple solution for drug delivery applications where more than a single therapeutic has to be administered in a sequential manner. (a) A confocal laser scanning microscopy image of particles made of polymers with different fluorescent molecules attached which are assembled with an electrified jet setting technique.[113] The distinct particle compartments made possible with this approach offer interesting possibilities for compartmentalization of therapeutics into nanoparticle domains that regulate their release or for organization of substances within a particle to create diverse microenvironments that may act synergistically towards a biomedical application. This is demonstrated in (b) with particles containing concentric compartments and housing different enzymes capable of a chain reaction.[114] The activity of each enzyme is visualized by substrate-derived fluorescence (red) and the delay between the activity of the first and last enzyme in the chain reaction was demonstrated to be tunable *via* the segregation of enzymes into nanoparticle compartments.

reasons, including an increased ability of platelets to 'sense' injured endothelium or tissue, and an increased ease with which platelets distribute from larger vessels into highly branched microvascular networks.[126] If drug delivery particles in the bloodstream also distributed in a similar manner, it could greatly promote their distribution and extravasation into target tissue.[19] Efforts are under way to examine the relationship between particle geometry and their resultant flow position within vessels.[19] Theoretical modeling of these interactions demonstrated that whereas spherical particles require an external force to sustain their trajectories along the vessel walls, non-spherical particles move with complex flow patterns that are dependent on their aspect ratio. These complex flow trajectories can potentially enhance targeting to injured endothelium or tissue, because they allow particles to come into closer contact with target cells.[127] *In vitro* experiments conducted with spherical, discoidal and quasi-hemispherical particles of the same weight in parallel flow chambers show a greater margination effect with discoidal and quasi-hemispherical particles than spherical particles. A smaller number of non-spherical particles make contact with the flow chamber walls across all shear rates examined. Further understanding of these relationships has been used to guide the design of drug delivery particles capable of exploiting the margination effect. For example, Muzykantov and co-workers demonstrated the targeting of endothelial cells with the peptide ICAM-1 using elongated nanoparticle shapes and used hydrodynamics of the nanoparticle to enhance its accumulation at or near the blood vessel walls.[128]

Other studies have also demonstrated the importance of nanoparticle shape in affecting processes of phagocytosis and endocytosis.[19,20,22,23,128–130] Circulating and tissue-specific macrophages constitute a major component of the RES system; therefore, circumventing phagocytosis of drug delivery particles by macrophages would increase the number of particles and the time they stay in the blood so that they may reach their intended targets in the body. The study of the dependence of particle shape on endocytosis is also relevant to the final stages of drug delivery, as certain therapeutics are required to be transported into the cell by the carrier via endocytosis for its intended therapeutic action.

Anisotropy in particle shape exerts similar effects on cellular uptake on both the micro- and nano-scale. Owing to its relatively larger size, the phagocytosis of anisotropic microparticles is highly influenced by its orientation as it encounters a macrophage[130,131] (Figure 12.4a). The local microparticle shape at the point of contact with the cell membrane determines whether the cell initiates phagocytosis or simply spreads out on the microparticle surface; this is because broader local particle shapes prevent the required actin structures associated with phagocytosis from forming and instead cause the cell to spread out along the entire surface. Although smaller or narrower points of contact facilitate the formation of the required actin structures for phagocytosis to initiate, the completion of phagocytosis is ultimately determined by a particle's overall volume and cells do not internalize efficiently microparticles that have larger volumes.[131]

The shape and aspect ratios of nanoparticles are also significant parameters governing both the rate and extent of their internalization into cells.[21,133] Asymmetric high aspect ratio nanoparticles (long cylinders) showed greater and faster internalization in HeLa cells *via* energy-dependent clathrin- or

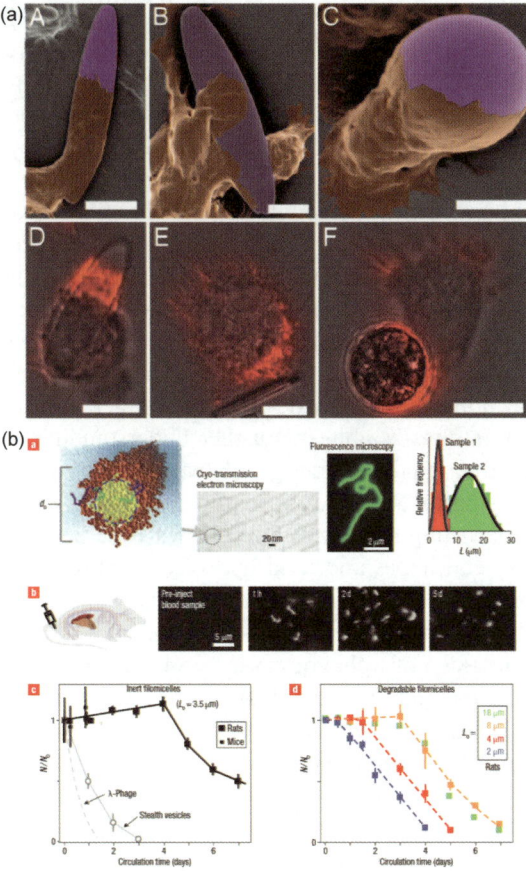

Figure 12.4 The target geometry of particles influences the rate and extent of their phagocytosis[131] (a) and informed geometrical designs can reduce nanoparticle RES clearance and increase their blood circulation times[132] (a). Scanning electron micrographs of macrophages in brown (top panel) and also with actin filament visualized *via* rhodamine phalloidin (red) demonstrate that particles (purple) with different aspect ratios are differentially taken up during phagocytosis. Narrower (left panels) or spherical (right panels) shaped surfaces enable the cell to spread and phagocytose the entire particle. When a broad particle surface is presented to macrophages, the cell actin polymerizes around the particle surface but no actin rings or cups are formed and subsequent internalization of the particle is unsuccessful (middle panels). This unique finding has been exploited to circumvent RES clearance by macrophages after systemic delivery. Filamentous particles (filomicelles) with high aspect ratios show persistence in systemic blood circulation compared with spherical counterparts. To account for these observed differences, *ex vivo* examinations of the uptake of particles were investigated under fluid flow conditions. Filamentous particles are readily extended by flow, presenting broad surfaces to macrophages that discourage their phagocytosis. In contrast, macrophages take up spherical nanoparticles and short filomicelles more readily.

caveolae-mediated processes compared with symmetric nanoparticles with low aspect ratios (spheres and cubes). Within a tested group of cylindrical nanoparticles, it was found that while HeLa cells eventually internalized all cylindrical nanoparticles to a similarly high degree, the kinetics of internalization varied with aspect ratio and size. When the total nanoparticle volume was kept constant, the higher aspect ratio cylindrical nanoparticles (d = 150 nm, h = 450 nm, volume = 7.9×10^{-3} μm^3, aspect ratio = 3) were four times faster than lower aspect ratio cylindrical nanoparticles (d = 200 nm, h = 200 nm, volume = 6.3×10^{-3} μm^3, aspect ratio = 1) at internalization. However, cylindrical particles having a diameter of 100 nm, an aspect ratio of 3 and a volume of 2.4×10^{-3} μm^3 were internalized to a lesser extent than the larger cylindrical particles having a diameter of 150 nm and the same aspect ratio. The internalization kinetics of cylindrical nanoparticles also appear to be similar to those of microparticles in that they depend not only on the effective relative length (aspect ratio), but also on the absolute size of the particle. This is likely due to the interplay between nanoparticle physical parameters in cell uptake and an optimum size and aspect ratio are required to maximize multivalent binding to cells and facilitate the subsequent processes of membrane invagination for uptake to occur.

Taking the needs for the entire systemic delivery process into consideration, it is conceivable to speculate that rod-like or high aspect ratio nanoparticles will be most suitable for increasing nanoparticle circulation and RES evasion while maintaining their capability to target and internalize into cells. When a comparison[132,134] was made between filamentous nanoparticles (diameters less than 100 nm and lengths of up to 18 μm) and stealth vesicles typically used in these applications, filamentous nanoparticles were observed in the blood for up to 1 week after injection, whereas stealth vesicles lasted for 2 days (Figure 12.4b). The uptake of these nanoparticles by activated macrophages examined *in vitro* and under flow conditions show that when smaller stealth vesicles contact the cells they adhere and are subsequently taken up. Longer filamentous nanoparticles that contact activated macrophages remain associated for short periods, but hydrodynamic shears ultimately pull them off macrophages. These observations likely account for the dramatically enhanced circulation of filamentous nanoparticles *in vivo*, which translated well into an anti-tumor effect; filamentous nanoparticles reduced tumor growth in mice eightfold with a single injection dose of 8 mg kg^{-1} of paclitaxel encapsulated within them. These results were comparable to promising Phase I clinical trials with paclitaxel-loaded spherical micelles of PEG–poly(lactic acid) that were administered three times at substantially higher doses.[132,134]

12.5   Conclusion

Researchers have developed several means of generating complex particles that exhibit asymmetric surfaces and compartments with a great deal of flexibility and precision with approaches such as stop-flow lithography.[135] Although these methods are now focused on micron-scale systems, there remain several areas of delivery for which the size is relevant, including delivery and release to the endothelial cells for vascular injury or targeted delivery to macrophages to address the immune system. Should this technology continue to move towards smaller particles and feature sizes, it is feasible that nanoparticles relevant to tumor targeting may evolve. There have also been recent advances in the generation of Janus particles that contain both a hydrophobic and a hydrophilic component; such systems may yield new possibilities for achieving high drug loading and flexibility in dosing schedules and the introduction of synergistic drugs.[136] Developments in synthetic chemistry have also led to controlled Janus characteristics in dendrimers[137] and electrohydrodynamics presents additional opportunities for the synthesis of complex particle systems,[138] methods which should prove valuable when adapted to the various barriers of drug delivery for targeted nanoparticles.

References

1. S. M. Moghimi, A. C. Hunter and J. C. Murray, *Pharmacol. Rev.*, 2001, **53**, 283–318.
2. S. M. Moghimi, A. C. Hunter and T. L. Andresen, *Annu Rev. Pharmacol. Toxicol.*, 2012, **52**, 481–503.
3. F. Gu, L. Zhang, B. A. Teply, N. Mann, A. Wang, A. F. Radovic-Moren, R. Langer and O. Farkokhzad, *Proc. Natl. Acad. Sci. U. S. A.*, 2008, **105**, 2586–2591.
4. R. Haag and F. Kratz, *Angew. Chem. Int. Ed.*, 2006, **45**, 1198–1215.
5. R. Duncan, *Nat. Rev. Cancer*, 2006, **6**, 688–701.
6. M. E. Davis, Z. Chen and D. M. Shin, *Nat. Rev. Drug Discov.*, 2008, **7**, 771–782.
7. F. Alexis, E. Pridgen, L. K. Molnar and O. C. Farokhzad, *Mol. Pharm.*, 2008, **5**, 505–515.
8. V. H. L. Lee, *Adv. Drug Deliv. Rev.*, 2004, **56**, 1527–1528.
9. L. Balogh, S. S. Nigavekar, B. M. Nair, W. Lesniak, C. Zhang, L. Y. Sung, M. S. T. Kariapper, A. El-Jawahri, M. Llanes, B. Bolton, F. Mamou, W. Tan, A. Hutson, L. Minc and M. K. Khan, *Nanomedicine*, 2007, **3**, 281–296.
10. R. A. Petros and S. J. M. De, *Nat. Rev. Drug Discov.*, 2010, **9**, 615–627.
11. E. S. Lee, Z. Gao, D. Kim, K. Park, I. C. Kwon and Y. H. Bae, *J. Control. Release*, 2008, **129**, 228–236.
12. K. T. Oh, H. Yin, E. S. Lee and Y. H. Bae, *J. Mater. Chem.*, 2007, **17**, 3987–4001.
13. D. J. Irvine, *Nat. Mater.*, 2011, **10**, 342–343.
14. O. C. Farokhzad, J. Cheng, B. A. Teply, I. Sherifi, S. Jon, P. W. Kantoff, J. P. Richie and R. Langer, *Proc. Natl. Acad. Sci. U. S. A.*, 2006, **103**, 6315–6320.

15. J. A. Reddy, V. M. Allagadda and C. P. Leamon, *Curr. Pharm. Biotechnol.*, 2005, **6**, 131–150.

16. O. Lunov, T. Syrovets, C. Loos, J. Beil, M. Delecher, K. Tron, G. U. Nienhaus, A. Musyanovych, V. Mailaender, K. Landfester and T. Simmet, *ACS Nano*, 2011, **5**, 1657–1669.

17. C. He, Y. Hu, L. Yin, C. Tang and C. Yin, *Biomaterials*, 2010, **31**, 3657–3666.

18. H. Yue, W. Wei, Z. Yue, P. Lv, L. Wang, G. Ma and Z. Su, *Eur. J. Pharm. Sci.*, 2010, **41**, 650–657.

19. P. Decuzzi, R. Pasqualini, W. Arap and M. Ferrari, *Pharm. Res.*, 2009, **26**, 235–243.

20. S. Mitragotri, *Pharm. Res.*, 2009, **26**, 232–234.

21. S. E. A. Gratton, P. A. Ropp, P. D. Pohlhaus, J. C. Luft, V. J. Madden, M. E. Napier and J. M. DeSimone, *Proc. Natl. Acad. Sci. U. S. A.*, 2008, **105**, 11613–11618.

22. X. Huang, X. Teng, D. Chen, F. Tang and J. He, *Biomaterials*, 2010, **31**, 438–448.

23. T. J. Merkel, S. W. Jones, K. P. Herlihy, F. R. Kersey, A. R. Shields, M. Napier, J. C. Luft, H. Wu, W. C. Zamboni, A. Z. Wang, J. E. Bear and J. M. DeSimone, *Proc. Natl. Acad. Sci. U. S. A.*, 2011, **108**, 586–591.

24. B. C. Dash, G. Rethore, M. Monaghan, K. Fitzgerald, W. Gallagher and A. Pandit, *Biomaterials*, 2010, **31**, 8188–8197.

25. K. Xiao, Y. Li, J. Luo, J. S. Lee, W. Xiao, A. M. Gonik, R. G. Agarwal and K. S. Lam, *Biomaterials*, 2011, **32**, 3435–3446.

26. L. S. Zuckier, E. Z. Berkowitz, R. J. Sattenberg, Q. H. Zhao, H. F. Deng and M. D. Scharff, *Cancer Res.*, 2000, **60**, 7008–7013.

27. D. W. Bartlett, H. Su, I. J. Hildebrandt, W. A. Weber and M. E. Davis, *Proc. Natl. Acad. Sci. U. S. A.*, 2007, **104**, 15549–15554.

28. D. B. Kirpotin, D. C. Drummond, Y. Shao, M. R. Shalaby, K. Hong, U. B. Nielsen, J. D. Marks, C. C. Benz and J. W. Park, *Cancer Res.*, 2006, **66**, 6732–6740.

29. J. Du and R. K. O'Reilly, *Chem. Soc. Rev.*, 2011, **40**, 2402–2416.

30. K. J. Lee, J. Yoon and J. Lahann, *Current Opin. Colloid Interface Sci.*, 2011, **16**, 195–202.

31. A. B. Pawar and I. Kretzschmar, *Macromol. Rapid Commun.*, 2010, **31**, 150–168.

32. X. Banquy, F. Suarez, A. Argaw, J.-M. Rabanel, P. Grutter, J.-F. Bouchard, P. Hildgen and S. Giasson, *Soft Matter*, 2009, **5**, 3984–3991.

33. Y. Chen, C. Wang, Y. Li and Z. Tong, *Prog. Chem.*, 2009, **21**, 615–621.

34. K. H. Roh, D. C. Martin and J. Lahann, *Nat. Mater.*, 2005, **4**, 759–763.

35. K. H. Roh, M. Yoshida and J. Lahann, *Materialwiss. Werkstofftech.*, 2007, **38**, 1008–1011.

36. P. A. Suci, S. Kang, M. Young and T. Douglas, *J. Am. Chem. Soc.*, 2009, **131**, 9164–9165.

37. K. A. Whitehead, R. Langer and D. G. Anderson, *Nat. Rev. Drug Discov.*, **9**, 372.

38. Y. Doi, T. Ishida and H. Kiwada, In *Breaking the barriers to tumor-targeting via nanocarrier-based drug delivery to the tumor microenvironment*, Nova Science Publishers, Inc.: 2011; pp 141–160.

39. M. Rowland, L. Z. Benet and G. G. Graham, *J. Pharmacokinet. Biopharm.*, 1973, **1**, 123–136.

40. A. I. Minchinton and I. F. Tannock, *Nat. Rev. Cancer*, 2006, **6**, 583–592.

41. M. Dominska and D. M. Dykxhoorn, *J. Cell Sci.*, **123**, 1183–1189.

42. Y. Liu, J. Sun, J. Han and Z. He, *Curr. Nanosci.*, 2010, **6**, 347–354.

43. B. Romberg, W. E. Hennink and G. Storm, *Pharm. Res.*, 2008, **25**, 55–71.

44. R. Gref, Y. Minamitake, M. T. Peracchia, V. Trubetskoy, V. Torchilin and R. Langer, *Science*, 1994, **263**, 1600–1603.

45. K. Kataoka, G. S. Kwon, M. Yokoyama, T. Okano and Y. Sakurai, *J. Control. Release*, 1993, **24**, 119–132.

46. G. S. Kwon, *Adv. Drug Deliv. Rev.*, 2002, **54**, 167.

47. G. S. Kwon and T. Okano, *Adv. Drug Deliv. Rev.*, 1996, **21**, 107–116.

48. S. F. Chen, L. Y. Li, C. Zhao and J. Zheng, *Polymer*, 2010, **51**, 5283–5293.

49. S. Y. Jiang and Z. Q. Cao, *Adv. Mater.*, 2010, **22**, 920–932.

50. S. Martwiset, A. E. Koh and W. Chen, *Langmuir*, 2006, **22**, 8192–8196.

51. H. Maeda, J. Wu, T. Sawa, Y. Matsumura and K. Hori, *J. Control. Release*, 2000, **65**, 271–284.

52. T. Nobuyuki, *Saibo*, 2008, **40**, 504–507.

53. R. K. Jain, *Sci. Am.*, 1994, **271**, 58–65.

54. J. Denekamp and B. Hobson, *Br. J. Cancer*, 1982, **46**, 711–720.

55. R. H. Thomlinson and L. H. Gray, *Br. J. Cancer*, 1955, **9**, 539–549.

56. J. R. Less, T. C. Skalak, E. M. Sevick and R. K. Jain, *Cancer Res.*, 1991, **51**, 265–273.

57. J. M. Brown and A. J. Giaccia, *Cancer Res.*, 1998, **58**, 1408–1416.

58. D. J. Chaplin, P. L. Olive and R. E. Durand, *Cancer Res.*, 1987, **47**, 597–601.

59. T. P. Padera, B. R. Stoll, J. B. Tooredman, D. Capen, E. di Tomaso and R. K. Jain, *Nature*, 2004, **427**, 695.

60. A. J. Leu, D. A. Berk, A. Lymboussaki, K. Alitalo and R. K. Jain, *Cancer Res.*, 2000, **60**, 4324–4327.

61. C.-H. Heldin, K. Rubin, K. Pietras and A. Oestman, *Nat. Rev. Cancer*, 2004, **4**, 806–813.

62. R. K. Jain, *Adv. Drug Deliv. Rev.*, 2001, **46**, 149–168.

63. P. A. Netti, D. A. Berk, M. A. Swartz, A. J. Grodzinsky and R. K. Jain, *Cancer Res.*, 2000, **60**, 2497–2503.

64. E. Brown, T. McKee, E. di Tomaso, A. Pluen, B. Seed, Y. Boucher and R. K. Jain, *Nat. Med.*, 2003, **9**, 796–801.

65. S. H. Jang, M. G. Wientjes, D. Lu and J. L. S. Au, *Pharm. Res.*, 2003, **20**, 1337–1350.

66. A. Gabizon, A. T. Horowitz, D. Goren, D. Tzemach, H. Shmeeda and S. Zalipsky, *Clin. Cancer Res.*, 2003, **9**, 6551–6559.

67. A. David, P. Kopeckova, T. Minko, A. Rubinstein and J. Kopecek, *Eur. J. Cancer*, 2004, **40**, 148–157.

68. P. Midoux, C. Pichon, J.-J. Yaouanc and P.-A. Jaffres, *Br. J. Pharmacol.*, 2009, **157**, 166–178.

69. Cell Penetrating Peptides in Drug Delivery. Eric L. Snyder and Steven F. Dowdy. PHARMACEUTICAL RESEARCH Volume 21, Number 3 (2004), 389–393, DOI:10.1023/B:PHAM.0000019289.61978.f5

70. S. V. Burov, T. V. Yablokova, M. Y. Dorosh, E. V. Krivizyuk, A. M. Efremov and S. V. Orlov, *Russ. J. Bioorg. Chem.*, **36**, 581–588.

71. R. Duncan, *Nat. Rev. Drug Discov.*, 2003, **2**, 347–360.

72. T. M. Allen and P. R. Cullis, *Science*, 2004, **303**, 1818–1822.

73. M. L. Immordino, F. Dosio and L. Cattel, *Int. J. Nanomed.*, 2006, **1**, 297–315.

74. A. Lavasanifar, J. Samuel and G. S. Kwon, *Adv. Drug Deliv. Rev.*, 2002, **54**, 169–190.

75. T. Hamaguchi, K. Kato, H. Yasui, C. Morizane, M. Ikeda, H. Ueno, K. Muro, Y. Yamada, T. Okusaka, K. Shirao, Y. Shimada, H. Nakahama and Y. Matsumura, *Br. J. Cancer*, 2007, **97**, 170–176.

76. T.-Y. Kim, D.-W. Kim, J.-Y. Chung, S. G. Shin, S.-C. Kim, D. S. Heo, N. K. Kim and Y.-J. Bang, *Clin. Cancer Res.*, 2004, **10**, 3708–3716.

77. K. S. Lee, H. C. Chung, S. A. Im, Y. H. Park, C. S. Kim, S.-B. Kim, S. Y. Rha, M. Y. Lee and J. Ro, *Breast Cancer Res. Treat.*, 2008, **108**, 241–250.

78. E. H. Romond, E. A. Perez, J. Bryant, V. J. Suman, C. E. Geyer Jr, N. E. Davidson, E. Tan-Chiu, S. Martino, S. Paik, P. A. Kaufman, S. M. Swain, T. M. Pisansky, L. Fehrenbacher, L. A. Kutteh, V. G. Vogel, D. W. Visscher, G. Yothers, R. B. Jenkins, A. M. Brown, S. R. Dakhil, E. P. Mamounas, W. L. Lingle, P. M. Klein, J. N. Ingle and N. Wolmark, *N. Engl. J. Med.*, 2005, **353**, 1673–1684.

79. M. Westphal, D. C. Hilt, E. Bortey, P. Delavault, R. Olivares, P. C. Warnke, I. R. Whittle, J. Jaeaeskelaeinen and Z. Ram, *Neuro-Oncology*, 2003, **5**, 79–88.

80. N. Wolmark and B. K. Dunn, *Ann. N. Y. Acad. Sci.*, 2001, **949**, 99–108.

81. J. Cheng, B. A. Teply, I. Sherifi, J. Sung, G. Luther, F. X. Gu, E. Levy-Nissenbaum, A. F. Radovic-Moreno, R. Langer and O. C. Farokhzad, *Biomaterials*, 2007, **28**, 869–876.

82. F. Gu, L. Zhang, B. A. Teply, N. Mann, A. Wang, A. F. Radovic-Moreno, R. Langer and O. C. Farokhzad, *Proc. Natl. Acad. Sci. U. S. A.*, 2008, **105**, 2586–2591.

83. C. C. Lee, J. A. MacKay, J. M. J. Frechet and F. C. Szoka, *Nat. Biotechnol.*, 2005, **23**, 1517–1526.

84. W. R. Dichtel, S. Hecht and J. M. J. Frechet, *Org. Lett.*, 2005, **7**, 4451–4454.

85. E. R. Gillies and J. M. J. Frechet, *Drug Discov. Today*, 2005, **10**, 35–43.

86. S. J. Guillaudeu, M. E. Fox, Y. M. Haidar, E. E. Dy, F. C. Szoka and J. M. J. Frechet, *Bioconjug. Chem.*, 2008, **19**, 461–469.

87. Y. Sheng, C. Liu, Y. Yuan, X. Tao, F. Yang, X. Shan, H. Zhou and F. Xu, *Biomaterials*, 2009, **30**, 2340–2348.

88. C. W. Cairo, J. E. Gestwicki, M. Kanai and L. L. Kiessling, *J. Am. Chem. Soc.*, 2002, **124**, 1615–1619.

89. H. Abe, A. Kenmoku, N. Yamaguch and K. Hattori, *J. Inclus. Phenom. Macrocycl. Chem.*, 2003, **44**, 39–47.

90. M. Mammen, S.-K. Chio and G. M. Whitesides, *Angew. Chem. Int. Ed.*, 1998, **37**, 2755–2794.

91. M. Mammen, G. Dahmann and G. M. Whitesides, *J. Med. Chem.*, 1995, **38**, 4179–4190.

92. J. Wang, S. Tian, R. A. Petros, M. E. Napier and J. M. DeSimone, *J. Am. Chem. Soc.*, 2011, **132**, 11306–11313.

93. M. Mourez, R. S. Kane, J. Mogridge, S. Metallo, P. Deschatelets, B. R. Sellman, G. M. Whitesides and R. J. Collier, *Nat. Biotechnol.*, 2001, **19**, 958–961.

94. P. I. Kitov, J. M. Sadowska, G. Mulvey, G. D. Armstrong, H. Ling, N. S. Pannu, R. J. Read and D. R. Bundle, *Nature*, 2000, **403**, 669–672.

95. R. T. Lee and Y. C. Lee, *Glycoconjug. J*, 2000, **17**, 543–551.

96. C. H. J. Choi, C. A. Alabi, P. Webster and M. E. Davis, *Proc. Natl. Acad. Sci. U. S. A.*, 2010, **107**, 1235–1240.

97. S. Hong, P. R. Leroueil, I. J. Majoros, B. G. Orr, J. R. Baker Jr and M. M. B. Holl, *Chem. Biol.*, 2007, **14**, 107–115.

98. E. K. Woller, E. D. Walter, J. R. Morgan, D. J. Singel and M. J. Cloninger, *J. Am. Chem. Soc.*, 2003, **125**, 8820–8826.

99. E. A. L. Biessen, F. Noorman, M. E. van Teijlingen, J. Kuiper, M. Barrett-Bergshoeff, M. K. Bijsterbosch, D. C. Rijken and T. J. van Berkel, *J. Biol. Chem.*, 1996, **271**, 28024–28030.

100. D. J. Irvine, A. V. G. Ruzette, A. M. Mayes and L. G. Griffith, *Biomacromolecules*, 2002, **2**, 545–556.

101. G. Maheshwari, G. L. Brown, D. A. Lauffenburger, A. Wells and L. G. Griffith, *J. Cell Sci.*, 2000, **113**, 1677–1686.

102. J. Z. Rappoport and S. M. Simon, *J. Cell Sci.*, 2009, **122**, 1301–1305.

103. X. R. Zhang, Y. X. Li, S. D. Chu, N. Ding, C. X. Li and H. S. Guan, *Chin. J. Chem.*, 2004, **22**, 482–486.

104. H. Djohari and E. E. Dormidontova, *Biomacromolecules*, 2009, **10**, 3089–3097.

105. S. Wang and E. E. Dormidontova, *Biomacromolecules*, 2010, **11**, 1785–1795.

106. Z. Poon, S. Chen, A. C. Engler, H.-i. Lee, E. Atas, G. von Maltzahn, S. N. Bhatia and P. T. Hammond, *Angew. Chem. Int. Ed.*, 2010, **49**, 7266–7270.

107. C. W. Cairo, J. E. Gestwicki, M. Kanai and L. L. Kiessling, *J. Am. Chem. Soc.*, 2002, **124**, 1615–1619.

108. S. K. Choi, M. Mammen and G. M. Whitesides, *J. Am. Chem. Soc.*, 1997, **119**, 4103–4111.

109. S. Hong, P. R. Leroueil, I. J. Majoros, B. G. Orr, J. R. Baker and M. M. Banaszak Holl, *Chem. Biol.*, 2007, **14**, 107–115.

110. A. Verma, O. Uzun, Y. Hu, Y. Hu, H.-S. Han, N. Watson, S. Chen, D. J. Irvine and F. Stellacci, *Nat. Mater.*, 2008, **7**, 588–595.

111. J. B. Spangler, J. R. Neil, S. Abramovitch, Y. Yarden, F. M. White, D. A. Lauffenburger and K. D. Wittrup, *Proc. Natl. Acad. Sci. U. S. A.*, 2011, **107**, 13252–13257.

112. Q. K. T. Ng, M. K. Sutton, P. Soonsawad, L. Xing, H. Cheng and T. Segura, *Mol. Ther.*, 2009, **17**, 828–836.
113. K. H. Roh, D. C. Martin and J. Lahann, *J. Am. Chem. Soc.*, 2006, **128**, 6796–6797.
114. H. Baeumler and R. Georgieva, *Biomacromolecules*, 2010, **11**, 1480–1487.
115. A. P. R. Johnston, C. Cortez, A. S. Angelatos and F. Caruso, *Curr. Opin. Colloid Interface Sci.*, 2006, **11**, 203–209.
116. P. Tanner, S. Egli, V. Balasubramanian, O. Onaca, C. G. Palivan and W. Meier, *FEBS Lett.*, 2011, **585**, 1699–1706.
117. Y. Hong, D. Velegol, N. Chaturvedi and A. Sen, *Phys. Chem. Chem. Phys.*, 2010, **12**, 1423–1435.
118. N. P. Kamat, J. S. Katz and D. A. Hammer, *J. Phys. Chem. Lett.*, 2011, **2**, 1612–1623.
119. K. T. Kim, S. A. Meeuwissen, R. J. M. Nolte and J. C. M. van Hest, *Nanoscale*, 2010, **2**, 844–858.
120. K. Renggli, P. Baumann, K. Langowska, O. Onaca, N. Bruns and W. Meier, *Adv. Funct. Mater.*, 2011, **21**, 1241–1259.
121. H. J. Choi and C. D. Montemagno, *Nano Lett.*, 2005, **5**, 2538–2542.
122. S. M. Kuiper, M. Nallani, D. M. Vriezema, J. J. L. M. Cornelissen, J. C. M. van Hest, R. J. M. Nolte and A. E. Rowan, *Org. Biomol. Chem.*, 2008, **6**, 4315–4318.
123. S. F. M. van Dongen, M. Nallani, J. L. L. M. Cornelissen, R. J. M. Nolte and J. C. M. van Hest, *Chem. Eur. J.*, 2009, **15**, 1107–1114.
124. G. Schneider and G. Decher, *Nano Lett.*, 2004, **4**, 1833–1839.
125. S. F. M. van Dongen, W. P. R. Verdurmen, R. J. R. W. Peters, R. J. M. Nolte, R. Brock and J. C. M. van Hest, *Angew. Chem. Int. Ed.*, 2010, **49**, 7213–7216.
126. J. Perkkio, L. J. Wurzinger and H. Schmidschonbein, *Thromb. Res.*, 1988, **50**, 357–364.
127. P. Decuzzi, S. Lee, B. Bhushan and M. Ferrari, *Ann. Biomed. Eng.*, 2005, **33**, 179–190.
128. S. Muro, C. Garnacho, J. A. Champion, J. Leferovich, C. Gajewski, E. H. Schuchman, S. Mitragotri and V. R. Muzykantov, *Mol. Ther.*, 2008, **16**, 1450–1458.
129. N. Doshi and S. Mitragotri, *J. R. Soc. Interface*, 2010, **7**, S403–S410.
130. N. Doshi and S. Mitragotri, *PLoS One*, 2010, **5**(4), e10051.
131. J. A. Champion and S. Mitragotri, *Proc. Natl. Acad. Sci. U. S. A.*, 2006, **103**, 4930–4934.
132. Y. Geng, P. Dalhaimer, S. Cai, R. Tsai, M. Tewari, T. Minko and D. E. Discher, *Nat. Nanotechnol.*, 2007, **2**, 249–255.
133. B. D. Chithrani, A. A. Ghazani and W. C. W. Chan, *Nano Lett.*, 2006, **6**, 662–668.
134. D. E. Discher, in *Proceedings of the ASME Summer Bioengineering Conference 2008, Parts A and B*, 2009, p. 739.

135. M. E. Helgeson, S. C. Chapin and P. S. Doyle, *Curr. Opin. Colloid Interface Sci.*, 2011, **16**, 106–117.
136. H. Xie, Z. G. She, S. Wang, G. Sharma and J. W. Smith, *Langmuir*, 2012, **28**, 4459–4463.
137. J. A. Chute, C. J. Hawker, K. O. Rasmussen and P. M. Welch, *Macromolecules*, 2011, **44**, 1046–1052.
138. J. Lahann, *Small*, 2011, **7**, 1149–1156.

Subject Index

Page numbers in *italics* refer to entries in figures or tables.

2D COMSOL simulation 179, *185,* 187, *192,* 193
3D bundles *188,* 190
3D DNA crystals 219–20

AB-diblock brush copolymers 7–8
acetylene–PLGA 66, *68, 70*
acrylic acid 78
actin *270*
adsorption
  emulsions 245, 246, 255
  and particle size 255
adsorption energy 247–8, 249–50, 251, *252*
aggregates/aggregation *104, 105*
  self-assembly 138
  see also clusters
amphipathic colloids *162*
amphiphilic Janus particles 138–41, 164–6
  emulsions stabilized by 253
  experimental methods 148–9
  experiments and simulations 154–61
  numerical methods 145–7
  off-balance 161–4
angle of variation *ω 228,* 229
angular dependence 111, 117, 123, 141–2, *143,* 145–6, 154, 156
angular dependent factor 111
angular dependent potential 123
angular localization 227–30

angular mean square displacement *241*
anisotropic particles
  composition 54, 55, 62
  drug delivery 258, 262–7
  geometries 267–71
  self-assembly 168
anisotropic potentials 117, 118
anisotropic surface modification 66
anticancer drugs 260, 261
aspect ratios 269, *270,* 271
asymmetrically modified gold nano-particles 216–17
atom-transfer radical polymerization (ATRP) initiators 67, 76
atomic force microscopy (AFM)
  AB-diblock brushes 7, *8*
  cylindrical brush polymers 6
  heteroarm star polymers 4
ATP hydrolysis 210
attractive part 111, *128*
auxiliary function $\gamma(12)$ 117, 118
axial coefficients 119
'axial' frames 118
azimuthal angle 225, *226,* 229, 230, 233, 234, 239
azobisisobutyronitrile 8

Barker–Henderson perturbation theory 121–3, 129, *131, 132*
base pairs 205, 206, 207, 210
biaxial electric fields 174–6

Janus particles in 195–7
bicompartmental particles, EHD co-
    jetting 59, *60, 62, 65, 69, 70*
binding affinity 263, *264,* 265–6
biodegradable materials 58, 71
biomedical applications
    DBNPs 41–8
    EHD co-jetting 63
    Janus particles 166
    PRINT process 92
    see also cancer therapy; drug
        delivery
biotin 66, *67, 69*
biphasic droplets 56, *57*
biphasic grafting *79*
black hole quencher-1 (BHQ-1) 207,
    208, 210
block copolymers
    AB-diblock brush 7–8
    anticancer drugs 261
    self-assembly 1, 2, 261
blood circulation profile 258, 259,
    270
body-centered cubic (BCC) 215
bottom-up approach 29, 204, 205
branching DNA units *205,* 206
bridge function $B(r)$ 116, 117, 119,
    120
brightness, Janus spheres 227–9
Brownian rotational dynamics *158,*
    199
brush polymers *3,* 6–8
bulk anisotropy 55
bundles, 3D *188,* 190

C-rich DNA strand 206–7
cadmium selenide 197
cadmium sulfide 37
cancer therapy
    anticancer drugs 260, 261
    drug delivery 48, *49*
    EGFR grafting 45
    field assembly 197, *198*
    tumor cells 259–60, 265

canonical *NVT* and *NPT* methods
    113, 145, 146
capillary needle configuration 60–2
capped trigonal bipyramid (CTBP)
    structures 158, *159*
carbodiimide coupling 78
carbon monoxide (CO) oxidation 40,
    41
cavity function $y(12)$ 121
cavity size, DNA crystals *219*
CCD camera *235,* 236
cell set-up *177,* 178
cell surface receptors 263, *264*
center of mass, Janus spheres *226,*
    227, 229
chain structures 175–6, 181, 187–9,
    191–6
chaining force 172–3
charge density *157*
charge distributions 142, 147
    dipolar Janus particles *151,* 152
chemical grafting 100–1, *102*
chemical potential 115, 121, 122, 124
chemical vapor deposition *69*
chloroform 4, *5*
cisplatin 48, *49*
Clausius–Mossotti function K 172,
    185
clearance 258, *270*
Clebsch–Gordan (CG) transforma-
    tions 118, 119
closed state *208, 210,* 211
clotting 267
clusters
    amphiphilic Janus particles 154,
        155, 157–8, *159,* 160–1, *162,* 163
    dipolar Janus particles 150, *151,*
        152–3
    folate 266
    ligands 262–3, *264,* 265
cobalt caps 191
'coffee-ring' effect 171, 179
colloids
    amphipathic *162*
    emulsion interface 244

Janus *79*
  localization and tracking 232
  molecules 140
  multifunctional 63–4
  potential 144
  from preformed particles *84, 85*
  rotational dynamics 223–4, 240
  synthesis 139–40, 147–9
  tracking 232
  triblock Janus 129, 133
  see also patchy colloids
compartmentalization
  interior particle 267, *268*
  NPs and MPs 55–62
compartmentalized fibers
  microsectioning 63–4
  multicompartmental *61, 62,* 64
computed tomography (CT) 46
2D COMSOL simulation 179, *185,*
    187, *192,* 193
conductivity 172
confocal laser scanning microscopy
  DBNPs 46
  drug delivery *268*
  EHD co-jetting *59, 62, 63, 69*
  field assembly *175*
confocal Raman spectromicroscopy
    66
consumer electronics 165
contact angle *247,* 248–9, 250, 251,
    252
contact line pinning 169, 180
contour plot 12, *235,* 236, 237, *238,*
    239
contrast agents 30, 45
convective flow fields 169, 171–4
  Janus particles in 179–83
  particle–field interaction *170*
coordinates, Janus spheres 227, *228*
coverage (χ) 111, 126–7
  half-coverage 127–9
  two-patch case 129, *130*
cross-sectional area, DNA crystals
    219
crossover frequency 172

cryo-transmission electron micro-
    scopy 139
cryosectioning *63,* 64, 69
crystalline structures
  3D DNA 219–20
  DNA-gold nanoparticles 214–15
  DNA self-assembly 217–18
  see also face-centered cubic (FCC)
    crystal; gold nanocrystals
    (AuNCs)
crystallization 133, *134*
  2D *188,* 189
cyanoacrylate *94, 97, 99, 102*
cylinders
  brush polymers *3,* 6–8
  nanoparticles 269, 271
  see also microcylinders
cytosine 206, 207
cytosol *264,* 266

DBNPs *see* dumbbell-like nanoparti-
    cles (DBNPs)
Debye screening length 146, 148, *157,*
    158, 171
Debye–Stokes–Einstein rotational
    diffusion constant 230, 232
degree of cooperativity 266
denaturing 213
dendrimers 260–2, 265
diblock brush copolymers 7–8
α,ω-dibromoalkane 5, *6*
dielectric permittivity 171, 172, 186
dielectrophoresis *170,* 183–6
dielectrophoretic (DEP) force 172,
    173, 185–6, *188,* 190
diffraction 236
  electron diffraction *37*
diphasic particles *97,* 98, *99,* 103, *105*
diphenyl ether solvent 33
dipolar Janus particles 138–41, 164–6
  experimental methods 147–8
  experiments and simulations 150–4
  numerical methods 141–5
dipole–dipole interactions 173, 174,
    176, 191–3

dipole strength parameter λ 174
direct correlation function $c(r)$ 115–
        16, 117
dispersants 42–3
dissociation rate constant 265, 266
distribution function $g(12)$ 120
divinylbenzene 8, *9*
DNA 204–6
    B–Z transition 206, *207*
    enzymatic activity 210–11
    macromolecular engineering
        217–20
    self-assembly 211–17
    strand displacement 208–10
DNA crystals 219–20
DNA duplex 206, 207, 211
DNA nanomachines 206–11
    applications 206, 214
    conformational changes 206–8
    from micro-engineering 217–20
DNA oligomers 148
DNA triplex 207, *208*
dodecamer cluster *155,* 157
dopamine 43, 44, 48
double chain *192,* 193, 194, 195, *196*
Doxil 260
doxorubicin 260, 261
droplets
    biphasic 56, *57*
    evaporation from 169, 179–80,
        181–2
    synthesis in 74–5, 79–82
drug delivery 257–8, 271–2
    anisotropic, patchy and Janus
        particles 258, 262–71
    DBNPs 47–8, *49*
    nanoparticle design 258–60
    nanoparticle examples 260–2
    see also cancer therapy
dumbbell-like nanoparticles
        (DBNPs) 29–30, 48–50
    biomedical applications 41–8
    components *30*
    containing more than 2 particles
        38–40

from emulsion polymerization *83*
heterogeneous catalysts 40–1
morphology 44
PRINT technique *99*
stability 48
synthesis 30–40
dynamic light scattering *9*

EC_{50} 265
edge lengths, DNA crystals *219*
EHD *see* electrohydrodynamic
        (EHD) co-jetting
electric fields
    biaxial 174–6, 195–7
    cell set-up *177,* 178
    frequency and strength *188,* 189
    Janus particles in 183–91
    particle–field interaction *170*
    uniaxial 169, 171–4
electrocatalysis 41, *42*
electrochemical materials 49
electrohydrodynamic (EHD)
        co-jetting 54–71
    microsectioning 63–4
    NP and MP compartmentalization
        55–62
    schematic description *57*
electron diffraction *37*
electron microscopy *see* cryo-
        transmission electron micro-
        scopy; scanning electron
        microscopy (SEM); transmis-
        sion electron microscopy
        (TEM)
electronic devices 49
electronic paper 197
electroosmosis 183–6
electrophoresis 171
    induced-charge 183–6, 189
electrophoretic deposition (EDP)
        170, 171
electrophoretic mobility μ 171
electrospinning 56, *57,* 63, 64
electrospraying 56, *57,* 63
electrostatic energy *157, 158*

electrostatic interactions 154
electrostatic screening length 154, 165
emulsion-based methods 74–88
emulsion polymerization 76, *77*, 78
    preformed particles 83–7
emulsions
    adsorption 245, 246, 255
    Janus particle-stabilized 244–5,
        253–5
    see also Pickering emulsions
endocytosis 260, 266, 269
endosomes *264*, 266
energy-dispersive X-ray (EDX) spec-
        troscopy *65*, 101, *103*
enhanced permeation and retention
        (EPR) 259
environment
    DNA nanomachines 206–8
    tumor 259–60
enzymes 267, *268*
    DNA 210–11
epidermal growth factor receptor
        (EGFR) antibody 43, 44–5, 46
epifluorescence microscopy 148, 150,
        151, 155, 230
epitaxial growth 32
equilibration process 113
erythrocytes 267
evaporation 169, *177,* 179–80, 181–2
    liquid droplet synthesis 79–80
evaporative flux J_s 180
excess free energy 120
extracellular matrix 259

face-centered cubic (FCC) crystal
        132, 161, *162,* 163, 215
face-centered cubic (FCC–FCC)
        transition 132
field assembly 168–9
    applications 197–9
    biaxial electric and magnetic fields
        174–6, 195–7
    convective flow/electric/magnetic
        fields 169, 171–4

Janus particle preparation and cell
        set-up 176–8
Janus particles 178–97
    particle–field interaction *170*
filamentous particles *270,* 271
five degrees of freedom 198, 199
flow cytometry *67*
flow rate, polymer solution 59, *60*
fluid–fluid coexistence curves 124–7
    thermodynamic perturbation the-
        ory 129, 131
    two-patch particles *130*
fluid–solid coexistence 115, 131–2
fluorescein *o*-acrylate 100, 101
fluorescence, DNA nanomachine
        207–8
fluorescence microscopy
    EHD co-jetting *67*
    PRINT technique *99, 100, 102,
        104, 105*
    see also epifluorescence microscopy
fluorescence resonance energy trans-
        fer (FRET) spectroscopy *207*
folate *264,* 265–6
Fourier transform infrared spectro-
        scopy (FTIR) 57, *58*
Fourier transforms 116, *117,* 118, 119
free energy *F* 120, 121, 122, 249
    Gibbs free energy 31, 32
Frenkel–Ladd procedure 115
full width at half-maximum
        (FWHM) 213

gel trapping technique 102, 246
General Polygon Clipping Library
        237
Genexol 261
Gibbs–Duhem integration 115
Gibbs Ensemble Monte Carlo
        (GEMC) method 114, *125,
        126, 127*
Gibbs free energy (ΔG_s) 31, 32
glassy system 240
gold-capped Janus particles 177, 178,
        179, 181, 184–91

gold-cappedJanus particles *see* gold-
    capped Janus particles
gold–gold–iron oxide (Au–Au–
    Fe_3O_4) NPs 38, *39*
gold–iron oxide (Au–Fe_3O_4) NPs 30,
    32–3, 38, *45*
  CO oxidation 40, *41*
  drug delivery 48, *49*
  molecular imaging 44–6
  surface modification 43
gold nanocrystals (AuNCs) 64, *65*
  at a planar interface 246–7
gold nanoparticles (AuNPs) 32–3
  asymmetrically-modified 216–17
  DNA-modified 212–17
  drug delivery *264*, 266–7
  emulsion polymerization 86, *87*
  optical properties 212
  surface plasmon resonance 46, 212,
    213
gold–oxide catalysts 30
gold silver–iron oxide (AuAg–Fe_3O_4)
    NPs *36*
grand-canonical μVT ensemble 114,
    122
green fluorescent protein 47
Gyricon balls 197, *198*

half-coverage (χ) 127–9
Hankel transform 118, 119
hard-sphere model 109, 110–12
HEK293T cell 46, *47*
HeLa cells 269, 271
Helmholtz free energy 122
Helmholtz–Smoluchowski equation
    171
hemispherical Janus particles 224–5,
    226, 227, 230, 233–9
heptamer clusters *152*, 153, 157, *160*,
    161
Herceptin 48, *49*
heteroarm star polymers *3*, 4, *5*
heterogeneous catalysts 40–1
heterogeneous nucleation 31–2
hexagonal lattice *133, 134*

hexagonally close packed (hcp)
    structure 175, 181
hexamer clusters *152*, 153, *155*, *159*,
    160, 161
high-resolution TEM (HRTEM) *33*,
    34, *37*, *39*, *40*
Holliday junction 205
homogeneous nucleation 31
horizontal stepwise mold filling 95–9,
    100
Hückel equation 171
hybrid Janus particles 64–6
hybridization, DNA-directed *209*,
    213, 215, 216–17
hydrodynamic diameter 44
hydrophilic–lipophilic balance (HLB)
    76, 245, 251, 252–3
hydrophilic particles/side
  amphiphilic Janus 161, *162*
  block polymers 261
  contact angle *248*, 249
  Janus balance 250–1, *252*
  PRINT technique *97*, 102–3, *104*
hydrophobic particles/side
  amphiphilic Janus 145, 146, 161
  block polymers 261
  contact angle *248*, 249
  Janus balance 250–1, *252*
  PRINT technique *97*, 103, *104*
hydrophobicity 248, 250

i-motif 206
ICAM-1 269
icosahedral structures *155*, 161
image preprocessing 235–6
inactive state *210*
induced-charge electroosmosis
    (ICEO) 183–6
induced-charge electrophoresis
    (ICEP) 183–6, 189
integral equation theories
  fluid–fluid coexistence curve 124
  general scheme 115–17
  iterative procedure 117–20
  RHNC 124–7

thermodynamics 120–1
inter-junction distances, DNA crystals *219*
interaction space 150
interfacial energies 31, 32, 248
interior particle compartmentalization 267, *268*
interparticle potential 110
interstitial fluid pressure 259
ionic strength 154
iron-capped Janus particles 178, 191, *194,* 195
iron oxide-capped Janus particles 178, 191, 193–7
iron oxide (Fe_2O_3) NPs *37, 38*
iron oxide (Fe_3O_4) 32–3, 34–5, *39*
  field asembly *170,* 174
  see also gold–iron oxide ($Au–Fe_3O_4$) NPs
iron oxide–cadmium selenide ($Fe_2O_3–CdSe$) DBNPs 46, *47*
iron oxide–cadmium sulfide ($Fe_2O_3–CdS$) DBNPs *37,* 38
iron oxide–gold–iron oxide ($Fe_3O_4–Au–Fe_3O_4$) NPs 38, *40*
iron oxide–gold–lead selenide ($Fe_3O_4–Au–PbSe$) NPs 38
iron oxide–gold–lead sulfide ($Fe_3O_4–Au–PbS$) NPs 38
iron pentacarbonyl ($Fe(CO)_5$) 32–3
iron platinum–cadmium sulfide ($FePt–CdS$) DBNPs 36–7
isolated particle localization 225–32
  angular 227–30
  experimental validation 230, *231,* 232
  spatial 225–7
isomerization *159*
isothermal–isobaric method *(NPT)* 113
isothermal–isochoric method *(NVT)* 113, 145, 146
iterative subtraction process 237, *238*

Janus balance 75–6, 140, 149, 161, 165, 247–53
  definition 245
  example 251, *252*
  outlook and implications 252–3
  quantification 250–1
Janus limit 127–9
Janus nanosheets 81–2
Janus particles 1, 140
  amphiphilic *see* amphiphilic Janus particles
  contact angle at an interface 248–9
  convective flow fields 179–83
  dipolar *see* dipolar Janus particles
  from direct self-assembly/transformations in solution 9–15
  drug delivery 262–7
  electrohydrodynamic co-jetting 54–71
  emulsion-based methods 74–88
  emulsions stabilized by 244–5, 253–5
  field assembly *see* field assembly
  geometry at oil/water interface *248*
  hemispherical *see* hemispherical Janus particles
  hollow spheres 81, *82*
  hybrid 64–6
  iron-capped *see* iron-capped Janus particles
  iron oxide-capped *see* iron oxide-capped Janus particles
  localization *see* particle localization
  from macromolecular engineering 1, 3–9
  metallodielectric 186–90
  at a planar interface 246–7
  polymer-based 2
  PRINT technique 93–100, 102–5
  topologies *2*
Janus spheres *see* hemispherical Janus particles
jetting parameters 58–9, 68

kagome lattice 132–4
Kaposi's sarcoma 260
Kern–Frenkel (KF) model 110–12,
    118, 122, 123, 135
  amphiphilic Janus particles 146,
    147
  fluid–fluid coexistence curves from
    125, *126, 127*
  two-patch particles *130,* 133
Kern–Frenkel (KF) particles 114
kidneys 258
kinetic pathways, amphiphilic Janus
    particles 158, *159,* 160

latex spheres *170,* 173
lattice mismatch 32
lead sulfide ($PbS–Au–Fe_3O_4$) NPs 38,
    39
Lennard-Jones interactions 145, 153,
    179, 181
ligands
  addition 42–3
  exchange 43, 44
  functionalization 262–3, *264,*
    265–7
linear superposition of image overlap
    236–7
linker molecules 263
liposomes 260–2
liquid droplets *see* droplets
liquid/liquid interface
  DBNP formation *34*
  Pickering emulsion 78, *79*
lithography
  photolithography 56, 69
  PRINT technique 91–106

macromolecular engineering
  DNA 217–20
  Janus particles 1, 3–9
macrophages 269, *270,* 271
magnetic control
  DBNPs 46–7
  hybrid Janus particles 65
magnetic dipole moment μ 173, 174

magnetic energy density *192,* 194
magnetic field assembly
  iron/iron oxide Janus particles 195
  simulation 193–4
magnetic fields
  biaxial 174–6, 195–7
  cell set-up *177,* 178
  Janus particles 191–5
  particle–field interaction *170*
  uniaxial 169, 171–4
magnetic particles
  bicompartmental 65
  Janus composite 79, *80,* 82
magnetite *65,* 79, 80, 83
margination 267, 269
Markov chain in configuration space
    113
Maxwell–Wagner charge relaxation
    time 172, 185
mean square displacement (MSD)
    230, *231, 241*
melting transition 213
1-mercapto-1-undecanesulfonate 266
mesoparticles, amphiphilic Janus 161
metal deposition 101–2, *103*
metallodielectric Janus particles
    186–90
micelles *9,* 260–2, 265, 266
  Janus 13–17, 19–21
  Janus limit *128,* 129
  multicompartment 1, *6*
  self-assembly *11*
microcylinders
  from cryosectioning *63,* 64
  PEGMA surface layer *70*
  PLGA 67, *68,* 69
  see also cylinders
microfluidics 56, 197
micromolecular engineering 217–20
microparticles (MPs), compartmen-
    talization 55–62
microsectioning 63–4
modulated optical nanoprobes 224
mold filling
  horizontal stepwise 95–9, *100*

vertical stepwise 93–5
molecular dynamics, field assembly
    179, 181–2
molecular resonance imaging (MRI)
    44, 45, 46
Monte Carlo simulations 112–15,
    125, 126, 127
    amphiphilic Janus particles 146,
        154, *155, 160*
    cylindrical brush polymers *7*
    dipolar Janus particles 145, *151*
    field assembly 175
moon-like appearance 224, *225, 234*
multicompartment micelles (MCMs)
    1, *6*
multicompartmental microfibers *61,
    62, 64*
multicompartmental particles
    EHD co-jetting 70
    properties *55*
multifunctional platform, DBNPs
    41–8
multiphase particles, PRINT techni-
    que *97,* 102, 106

N-hydroxysuccinimide (NHS)-
    rhodamine 100, *101*
nano-robots 82
nanomachines, DNA 206–11
nanoparticles (NPs)
    binding affinity 263, *264*
    compartmentalization 55–62
    cylindrical 269, 271
    drug delivery 258–60, *264,* 265–7
    examples 260–2
    material properties 257
    seed-mediated growth 31, 32–4, 38
    self-assembly 211
    virus-like protein NPs *209,* 215
    see also dumbbell-like nanoparti-
        cles (DBNPs); gold nanoparti-
        cles (AuNPs)
nanosheets 81–2
nanotechnology 204
nickel caps 191

noble metal–magnetic NPs *36*
noble metal oxide NPs 32–6
    see also gold–iron oxide
        (Au–Fe$_3$O$_4$) NPs
non-spherical particles 63
    synthesis 76, *77,* 81
non-Watson–Crick base pairs 206,
    207
NPT (isothermal–isobaric) method
    113
nuclear Overhauser effect (NOESY)
    NMR 4
nucleation theory 31
NVT (isothermal–isochoric) method
    113, 145, 146

octadecanethiol 246
octadene solvent 32, 33
octahedral clusters *159,* 160
1-octanethiol 266
off-balance amphiphilic Janus
    particles 161–4
oil-in-water emulsions 244, 245
    magnetic *80*
oil/water interface
    contact angle *248, 249,* 251
    PRINT technique 102–5
open state *208, 210,* 211
opsonins 262
optical devices 49
optical imaging
    DBNPs 44, 46
    tensegrity triangle *218*
optical microscopy
    field assembly *179,* 181, *194*
    PRINT process *94*
optical trap *198,* 199
optically overlapping particle locali-
    zation 232–9
organic–inorganic nanocomposites
    64
organic-soluble polymers 58
Ornstein–Zernike (OZ) equation
    115–16, 117, 119–20
    schematic flow chart *118*

Ostwald ripening process *32*
ovarian cancer 260
overlapping Janus spheres, separation 236–7, *238*
overlapping object recognition 236
oxygen reduction reaction 41, *42*

p53 expression 48, *49*
packing fraction 111
paclitaxel 261, 271
'Pacman'-like particles 62
pair potentials 109, 149–50
  amphiphilic Janus particles 145, 154, 156
  dipolar Janus particles 141–2, *143*
paramagnetic particles 175
particle–cell interactions 262
particle–field interactions *170, 171*
particle localization 223–5, 242
  isolated particle 225–32
  optically overlapping 232–9
particle–particle interactions 169, *215*, 255, 262
particle replication in non-wetting templates (PRINT) 90–106
  advantages of 92
  Janus particle fabrication 93–100
  patchy PRINT particles 100–2, *103*
  schematic illustration *92*
  self-assembly 102–5
particle size 25, 255, 258
patchy colloids 108–35
  field assembly *177*, 190–1
  single-patch particles *111*
patchy particles, drug delivery 262–7
patchy PRINT particles 100–1, *102, 103*
pentagonal structure *160*
pentamer clusters *160*, 161
perfluorodecalin 102–5
perfluoropolyether (PFPE) 91–2, 95, 102, *104*
perturbation theory 121–3
  fluid–fluid coexistence curves 124, 129, 131
    fluid–solid coexistence *132*
pH, DNA nanomachine 207, *208*
phagocytosis 269, *270*
phase diagrams 108–35, 140
  amphiphilic Janus particles *157, 158*, 161
  field assembly *188*
  fluid–fluid *131*
  fluid–solid *132*
  Janus limit *128*
  kagome lattice *134*
  one- and two-patch coverage *130*
phase separation
  cylindrical brush polymers 6
  polymer arms 3–4, *5*
photo-curing 93, *94*, 95, *96*, 100
photolithography 56, 69
physiological barriers 258
Pickering emulsions 74, 75–9, 148, *149*, 244
planar interface, Janus particles at 246–7
platelets 267–8
platinum–gold–iron oxide (Pt–Au–Fe$_3$O$_4$) DBNPs 48, *49*
platinum–iron oxide (Pt–Fe$_3$O$_4$) DBNPs *36*
  CO oxidation 40, *41*
  electrocatalysis 41, *42*
PLGA *see* poly(lactic-*co*-glycolic acid) (PLGA)
PMMA *see* poly(methyl methacrylate) (PMMA)
poly-*N*-isopropylacrylamide (PNIPAM) 62, 83
poly(2-(diethylamino)ethyl methacrylate) (PDEAMA) 4
poly(2-(dimethylamino)ethyl methacrylate) (PDMAEMA) 6
poly(2-vinylpyridine) (P2VP) 4, 6
poly(4-vinylpyridine) (P4VP) 8
poly(acrylamide-*co*-acrylic acid) [P(AAm-*co*-AA)] 57, *58*, 60, 64
  biotin- and acetylene-modified 66

poly(acrylic acid) (PAA) 56, 79
polyacrylonitrile (PAN) 83, 85
poly(ε-caprolactone) (PCL) 6, 7, *8*
polydimethylsiloxane (PDMS) 91, 246–7
polydivinylbenzene (PDVB) 8, *9*
poly(ethylene glycol) methacrylate (PEGMA) 67, *70*
poly(ethylene glycol) (PEG) 56, 66, *67*, 259, 262, 263
[poly(ethylene oxide)-*b*-poly(benzyl L-aspartate)] (PEO-*b*-PBLA) 261
[poly(ethylene oxide)-*b*-poly(lactic acid)] (PEO-*b*-PLA) 261
poly(ethylene oxide) (PEO) 6, 8
poly(ethylene terephthalate) (PET) 91
poly(lactic-*co*-glycolic acid) (PLGA) 58–60, 62, *63*, 64, 69
 acetylene functionalized 66, *68*, *70*
 microcylinders 67, *68*, 69
polymer arms 3–4, *5*
polymer-based Janus particles 2
polymerase chain reaction (PCR) 212
poly(methyl methacrylate) (PMMA) 6, 83, *84*, 240
 field assembly 174–5
poly(*n*-butyl acrylate) (PnBA) 7, *8*
poly(*N*-isopropylacrylamide)-(PNIPAAm) 4, 8–9
poly[oligo(ethylene glycol) methyl ether methacrylate] (OEGMA) 67, *70*
polystyrene (PS) 6
 emulsion-based methods 76, *77*, 78, 83–7
 field assembly *170*, 171, 176, *179*, 181, 184–6, 188–91, 195
polystyrene-*block*-poly(2-vinylpyridine)-*block*-poly-(ethylene oxide) (PS-*b*-P2VP-*b*-PEO) 4–5, *6*
polystyrene-*block*-poly(acrylic acid) (PS-*b*-PAA) 79
poly(vinyl alcohol) (PVA) 198

poly(vinyl cinnamate) (PVCi) 62
polyvinylpyrrolidone (PVP) 64
positional mean square displacement *241*
potential difference *185*, 187
preformed particles, synthesis on 75, 83–7
pressure *P* 124
PRINT *see* particle replication in non-wetting templates (PRINT)
probing translational and rotational dynamics 239–42

quantum dots (QDs) 198–9
 DBNPs containing 36–8

radial distribution function *g(r)* 113, 115, 117
receptor-mediated drug delivery 260, 263, *264*
reference hyper-netted chain (RHNC) closure 116, 117, 119, 120, 121
 fluid–fluid coexistence curves from 124–7
reflection imaging, DBNPs *45*, 48, *49*
repulsive part *111*
reticuloendothelial system (RES) 44, 258, 269, *270*
RHNC *see* reference hyper-netted chain (RHNC) closure
rhodamine dye 100, 101
rhodamine green 207, *208*, *210*
rhombohedral crystals *218*, 219–20
rotating disk electrode (RDE) measurements *42*
rotation, horizontal mold filling *98*, *99*
rotational diffusion constant *D*r 229–30, *231*, 232
rotational dynamics 223–4, 225, *228*, 230, *231*
 Brownian *158*, 199
 probing 239–42

rotational orientation 225

salt concentration 140, 141, 146, 154,
        155, 156
    DNA self-assembly 212
    NaCl 159
scanning electron microscopy (SEM)
    drug delivery 270
    EHD co-jetting 59
    emulsion-based methods 85, 86
    field assembly 170, 171, 181
    Janus nanosheets 82
    at a planar interface 246, 247
    PRINT process 94, 101, 103, 104
seed-mediated growth
    dumbbell particles 83
    nanoparticles 31, 32–4, 38
segmentation 236
self-assembly 138–41, 164–6
    amphiphilic and dipolar Janus
        particles 138–66
    block copolymers 1, 2, 261
    branching DNA units 205
    DNA, micro- to macro-engineering
        217–20
    DNA-enabled 211–17
    EHD co-jetting 66–8, 69, 70
    experimental methods 139, 147–9
    experiments and simulations
        149–61
    Janus particles from 9–15, 91
    Janus PRINT particles 102–5
    numerical methods 141–7
    patchy colloids 108–35
    predefined kagome lattice 132–4
    see also field assembly
semiconductor NPs 36–8
semipermeable compartments 267
sensors 198
separation, overlapping Janus
        spheres 236–7, 238
sheets
    field assembly 175
    nanosheets 81–2
silanes 75, 76, 82

silica particles 77, 78, 80
    Janus balance 251
    Janus nanosheets 81, 82
    Pickering interface 75, 76, 148, 149
    planar interface 246–7
    preformed synthesis 83–5
silver–iron oxide (Ag–Fe$_3$O$_4$) NPs 32,
        34–5, 36
silver–iron platinum (Ag–FePt) NPs
        34
silver nitrate (AgNO$_3$) 34
single-patch particles 111, 125, 127
    Janus limit 127–8
    vs two-patches 129, 130
single-phase particles 102, 103, 104
SkBr3 cells 48, 49
small-angle X-ray scattering 215
smart materials 82
sol–gel processing 82
solutions
    polymer 59, 60
    transformations in 9–15
solvent evaporation 79–80
space coefficients 119
space group, DNA crystals 219
spatial localization 225–7
spatioselective surface modification
        66, 70
spleen 258
square-well case 109, 111, 121, 124–6
    fluid–solid coexistence 131
    two-patches 129, 130
staggered chains 187–9, 192, 193–5,
        196
statistical physics 109
    tools 112
stealth coatings 259, 262, 271
sticky ends 205, 206
    cohesion 219
strand displacement 208–10
streptavidin 67, 69, 170, 171
sulfate-terminated polystyrene (PS)
        176
supraballs 197, 198
surface anisotropy 55

surface immobilization 212
surface modification
  chemical grafting 100–1, *102*
  DBNPs 42–4
  EHD co-jetting 66–8, *69, 70*
  Janus particles 91
  metal deposition 101–2, *103*
surface plasmon resonance 46, 212, 213
surface-to-surface distance 144
surfactants *9*, 42–3, 76
  amphiphilic particles 165
  emulsion stabilization and 244
  hydrophilic–lipophilic balance 245, 251, 252–3
  self-assembly 140
systemic barriers 258

Taylor cone 56, *57, 65,* 68
TEM *see* transmission electron microscopy (TEM)
tensegrity triangle 218–20
ternary DBNPs 38, *39, 40*
tetrahydrofuran 4, *5*
tetramer clusters *152, 153, 155,* 157, *160*
thermal cross-linking, EHD co-jetting 57, *58*
thermal evaporation *177*
thermodynamic integration 115
thiol-modification 190, 212
three-particle angle 193, 194
tilt angles *156, 158*
time-resolved fluorescence anisotropy 223
titanium dioxide (TiO$_2$) 64, *65*
toluene/water 246, 247, 253, *254, 255*
top-down approach 91, 204
total correlation function $h(r)$ 115–116, 117, 120
total potential U 122
tracking
  colloids 232
  Janus spheres *235*
trans-illumination *234,* 235, 239

transformations in solution 9–15
transition metal oxide NPs 32–6
translational diffusion constant *231,* 232
translational dynamics 230, *231*
  probing 239–42
transmission electron microscopy (TEM) 4, *6, 9*
  Ag–Fe$_3$O$_4$ NPs *34, 35*
  Au–Fe$_3$O$_4$ NPs *33, 45*
  DNA-modified gold nanoparticles *217*
  FePt–CdS NPs *37*
  hybrid Janus particles *65*
  Pt–Fe$_3$O$_4$ NPs *42*
  ternary DBNPs *39, 40*
triblock Janus colloids 129, 133
tricompartmental microfibers 60, *61*
trimer clusters *160*
triphasic particles 95, *96, 97,* 103–4, *105*
tumor cells 259–60, 265
tweezers, DNA 208–10
two-body potential $\varphi(r)$ 116
two-patch particles 129, *130,* 133

uniaxial electric fields 169, 171–4
unit vectors 110–11
unperturbed images 237
UV irradiation 93, 95

van der Waals interactions 153
vapor deposition 101–2, *103*
Verlet–Weiss expression 116, 120
vertical stepwise mold filling 93–5
vesicles, Janus limit *128,* 129
virial pressure P 121
virus-like protein nanoparticles (VLPs) *209,* 215
viscosity, polymer solution 59, *60*

water-in-oil emulsions 244, 245, *254,* 255
water-repellent fibers 165

water-soluble polymers, EHD
    co-jetting 56–7
Watson–Crick base pairs 205, 207,
    210
wax 75, 76, *77,* 148, *149*
wet etching 76, *77, 85*
wettability 245
'worm-like' clusters 154, *155,* 158,
    160, 161, *162,* 163

X-ray absorption 46

X-ray crystallography 139

yttrium hydroxide nanotubes (YNTs)
    8, *9*
Yukawa potential 146

zenith angle 224, 229, 230, 234, 239
zeta potential 147, *151,* 171, 183, 190
zwitterionic particles 147, 148, 150,
    153